Laser Analytical Spectrochemistry

The Adam Hilger Series on Optics and Optoelectronics

Series Editors: **E R Pike** FRS and **W T Welford** FRS

Other books in the series

Aberrations in Optical Systems
W T Welford

Laser Damage in Optical Materials
R M Wood

Waves in Focal Regions
J J Stamnes

Other titles of related interest

Infrared Optical Materials and their Antireflection Coatings
J A Savage

Principles of Optical Disc Systems
G Bouwhuis, J Braat, A Huijser, J Pasman, G van Rosmalen and
K Schouhamer Immink

Thin-film Optical Filters
H A Macleod

Aberration and Optical Design Theory
G G Slyusarev

Lasers in Applied and Fundamental Research
S Stenholm

The Adam Hilger Series on Optics and Optoelectronics

Laser Analytical Spectrochemistry

edited by

V S Letokhov

USSR Academy of Sciences,
Institute of Spectroscopy

Adam Hilger, Bristol and Boston

© V S Letokhov 1985

English edition © IOP Publishing Ltd 1986

British Library Cataloguing in Publication Data
Laser analytical spectrochemistry.
  1. Laser spectroscopy
  I. Letokhov, V S
  543′.0858   QD96.L3

  ISBN 0-85274-568-0

Published under the Adam Hilger imprint by IOP Publishing Ltd
Techno House, Redcliffe Way, Bristol BS1 6NX, England
PO Box 230, Accord, MA 02018, USA

Typeset by Mid-County Press, 2a Merivale Road London, SW15 2NW
Printed in Great Britain by J W Arrowsmith Ltd, Bristol

Contents

Series Editors' Preface

Optics has been a major field of pure and applied physics since the mid 1960s. Lasers have transformed the work of, for example, spectroscopists, metrologists, communication engineers and instrument designers in addition to leading to many detailed developments in the quantum theory of light. Computers have revolutionised the subject of optical design and at the same time new requirements such as laser scanners, very large telescopes and diffractive optical systems have stimulated developments in aberration theory. The increasing use of what were previously not very familiar regions of the spectrum, e.g. the thermal infrared band, has led to the development of new optical materials as well as new optical designs. New detectors have led to better methods of extracting the information from the available signals. These are only some of the reasons for having an *Adam Hilger Series on Optics and Optoelectronics*.

The name Adam Hilger, in fact, is that of one of the most famous precision optical instrument companies in the UK; the company existed as a separate entity until the mid 1940s. As an optical instrument firm Adam Hilger had always published books on optics, perhaps the most notable being Frank Twyman's *Prism and Lens Making*.

Since the purchase of the book publishing company by The Institute of Physics in 1976 their list has been expanded into all areas of physics and related subjects. Books on optics and quantum optics have continued to comprise a significant part of Adam Hilger's output, however, and the present series has some twenty titles in print or to be published shortly. These constitute an essential library for all who work in the optical field.

Preface

The progress of tunable laser technology has led to profound changes in the techniques of optical spectroscopy, making it a still more effective tool for studying atoms, molecules, and solids. Tunable lasers have breathed new life into every known spectroscopic technique and given rise to many essentially novel methods of nonlinear spectroscopy.

Almost from the time of its establishment at the end of 1968, the Institute of Spectroscopy of the USSR Academy of Sciences has been engaged in intensive research in laser spectroscopy, particularly in ultra-high-resolution nonlinear spectroscopy, ultra-sensitive photoionisation spectroscopy of multistep atomic and molecular excitation, spectroscopy of electronic transitions of molecules in low-temperature matrices, spectroscopy of highly excited vibrational states in molecules, picosecond spectroscopy of biomolecules, and so on.

During the period that has elapsed since then, important results have been achieved in each of the above fields. Suffice it to mention such techniques as laser detection of single atoms and molecules, laser burning of narrow holes in electronic spectra of complex molecules, and laser cooling of atoms.

As fundamental investigations with the aid of new laser spectroscopy techniques got under way, it became clear that they were of considerable practical interest as they allowed for spectral analyses of substances at a level unattainable with the former, pre-laser methods. During the last ten years, extensive investigations have therefore been undertaken at a number of the Institute's laboratories into the analytical capabilities of the techniques of laser spectroscopy. The publications in scientific journals and conference reports by the Institute's investigators engaged in analytical laser spectroscopy have attracted a stream of visitors to the Institute, seeking detailed information on the methods being developed.

Research workers and specialists in allied fields are in great need of qualified assistance and information on problems of analytical laser spectroscopy. For this reason we have decided to systematise the results of the research done at the Institute in the form of reviews by the Institute's leading scientists actively concerned with the development of the new techniques and to publish them in a collective monograph furnished with a general introduction to the problem. The result is the present book.

The book deals with most of the atomic and molecular laser spectroscopy techniques which are now finding analytical applications. As regards atomic spectral analysis, it describes methods of laser induced fluorescence (Dr M A Bol'shov) and multistep resonance ionisation (Dr G I Bekov and Professor V S Letokhov). In the field of molecular spectral analysis, consideration is given to the methods of IR absorption spectroscopy using injection diode lasers (Dr Yu A Kuritsyn), laser excitation of fine-structure fluorescence spectra of complex molecules in a low-temperature matrix (Professor R I Personov), and multiphoton ionisation of molecules (Dr V S Antonov and Professor V S Letokhov). Special attention is devoted to laser techniques that can be used in combination with well known analytical methods, such as opto-acoustic spectroscopy combined with gas chromatography (Dr V P Zharov) and molecular laser ionisation combined with mass spectrometry (Dr A N Shibanov). Thus the reader is provided with adequate information on a fairly wide range of analytical applications of laser spectroscopy currently in a state of explosive development. The authors hope that the publication of this monograph will contribute to further progress in the field of laser application.

It is only natural that a book of this kind, one of the first in the field and written by authors working in different lines of research and having different tastes and temperaments, should have certain shortcomings. The authors would be grateful for any criticism and suggestions for improvement coming from their readers.

The Russian version of this monograph is being published simultaneously by the Nauka publishing house of the USSR Academy of Sciences. The authors wish to express their gratitude to Adam Hilger and Mr J A Revill personally for publication of the present English version of the book.

Professor V S Letokhov
142092 Troitsk, Moscow Region, USSR, November 1985

1 Introduction to Laser Analytics

V S Letokhov

The lasers created a quarter of a century ago have received much recognition today in various fields of science and technology. In some fields (medicine, material processing, isotope separation etc) the application of laser light has proved unexpectedly successful. In other fields which have been related to light for a long time (spectroscopy, metrology, photochemistry) the use of lasers seemed to be very promising from the very beginning. But it took a decade in the 1960s to realise successfully in experiments even the more obvious applications of lasers in spectroscopy. This was due to the fact that the first pulsed and continuous-wave lasers operated at a limited number of discrete frequencies of visible and near infrared ranges. At the same time, most of the methods of atomic and molecular spectroscopy require that the laser radiation frequency should coincide with the frequency of a definite quantum transition between two energy levels of the studied atom or molecule. Late in the 1960s various methods of generating coherent light with a tunable frequency began to develop rapidly. So it was in that period that the methods of laser spectroscopy began to progress.

In the 1970s thousands of researchers got involved in the studies on laser spectroscopy making use of synthesis of the methods and means of quantum electronics. In that period international conferences on laser spectroscopy [1.1–1.6] and tunable lasers [1.7–1.9] were held regularly. Several monographs on general principles of laser spectroscopy [1.10–1.12] and many monographs on different trends in laser spectroscopy were published over a short period of time. These are quoted below. The analytical applications of laser spectroscopy, of course, are highly effective and rapidly developing. The field of laser spectroscopy has been covered in several collective monographs

1

[1.13–1.15] concerned with most of the methods which are of interest for analytics. It must be noted, however, that progress in this field is very rapid, and many new results have been obtained in the last two or three years. Hence there was a need for another monograph on analytical laser spectroscopy which would complement the monographs [1.13–1.15] and would include the experiments carried out at the Institute of Spectroscopy, Academy of Sciences USSR.

The purpose of this chapter is to consider in general the methods of laser analytics and the role of these methods which are considered in more detail in the subsequent chapters.

1.1   Properties of laser light

The revolutionary influence of the laser on optical spectroscopy is based on a combination of unique characteristics of laser source radiation. In this short introduction it is impossible to examine in great detail the tunable-frequency lasers which are the main instrument of almost all methods of laser spectroscopy. So here we are going to restrict ourselves just to listing the most important laser radiation characteristics which make the laser an extremely valuable and effective instrument in laser analytical spectroscopy.

(1) The *wavelength tunability* enables radiation at any wavelength in the IR, visible, UV and, recently, vacuum ultraviolet (VUV) spectral regions to be obtained. Figure 1.1 shows the principal regions of operation of different tunable lasers including the methods of nonlinear frequency conversion. The very fact that laser radiation can be attained at any wavelength over a wide range from 0.2 to 20 μm makes it possible to study almost any quantum transition of atoms and molecules.

(2) The *high intensity* of laser radiation allows easy realisation of nonlinear interaction of light with atoms and molecules in both gas and condensed media. In one-quantum resonance interaction the nonlinearity manifests itself in the absorption saturation effect when a considerable fraction of particles are transferred to an excited state. At a higher intensity forbidden one-quantum and multiquantum resonance transitions between atomic (or molecular) levels, which cannot be observed at low light intensities, become probable.

(3) The *short (controllable) duration* of radiation that can be made shorter than the lifetime of any excited state of an atom or molecule provides a means for the multistep excitation of high-lying energy levels ($E_{exc} = 5$–20 eV) in a time interval shorter than the relaxation time of any intermediate quantum state. This essentially widens the scope of optical spectroscopy, the spectral transitions between highly excited quantum states being accessible to measurement.

(4) The *monochromaticity* or *high spectral intensity* of laser radiation allows,

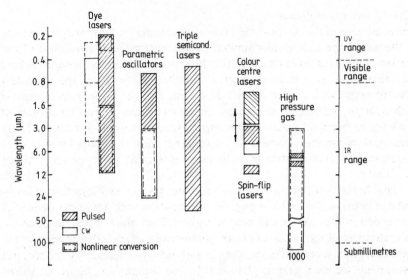

Figure 1.1 Tuning ranges of main types of pulsed (hatched areas) and CW (white areas) lasers for dye lasers, opitcal parametric oscillators, diode lasers on triple compounds, colour centre lasers, spin-flip lasers and high-pressure molecular gas lasers, including expansion of the tuning range by nonlinear conversion of frequency.

first, measurement of spectra with almost any required resolution and, second, selective excitation of atoms or molecules of definite sorts in mixtures. This latter consideration is particularly important for analytical applications.

(5) The *spatial coherence* of radiation makes it possible to form highly collimated light beams for spectral probing of regions or to focus the radiation onto a small area for local analysis.

A compromise between all these valuable properties in a very efficient laser source of coherent optical radiation offers remarkable possibilities for the development of methods for analysis of the composition of matter on the atomic–molecular level with extremely high sensitivity and selectivity.

1.2 The problems of laser spectroscopy

The above-listed unique characteristics of laser radiation have made it possible to solve, or have provided a potential possibility of solving, the basic problems of spectroscopy; these are to obtain necessary spectral and time resolution, sensitivity and selectivity. Below we shall consider briefly the progress achieved here from the standpoint of analytical requirements.

1.2.1 Spectral resolution

The resolution of the classical methods of atomic and molecular spectroscopy in the gas phase was usually limited by the instrumental resolution of the spectrometer. This limitation is particularly noticeable in the IR region where the resolution of only a limited number of unique spectrometers (spectrometers with a large diffraction grating and the Fourier spectrometer with a large pathlength difference) reaches $0.01\ cm^{-1}$. The resolution of ordinary commercial spectrometers approximates to only $0.1\ cm^{-1}$. These numerical parameters should be compared with the Doppler width of vibrational–rotational absorption lines of molecular gases which is about $10^{-3}\ cm^{-1}$.

Tunable lasers with a narrow radiation line, particularly injection IR lasers and dye lasers in the visible region (in combination with nonlinear frequency conversion in the near UV and near IR regions), have allowed realisation of the ultimate spectral resolution of linear spectroscopy which is determined by the real absorption spectrum of the sample without any influence by the spectral instrument (see monographs [1.10–1.12]). For comparison, figure 1.2 shows the ammonia absorption line sP (1;0) of the band v_2 near 10.6 μm measured with a tunable diode laser on the three-component compound $Pb_{0.88}Sn_{0.12}Te$ and with a high-quality spectrometer with a diffraction grating and a resolution of $0.1\ cm^{-1}$. This example is taken from [1.16] and vividly demonstrates the advantages of tunable lasers in high-resolution spectroscopy.

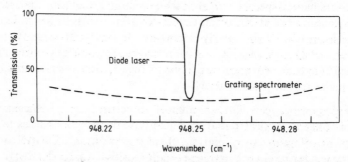

Figure 1.2   Experimental comparison between a tunable $Pb_{0.88}Sn_{0.12}Te$ diode laser scan and diffraction grating scan of the sP(1; 0) line of the v_2 band of NH_3 at room temperature (from reference [1.16]).

The high monochromaticity has made it possible to elaborate quite new methods of Doppler-free nonlinear spectroscopy of atoms and molecules in the gas phase. This trend of laser spectroscopy is covered in some special monographs [1.17–1.18] (also, see the corresponding sections in [1.10–1.12]). Here it should be emphasised that the narrowing of Doppler-broadened spectral lines by the methods of nonlinear laser spectroscopy is of considerable interest for analytics since in this way one can considerably increase the information capacity of the optical spectrum of the sample under study.

1.2.2 Temporal resolution

The temporal resolution in pre-laser kinetic spectroscopy was, at best, about 10^{-8} s due to the use of pulsed flash-lamps and a Kerr-cell shutter. The creation in the late 1960s of ultrashort pulse lasers which employ the effect of synchronisation of a great number of modes of wide spectral range laser [1.19] has culminated in a rapid progress in the methods of laser picosecond resolution spectroscopy. In recent years further progress has been made in generating subpicosecond pulses (up to 3×10^{-14} s = 30 fs [1.20]). Ultrashort laser pulses with a tunable frequency enable selective excitation of a considerable fraction of molecules to a definite quantum state and picosecond resolution of primary photophysical and photochemical processes with excited molecules.

The methods of picosecond laser spectroscopy are of great importance for the analysis of short-lived particles (radicals, complexes, etc).

1.2.3 Sensitivity

Although all the spectral information is contained in one atom or one molecule, real measurements of spectra are possible only with a great number of particles in the sample (approximately from 10^{10} to 10^{20} for different methods and objects) because the sensitivity of all the spectral methods is very limited. So the problem of increasing the sensitivity of spectroscopic methods has always been very important for analytical applications.

The development of tunable lasers and their use in spectroscopy has materially increased the sensitivity of all the known methods of spectroscopy (transmission, absorption, fluorescence and other methods) both for atoms and for molecules. At the same time quite new methods have been elaborated on the basis of lasers; for example, multistep photoionisation spectroscopy of atoms and molecules [1.21], intracavity absorption [1.22] and coherent anti-Stokes Raman scattering [1.23]. The sensitivity of these new methods is many orders higher than that of the known methods of spectroscopy even when laser radiation is used. For example, the method of photoionisation laser spectroscopy based on resonance stepwise photoionisation of atoms and molecules enables detection of single atoms and molecules [1.24].

Thus, using the laser as a basis, it has become possible to develop some methods of ultrasensitive spectroscopy of atoms and molecules which permit operation with a much smaller number of particles (from 1 to 10^{10}) in the sample than in any known method of conventional spectroscopy. This impressive progress in the sensitivity of optical spectroscopy with the use of lasers offers excellent possibilities for analytics.

1.2.4   Selectivity

In analysing a real sample selectivity is also of critical importance; that is the ability to detect the presence of atoms or molecules of definite sorts in their mixtures. The potentialities of optical spectroscopy in this respect have always been limited. For the most part, this holds true for the analysis of condensed media characterised by broad absorption lines which have low information capacity. The spectral analysis of molecules is then often performed after preliminary separation of these molecules by, for example, chromatography.

The use of laser radiation has, firstly, made it possible in some cases to simplify optical spectra of molecules, for example the fluorescence spectra of molecules in a low-temperature matrix [1.25], and has increased the selectivity of analysis. Secondly, the spectra of multistep laser excitation of atoms and molecules are characterised by a far higher selectivity than ordinary absorption spectra. Finally, higher sensitive laser methods can be easily combined with such well known methods of analysis as chromatography, mass spectrometry, etc. Therefore, the high selectivity of the methods of laser spectroscopy in combination with a high sensitivity is extremely valuable for analytical purposes.

1.2.5   Remote analysis

The spectra of light from remote sources have long been used to study the composition of matter. In this respect the spectral analysis is remote but only for radiative objects, such as flames, discharges or stars. Laser light has permitted this unique characteristic of optical spectroscopy to be applied to nonradiating objects. Indeed, a directed laser beam can induce fluorescence or scattering in a remote region, in the upper atmosphere for example [1.26]. So it is possible to analyse the atomic–molecular composition of a region irradiated at a distance from the observer. This principle is applied to a great number of methods of remote laser spectroscopy which are now under active elaboration for control over the environment (see monographs [1.27 and 1.28]).

1.2.6   Locality

The spatial coherence of laser radiation enables it to be focused onto an area with minimum dimensions of the order of light wavelengths, that is a fraction of a micrometre for the visible and uv ranges. This property is well known for 'point' nonlaser light sources but the very small radiation power of ordinary sources which can be focused onto an area of about λ^2 is not often sufficient for measurements. In the case of a laser source one can obtain, by means of focusing, giant powers which provide fast heating and evaporation of the irradiated local region. This laser property has formed the basis for microspectral emission analysis of atoms [1.29] and local mass spectral analysis of molecules [1.30].

This short analysis of the progress in optical spectroscopy achieved by the use of laser light apparently points to the high potential for analytics of the methods of laser spectroscopy. Of particular importance is a radical increase in sensitivity of the spectroscopic methods which form the basis for analytical applications.

In the next section we will consider briefly some different approaches to the application of lasers for analytical purposes based on the above-listed characteristics of laser light and optical spectroscopy.

1.3 Ways of using lasers in analytics

Analysis of matter is a complex process consisting of the following successive stages: preparation of a sample in the state under analysis; separation of the sample into different fractions if it is necessary for further detection; and measurement or detection of the content of definite particle components. The applications of laser light in analytics thus vary greatly since they can be, in principle, realised at each of the listed stages separately or together. In this connection we shall try to classify the fields of application of lasers in analytics following the sequence of the process illustrated in a simplified way in figure 1.3.

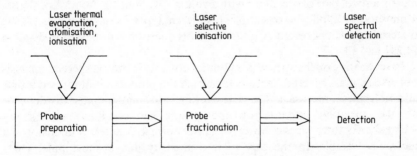

Figure 1.3 The application of lasers to the analytical process.

1.3.1 Preparation of the sample
In many methods of analysis of condensed media the substance to be analysed must first be converted to the gaseous or plasma state. For the method of atomic absorption spectroscopy, for example, the substance should be atomised, for the method of emission spectral analysis the excited atoms should be available, and for mass spectral analysis it is necessary to produce atomic or molecular ions. In analytics there are various methods of atomisation and ionisation of matter, for example, by means of contact heating, a discharge, an electron beam, etc. Before the laser was invented, optical radiation had never been used for these purposes because of its low intensity.

High-intensity laser radiation can easily heat the exposed substance to any desirable value over a range from fractions of a degree to 10^4–10^5 K depending on the laser radiation intensity; the heating can be continuous or pulsed.

The heating of substances by laser radiation has certain advantages over other methods. First, laser heating is contactless and in this respect can be compared with heating by an electron beam. But unlike the electron beam method, the laser method does not call for vacuum and enjoys the merit of simple optical deflection and propagation methods for the beam.

Secondly, in contrast to the method of heating by electric current that is commonly used, laser radiation can easily heat *nonconducting* materials. It is only necessary that the exposed substance should have some absorption at the laser wavelength. This applies to the case of heating at moderate laser powers when it is necessary to absorb a certain fraction of light since the absorption is linear. At a high radiation power ($> 10^{10}$ W cm^{-2}) any, even translucent, material is subjected to heating and destruction because of inevitable multiphoton light absorption.

Finally, a laser beam can be focused onto a small area and bring about local atomisation and ionisation of matter. The use of a focused laser pulse in microsample spectral analysis was demonstrated soon after the advent of the laser [1.31] and has found wide application since then. With the laser pulse energy and intensity properly chosen and controlled, it is possible to evaporate locally a small part of substance with a volume of about 10^{-8} cm^3. Local laser atomisation is applicable to samples of different types: metals, minerals, plastic materials, powders, ceramics, glass, etc. This question is discussed in detail in [1.29] and [1.32].

Laser heating opens up new possibilities not only in optical spectral analysis but also in mass spectral analysis for which the production of ionised atoms and molecules is necessary. The state of evaporated matter materially depends on the radiation intensity on the sample surface. Figure 1.4 shows schematically three types of action of a laser pulse with different intensity on a molecular sample. In the case of low intensity and a short pulse (10^4–10^6 W cm^{-2}) slight pulsed heating of the substance and thermal *desorption* of a negligible fraction of the molecules adsorbed on the surface, without their dissociation or fragmentation, take place. As the radiation intensity increases, *intense evaporation* of the substance occurs. In the dense heated vapour formed chemical reactions followed by chemi-ionisation take place. The resulting molecular and fragmentary ions can enter the mass spectrometer. These conditions are used to advantage in the mass spectral microsample analysis of organic and biological materials [1.30]. Finally, at a very high intensity ($> 10^9$ W cm^{-2}), stronger heating of the substance and the dense vapour comes about with the formation of a *laser plasma*. In such a plasma full atomisation of matter and a considerable degree of ionisation, including the formation of multicharged ions, can be realised. Such laser plasma formation is also used in mass spectral analysis [1.33].

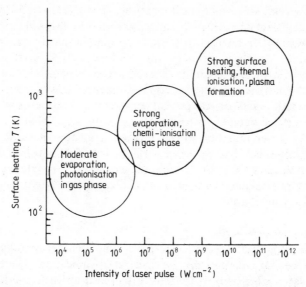

Figure 1.4 Various regimes of thermal action on the surface of a molecular sample by a laser pulse of differing intensity.

Thus, using laser radiation with different intensities one can convert the substance under analysis to a gas phase in different states: molecules, atoms and ions. When combined with the locality and short duration of pulse–substance interaction, this opens up new and interesting possibilities for analytics.

1.3.2 Separation of sample components

At present laser radiation is not often used for separating the components of the mixture under analysis since it cannot compete with well developed universal methods of separation, such as chromatography, mass separation, etc. Nevertheless, a highly effective and subtle application based on the ability of laser radiation to realise resonance-selective ionisation of atoms and molecules, as they absorb several photons, seems to be possible here. In this case the problem of mass separation either no longer arises or becomes much less troublesome since laser radiation can ionise only definite components (atoms, isotopes or molecules) in the mixture. In essence, the role of the mass spectrometer in this case can be reduced only to identifying the ionised fragments for the given component in the mixture. Such an approach to analysis of molecular mixtures was proposed rather a long time ago [1.21, 1.24] but has only recently found wide application. A number of reviews [1.34–1.36] are concerned with laser photoionisation mass spectroscopy. The latest achievements in this field are also considered in Chapter 7 of this monograph.

There are several ideas for optical mass-separation of atoms based on the use of resonance light pressure [1.37] (see also review [1.38]) which are still far from being practically realised. It is not improbable that this method will prove useful for analysis of very rare isotopes [1.39] particularly in combination with various methods of laser isotope separation [1.40].

Laser light can simultaneously act on a substance at two stages, both for sample preparation and for component separation. For an example we can point to the method of laser photoionisation spectroscopy of molecules on a surface [1.41]. Laser radiation in this method irradiates the surface of the substance. Slight pulsed heating of the surface results in desorption of a small number of molecules (10^{-6}–10^{-10} monolayers) which are later resonantly ionised either by the same laser pulse or an additional one. This method is also considered later in Chapter 7.

1.3.3 Spectral detection

The methods of optical spectroscopy have long been successfully used in analytics at the stage of detection for spectral identification. Therefore, the methods of laser spectroscopy are most often used at this stage for highly sensitive and highly selective spectral detection. The application of an expensive and complex tunable laser is, of course, justified in cases when the sensitivity and resolution (or selectivity) of the methods of conventional spectroscopy turn out to be insufficient. As an analyst is always interested in measuring a certain component in minimum concentration, with the influence of the rest of the components at a minimum, a laser is always of great use.

The unique potentialities of the methods of laser spectroscopy for detection have been already listed briefly in §1.2 in considering the main problems of optical spectroscopy (spectral and temporal resolution, sensitivity and selectivity, remote analysis and locality) which can be successfully solved using laser light. Here we want just to emphasise that the high sensitivity, spectral resolution, selectivity and other characteristics of the methods of laser spectroscopy are based on a great difference in spectral brightness between thermal and laser light sources. Here are several numerical examples. The effective brightness temperature of a low-power laser with an output of only 1 mW and spectral width 1 MHz equals 10^{14} K. For another numerical example we may mention a typical injection diode laser with a power of just 10^{-4}–10^{-5} W has a spectral power density of 10^{-9} W Hz^{-1} which corresponds to the radiation of a black body with a temperature of 2000 K and a diameter of several metres. This high spectral power density together with radiation spatial coherence provide high spectral brightness. For the above typical cw diode laser the spectral brightness of radiation may be as high as 10^{-5} W cm^{-2} sr^{-1} Hz^{-1}, that is many orders higher than the spectral brightness of thermal IR radiation sources. For example, the spectral radiation brightness of a black body at 2000 K at 10 μm is only 2×10^{-14} W cm^{-2} sr^{-1} Hz^{-1}.

All methods of laser spectroscopy can be divided into two classes in accordance with the type of interaction of the light with the analysed component or substance. The first class includes all the methods of *linear* laser spectroscopy based on single-quantum linear interaction of laser light with the substance, and the second class includes the methods of *nonlinear* laser spectroscopy based on nonlinear single-quantum or multiquantum interaction of laser light with the substance. Both classes consist of many specific and various methods which are of great interest for analytical applications. They are reviewed briefly below.

1.4 Methods of linear laser spectroscopy

More than ten different methods of linear laser spectroscopy can be used for analytical purposes. It is advisable to give a simple classification of them so that one is able to gain an easy understanding of them and, in addition, to mark out the methods to be described in the subsequent chapters and already described in other books.

1.4.1 *Classification of linear methods*

There are three methods that are most widely used in studying the spectrum of a substance with a tunable laser, that is without conventional spectral instruments. These methods are shown schematically in figure 1.5.

In the simplest case the dependence of the radiation intensity passed through the sample at a tunable laser wavelength is measured (figure 1.5(*a*)).

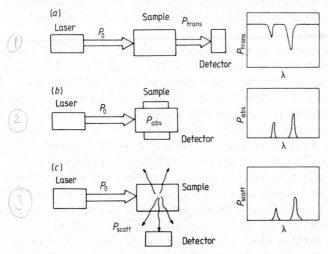

Figure 1.5   Principal methods of linear laser spectroscopy using a tunable laser: (*a*) absorption method; (*b*) opto-acoustic method; (*c*) fluorescence method.

This method for measuring *transmission* is particularly useful in the case of lines which are sufficiently intense when the absorbed intensity is a measurable fraction of the incident one. If the radiation absorption per single pass of the sample is very small, it is quite possible, of course, to use multiple passage of laser radiation through the sample. The number of possible passes is usually small due to inevitable losses in the optical windows of the cell and other elements.

A radical increase of the number of passes can be achieved by using the method of *intracavity absorption* [1.22] when the sample is placed inside the cavity of a laser with a wide amplification line. In this case the amplification of the laser medium compensates for the nonresonant light losses in the cell which allows a large increase ($\gtrsim 10^6$) in the effective number of laser light passes in the sample and thus, respectively, an increase in sensitivity of the absorption method.

When the density of absorbing particles is small, but it is still possible to modulate the position of their absorption line frequency, a small fraction of modulated absorption can be discriminated against the background. This approach to increasing the sensitivity of the transmission method can be applied with advantage, for example, in *laser spectroscopy of magnetic resonance* [1.42].

All the above-listed methods of absorption spectroscopy based on measuring the transmission spectrum of samples can be generally called *absorption–transmission methods*. Their common feature is universality, or nonsensitivity to the subsequent state of excited particles. Figure 1.6 presents a general classification of laser spectroscopy methods from the standpoint of the various pathways of excited state relaxation. This classification is especially convenient for systematising the numerous methods of laser spectroscopy.

If the absorbed energy is a small fraction relative to the incident energy it is often more convenient to record directly the energy absorbed in the sample using an effect arising in the sample due to absorbed radiation (figure 1.5(*b*)). For example, as the sample is irradiated, the radiation absorption is followed by variations in temperature and pressure of the medium. As a result, acoustic vibrations are generated. This effect forms the basis for *opto-acoustic spectroscopy* which received much attention even before the advent of the laser on the basis of incoherent radiation sources (see monograph [1.43]). The application of tunable lasers has made this method very simple and sensitive for spectroscopy of molecules both in a gas medium and in a condensed state (see reviews [1.44, 1.45] and monograph [1.46]). In some cases the effect of increase in temperature can be recorded from the variation in the refractive index of the medium which, among other things, gives rise to the lens effect. This 'thermal lens' method, termed the *opto-refractometric* method, has been the subject of several reviews [1.47, 1.48]. Sometimes it is possible to measure the energy absorbed by the molecules from the laser beam during their

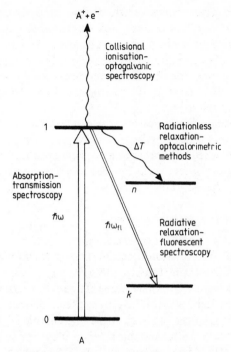

Figure 1.6 Classification of laser spectroscopy methods from the point of view of the various pathways of excited state relaxation.

interaction with the thermal detector (bolometer or pyroelectric [1.49, 1.50]). This method of absorption spectroscopy is called *optothermal*.

This entire group of spectroscopic methods based on direct measurement of absorbed light energy by means of its degradation to heat can be generally termed *optocalorimetric* methods (figure 1.6).

Finally, during laser excitation of atoms or molecules in a low-pressure gas discharge or in a flame the conduction of the medium may change. This is called the *optogalvanic* effect [1.51] and may be successfully applied in spectroscopy for analytical purposes [1.52–1.54]. For this method it is necessary that the excitation of a particle should create the charged particles or cause their number to be changed (figure 1.6).

In many cases the resonant absorption of radiation by atoms and molecules is followed by its subsequent re-emission by excited particles (i.e. fluorescence occurs). The *fluorescence* method of laser spectroscopy is based on measuring fluorescence intensity as a function of the exciting laser wavelength (figures 1.5(*c*) and 1.6). This method has long and universally been used in laser methods of atomic and molecular spectroscopy, especially for analytical purposes. Atomic fluorescence spectroscopy, for example, is covered in [1.55, 1.56] and molecular fluorescence spectroscopy is reviewed in [1.57, 1.58].

Raman scattering can also be considered as a radiative method. Here the exciting radiation is off resonance with the quantum transition and the scattered photons have a lower energy relative to the incident ones due to transfer of the energy of an internal degree of freedom of the molecule. Raman scattering spectroscopy does not call for continuously tunable radiation and the successful use of lasers in this method is based on the high spectral brightness of laser radiation.

All these methods of laser spectroscopy can be called *linear* methods. Here linearity means that these methods, unlike nonlinear methods, are most effective when the laser radiation intensity is much lower than the intensity which brings about such nonlinear effects in substances as absorption saturation and multiquantum transitions.

1.4.2   *Absorption–transmission methods*

(a) *Measurement of transmission spectrum*
The advantages of transmission spectrum measurement using tunable lasers are especially evident when studying the spectrum of vibrational–rotational spectral lines of molecules in the IR range. The first experiments were carried out at the Lincoln Laboratory, USA, using semiconductor diode CW lasers [1.16]. Figure 1.2 shows the absorption line of ammonia sP(1;0) of the band v_2 near 10.6 μm measured with a tunable laser diode on the semiconductor alloy $Pb_{0.88}Sn_{0.12}Te$ [1.16]. The measurements were taken using a gas cell length $l = 10$ cm, at room temperature and at a pressure of 0.05 Torr. The linewidth of the diode laser was less than 0.1 MHz, i.e. the laser spectrometer resolution was better than 3.3×10^{-6} cm^{-1}. At a low pressure the absorption linewidth coincides with the Doppler broadening for NH_3, $\Delta v_{Dopp} = 85$ MHz. When a high-quality spectrometer has a diffraction grating with a resolution of 0.1 cm^{-1}, the absorption signal has the same amplitude as the laser spectrometer but at a pressure of 5 Torr and with $l = 200$ cm. This means that measuring the absorption spectrum with a good spectrometer with a diffraction grating requires an amount of substance 2000 times greater. Besides, the spectral resolution of even a very good spectrometer is much larger than the Doppler width of gas absorption lines at low pressures (less than several torrs).

Infrared tunable lasers make it possible to measure directly the shape and structure of molecular absorption lines. The measurements and identification of the complex structure of the vibrational–rotational spectrum of the SF_6 and UF_6 molecules carried out at the Los Alamos Laboratory are an excellent example of this kind of work [1.59].

Direct measurement of the spectral line Doppler width determined by the expression

$$\frac{\Delta v_{Dopp}}{v_0} = \left(\frac{8kT}{Mc^2}\right)^{1/2} \tag{1.1}$$

enables the mass of absorbing molecule M to be found if it is unknown. This gives additional information useful for identifying molecular impurities in low-pressure gases using methods of high-resolution laser spectroscopy.

To compare this method with other methods of laser spectroscopy it is useful to estimate the ultimate theoretical sensitivity of transmission spectrum measurement, i.e. a minimum detectable number of particles on the laser beam path. The resonance absorption factor at the transition 1–2 (figure 1.7(a)) per unit length is equal to

$$\kappa = \sigma(N_1 - N_2) = \sigma\frac{N_1^0 - N_2^0}{1 + (I/I_{sat})} = \frac{\kappa_0}{1 + (I/I_{sat})} \tag{1.2}$$

where σ is the absorption cross section at the transition between levels 1 and 2 and N_1 and N_2 denote the number of particles per unit volume at the lower (1) and upper (2) levels of transition. These numbers in the general case may differ from the populations of these levels N_1^0 and N_2^0 in the absence of laser radiation due to particle redistribution under the action of radiation with the intensity I, I_{sat} is the intensity that induces absorption saturation, i.e. decreases the absorption by half (figure 1.7(c)). The saturation intensity equals

$$I = \frac{\hbar\omega}{2\sigma\tau} \tag{1.3}$$

where τ is the relaxation time of excited particles to the ground state. In equation (1.2) it is assumed that the laser radiation interacts with all the

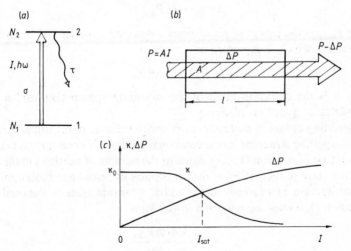

Figure 1.7 Estimation of the sensitivity of the absorption–transmissin method: (a) scheme of transitions; (b) geometry of measurement; (c) influence of absorption saturation.

molecules responsible for the absorption line at the transition 1–2 with the absorption factor per unit length for a weak field being κ_0.

As a laser beam passes through a gas cell length l, its power $P = IA$ will be reduced by the value

$$\Delta P = \kappa I A l \qquad \kappa l \ll 1 \tag{1.4}$$

where A is the beam cross section (figure 1.7(b)). From expressions (1.2)–(1.4) it can be seen that the detection of absorbing particles is optimal when the radiation intensity $I = I_{sat}$. When it is necessary to increase the absolute absorbed power ΔP, the radiation intensity I must be increased. But it is not advisable to increase I over the value I_{sat} since in this case the absorbing transition becomes bleached and the value of ΔP tends to be constant (figure 1.7(c)). Then from equations (1.2)–(1.4) it follows that the minimum number of detected particles will be [1.60]

$$n_{min} = (Al)(N_1^0 - N_2^0)_{min} = \frac{4\tau \, \Delta P_{min}}{\hbar\omega} \tag{1.5}$$

where ΔP_{min} is the minimum detectable variation of laser beam power the value of which depends on the method of detection and the efficiency of detectors in use. This point is considered in more detail in Chapter 4. Here we shall limit ourselves with simple estimations of the optimal case of heterodyne detection.

The sensitivity threshold of heterodyne detection in the ideal case, when the noise depends only on quantum fluctuations, is equal to [1.61]

$$\Delta P_{min} = 4(PP_{noise})^{1/2} \tag{1.6}$$

where P_{noise} is the power of photoelectric detector noise caused by quantum fluctuations which is expressed as

$$P_{noise} = \hbar\omega B / \eta \tag{1.7}$$

where B is the detection band of the recording system (Hz) and η is the photodetector quantum efficiency.

If detecting atoms or molecules are distributed over many initial quantum levels (hyperfine structure levels, rotational levels) it is necessary to take into account that estimation (1.5) is related to the number of particles in the initial quantum state participating in the absorption of resonance radiation. If the relative fraction of particles in the initial quantum state is denoted by q, estimation (1.5) must be rewritten in the form

$$n_{min} = \frac{1}{q} \frac{4\tau \, \Delta P_{min}}{\hbar\omega}. \tag{1.8}$$

For atoms and simple molecules q usually equals 0.1–1.0 and for polyatomic molecules q equals 0.001–0.1.

Table 1.1 Estimation of detection thresholds for atoms or molecules by laser absorption spectroscopy.

Parameters of absorption transition and laser†	Atoms	Molecules
λ	5000 Å	500 000 Å
σ	4×10^{-12} cm^2	4×10^{-16} cm^2
τ	10^{-8} s	2×10^{-4} s
A_{21}	10^8 s^{-1}	10 s^{-1}
B	1 Hz	1 Hz
η	0.1	0.2
P_{noise}	4×10^{-18} W	3×10^{-19} W
ΔP_{min} (for $P = 10^{-2}$ W)	8×10^{-10} W	2×10^{-10} W
I_{sat}	5 W cm^{-2}	0.5 W cm^{-2}
q	1	0.01
n_{min}	10^2	3×10^8

† Notations are given in the text.

As an illustration, table 1.1 presents numerical estimations for atomic and molecular absorption lines in the visible and IR ranges with the corresponding parameters of the photodetector. The laser beam power is taken on a real level, $P = 10^{-2}$ W, so that even at a small beam cross section ($A = 10^{-2}$ cm^2) the radiation intensity I is less than the absorption saturation power I_{sat}. In the case of allowed atomic transition the detection threshold equals 10^2 atoms in the irradiated volume with a detection time of about 1 s. In the case of allowed vibrational–rotational transitions, when an IR photodetector is used, all other things being equal, the detection threshold is equal to only 3×10^8 molecules. This is due to the slower relaxation of excited molecules and hence to a much smaller number of repeated radiation absorption cycles throughout the observation time as well as a far smaller ($q = 0.01$) relative population of the vibrational–rotational level participating in absorption.

(b) *Method of intracavity absorption*

Highly sensitive detection of weak absorption lines is possible using a multimode laser with a wide amplification line with intracavity weak absorption inside the laser amplification band [1.22]. The principle of this method is illustrated in figure 1.8 (see review [1.62]). Let a cell with weak absorption lines be placed outside the cavity of a multimode laser with a wide spectral band. Then, using a conventional spectral instrument with good resolution, one is able to detect a decrease of intensity at absorption frequencies provided that, for example, no less than 1% of intensity is absorbed. But if this cell is placed inside the cavity, the weak absorption at

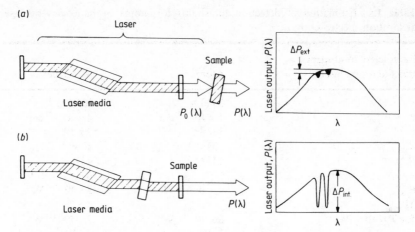

Figure 1.8   A comparison of the methods for detecting weak absorption lines when the absorption cell is (*a*) outside the cavity and (*b*) inside the cavity.

some frequencies leads to redistribution of radiation intensity in different axial modes. The modes falling within weak absorption lines are suppressed, i.e. selective mode quenching takes place. This gives rise to sharp holes in the laser radiation spectrum which can more easily be detected with a spectral instrument. This method is apparently realisable in lasers with wide amplification lines where many axial modes can coexist. In pulsed lasers a great number of axial modes develop almost independently. In cw lasers many axial modes coexist due either to inhomogeneous broadening of amplification lines or to spatially inhomogeneous interaction of modes with the amplifying medium (or to the simultaneous action of both effects). However, the ultimate sensitivity of the method of intracavity absorption is essentially different in these two types of laser operation.

To detect a weak absorption line using a pulsed laser it is necessary that throughout the laser oscillation time t_{osc} the light should make a sufficient number of passes through the sample inside the cavity. This multiple intensity attenuation of the modes interacting with a weak absorption line can be considered the criterion of sufficiency of the number of passes. Thus, the theoretically detectable ultimate absorption factor will be proportional to

$$\kappa_{int} = \frac{1}{ct_{osc}} \frac{L}{l} \qquad (1.9)$$

where c is the light velocity, l is the absorbing cell length and L is the cavity length. In the external cell such intensity attenuation can be attained with the absorption factor κ_{ext} proportional to $1/l$. Thus the increase of sensitivity obtained by using the method of intracavity absorption in comparison with the method of one-pass transmission of an external cell equals

$$S = \frac{\kappa_{ext}}{\kappa_{int}} = \frac{ct_{osc}}{L}.$$
(1.10)

Using a neodymium–glass laser operating in free regime ($t_{osc} \simeq 10^{-3}$ s) it is possible to detect extremely weak absorption lines of molecules (NH_3, HN_3, CO_2, C_2H_2, HCN, etc) at compound and overtone vibrational–rotational transitions in the region of 9380–480 cm^{-1} with absorption factors of about 10^{-7} cm^{-1}, i.e. to obtain sensitivity increases of $S = 10^2$–10^3 [1.63].

In continuous oscillation the sensitivity increase factor S, i.e. the increase of hole depth in the spectrum, for the case of an intracavity cell rises considerably due to an increased number of passes in comparison with the case of an external cell. Its maximum value however is limited by a number of effects. First, it is the contribution of spontaneous radiation to the cavity modes which makes it impossible to suppress completely the modes in the case of very slight absorption. Secondly, there may be 'a failure' in continuous oscillation at fixed modes caused by technical perturbations of the laser cavity. In the latter case the ultimate sensitivity is again determined by equation (1.9) in which t_{osc} represents the average oscillation time at fixed modes.

Experiments confirm that the intracavity method is characterised by high sensitivity in the continuous oscillation regime. For example, in the first experiments with a cw dye laser the sensitivity of detection of the absorption lines of I_2 was increased by 10^5 times [1.64]. Figure 1.9 shows the spectrum of weak absorption lines of air in the region 5800–6200 Å produced under laboratory conditions with a cw dye laser [1.65]. For comparison, it also shows the absorption spectrum of the atmosphere above sea level in the same region obtained in [1.66] on an optical path of 16 km. This gives conclusive evidence of the efficiency of the method of the intracavity absorbing cell.

At first sight it seems that the method of intracavity absorption can hardly be used for quantitative analysis. But this is not so. If the laser radiation spectrum is recorded at two subsequent instants of time t and $t + \Delta\tau_e$, which differ in hole depth in the spectrum by l times, it is possible to determine quite accurately the value of resonant absorption

$$\kappa_{res} = \frac{1}{c\,\Delta\tau_e}.$$
(1.11)

It should be noted that the method of intracavity absorption is very promising for recording weak absorption (and amplification) lines of short-lived (with a lifetime of 10^{-6}–10^{-3} s) products of chemical reactions, radicals and transient molecules [1.67].

Considerable progress has been achieved recently in developing lasers on colour centres in crystals with wide amplification lines in the near IR region (up to 3 μm). This enables the method of intracavity absorption to be applied to

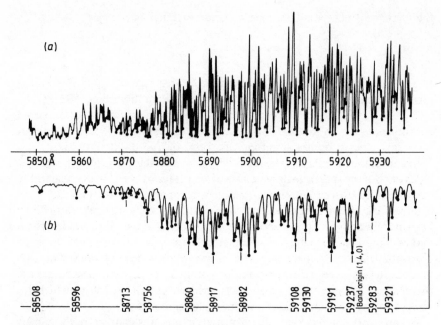

Figure 1.9 Spectrum of the very weak absorption lines of air in the red spectral region measured by (a) intracavity absorption method in laboratory conditions (from reference [1.65]) and (b) transmission measurements in sea air over a path of about 16 km (from reference [1.66]).

the region of vibrational–rotational molecular transitions, a very important region for analytical purposes [1.68].

1.4.3   *Absorption–calorimetric methods*

The principal limitation of sensitivity in the transmissiin method is connected with the unavoidable fluctuations in the photodetector, $P_{noise} \propto P^{1/2}$, caused by the shot noise of the photon beam with power P. Therefore it seems quite natural to apply the methods of measuring the ΔP power absorbed in the sample not against a background of a much higher power P but directly in the sample against 'a zero background'. In this case a higher sensitivity is expected if a sufficiently effective method of measuring is applied. This is, indeed, a way of increasing sensitivity which has given rise to new methods of detection of single atoms and molecules. This has not been achieved, however, with the use of the linear absorption–calorimetric methods of spectroscopy considered below. The sensitivity of these methods, although high, is nevertheless not sufficient for this limit (the detection of single atoms and molecules).

(a)   *The opto-acoustic method*

This method of linear spectroscopy is based on the so-called opto-acoustic effect discovered by Bell, Tyndal and Roentgen at the end of the last century (see its history in [1.43–1.46]). The effect manifests itself in gas pressure pulsations in a closed volume as an infrared radiation flux modulated at the sound frequency is absorbed. When the laser radiation frequency is coincident with the spectral line frequency of gas absorption, for example, the impurity molecules to be detected in the gas, the absorbed power increases resonantly (figure 1.10). The value of absorbed power ΔP is determined by equation (1.4), and its dependence on radiation power is given by relations (1.2) and (1.3).

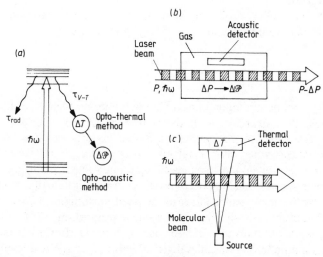

Figure 1.10 Opto-calorimetric methods of laser spectroscopy: (*a*) pathway of excitation energy conversion to heat and change of pressure; (*b*) geometry of opto-acoustic measurement in a gas; (*c*) geometry of optothermic measurement with a molecular beam.

The return of excited molecules to the initial state occurs in two channels (figure 1.10(*a*): through nonradiative vibrational–rotational relaxation with the time τ_{V-T} which is followed by gas heating; and radiative relaxation with the time τ_{rad}. For most of the molecules at pressures such that the heterogenous relaxation on the walls is inessential, the relation $\tau_{V-T} \ll \tau_{\text{rad}}$ holds true and hence the absorbed energy converts mainly to heat with an increase in the temperature ΔT and causes the gas pressure $\Delta \mathscr{P}$ to change. The gas pressure variation is recorded with a highly sensitive microphone (figure 1.10(*b*)). The opto-acoustic cell goes back to the initial state after the time τ_{rel} that is usually determined by the gas cooling time on the cell's walls. To obtain pressure variations with a maximum amplitude it is necessary that the modulation frequency f should not apparently exceed the inverse relaxation time τ_{rel}. When the transverse dimension of the gas cell is of the order of several centimetres, the characteristic time of cooling due to wall diffusion of molecules equals $\tau_{rel} \simeq 10^{-3}$ s. Therefore the radiation modulation frequency usually ranges from 10^2 to 10^3 Hz.

The sensitivity of a condenser microphone, for example, allows reliable detection of absorbed power ΔP_{min} of the order of 10^{-8}–10^{-9} W with the opto-acoustic cell length $l = 10$ cm and the recording system band $B = 1$ Hz [1.69]. The physical sensitivity threshold is determined by the Brownian noise of the microphone diaphragm due to the thermal motion of gas particles. In a well designed microphone ΔP_{min} exceeds this theoretical ultimate value by about one or two orders.

The least detectable absorption factor κ_{min} with the laser radiation power $P \ll AI_{sat}$, i.e. far from saturation, according to (1.4) equals

$$\kappa_{min} = \frac{1}{l} \frac{\Delta P_{min}}{P} \qquad (1.12)$$

where $\Delta P_{min}/l$ is the minimum detectable power per unit length of the opto-acoustic cell. For example, with the opto-acoustic cell length $l = 10$ cm, sensitivity $\Delta P = 10^{-9}$ W and quite accessible laser power $P = 1$ W, it is possible to detect rather weak absorption lines with $\kappa_{min} = 10^{-10}$ cm^{-1}. The laser radiation power should be increased up to the value $P_{opt} = AI_{sat}$ when the absorbing transition bleaching becomes observable. With further increase of power the useful signal can be increased twice at most (figure 1.7(b)) whereas the background spurious signal goes on rising in proportion to P. So the value $P_{opt} = AI_{sat}$ is the optimal value of laser radiation power. In this optimal case the minimum number of molecules detected within the laser beam volume in the cell will be

$$n_{min} = N_0^{min} Al = \frac{4\Delta P_{min}\tau}{\hbar\omega}. \qquad (1.13)$$

This value, of course, coincides with the expression given above (1.5) for the transmission method and differs from the latter only by the way of recording absorbed energy and hence by the numerical value of ΔP_{min}.

The value of n_{min} does not depend on the intensity (or the cross section) of the absorbing transition. Physically this is explained by the choice of optimal radiation power $P_{opt} = AI_{sat}$. For weak transitions the power required to achieve limit (1.13) is high and for strong absorption lines it is respectively lower. Every molecule is able to absorb and transform to heat the power $\hbar\omega/\tau$ provided that the laser radiation provides its excitation after relaxation faster than the time τ. The level of minimum detectable heat release in the opto-acoustic cell determines at once the minimum number of detected molecules n_{min}. Let a strong vibrational transition with $\sigma = 10^{-16}$ cm^2 at $\lambda = 10$ μm have the vibrational relaxation time $\tau = 10^{-7}$ s due to the presence of additional buffer gas. In an opto-acoustic cell with $\Delta P_{min} = 10^{-9}$ W the value $n_{min} =$

2×10^4 can be obtained theoretically with the laser radiation intensity $I = P/A = I_{sat} = 10^3$ W cm^{-2}. This corresponds to the detection of absorption at the level $\kappa_{min} = 2 \times 10^{-11}$ cm^{-1}.

The high sensitivity of the opto-acoustic method using IR laser radiation makes it possible to detect very low concentrations of molecules. This method is particularly effective in the detection of definite molecular impurities in a relatively simple environment when there is no need for high selectivity, for example, in detecting molecular pollutants in the air. But the selectivity of the opto-acoustic method, like any absorption method, is insufficient for analysis of complex molecular mixtures by IR absorption. Therefore, for analytical purposes it is most effective to combine the opto-acoustic method with chromatographic methods which permit preliminary separation of the mixture components and then to use a highly sensitive opto-acoustic spectrometer to detect molecules by IR absorption. It is this combination that is discussed in Chapter 5 from the standpoint of analytical applications.

(b) *Optothermal spectroscopy*
This is another method in the family of laser calorimetric methods that has recently found application in gas and condensed medium spectroscopy. In low-pressure gases it is possible to carry out direct detection of excited molecules by the use of their thermal action on the surface of a corresponding sensitive element (bolometer, pyroelectric, etc), as shown in figure 1.10(c). The first experiments on IR laser optothermal spectroscopy were performed with a molecular beam and a bolometer [1.49], using a low-pressure gas and a pyroelectric detector [1.50]. The ultimate sensitivity of this method is limited by the thermal noise of the detector and theoretically can be as high as 10^6–10^7 mol cm^{-3}. It is most advisable to apply this method at low gas pressures (fractions of a torr) when the opto-acoustic method does not work due to very slow relaxation of excitation energy to heat. As the pressure is increased, the homogeneous relaxation of excitation energy to heat begins to prevail. In principle it is also possible to measure gas temperature variations with a thermal detector. However, the fact that thermal detectors are imperfect compared with pressure (sound) detectors makes the optothermal method uncompetitive with the opto-acoustic one.

The optothermal method can be also applied to condensed media with a standard thermocouple used as a thermal detector. In this way it is possible to measure the absorption of transparent dielectrics at the level of 10^{-5} cm^{-1} [1.70]. The optothermal method is also effective for analysis of highly absorbing thin samples placed directly on a thermal detector. Finally, it is possible to detect the absorption in liquid media in a capillary tube from the variation of geometric dimensions of the liquid heated by laser radiation.

(c) *Optorefraction spectroscopy*
A variation of temperature in an absorbing medium changes its refractive

index in accordance with the distribution of beam intensity and absorption factor. Such a variation can be measured by different refraction methods and by the thermolens and interference methods.

In the thermolens method described in reviews [1.47, 1.48] the slight variations of refractive index are measured by recording the defocusing of an additional probe beam that passes through the excitation region. This method is most widely used in analysis of liquid substances. The sensitivity of this method for liquids exceeds even that of the opto-acoustic method. The main contribution to the refractive index gradient is made by the variation of medium density across the laser beam. Because of this gradient the probe beam wavefront varies and is subjected to defocusing which is easily recorded by a photodetector with a diaphragm. The theoretical sensitivity of the thermolens method is limited by a photodetector noise but in practice the sensitivity is restricted by intensity fluctuations of the laser beam.

The variation of refractive index can also be measured using the interference method [1.71] (see also review [1.72]). The sensitivity of this method is comparable with that of the thermolens method but it is more difficult to realise technically.

This short discussion of optocalorimetric methods (which are discussed in more detail in monograph [1.46]) is concluded with table 1.2 which presents the sensitivity of various calorimetric methods of laser spectroscopy in terms of absorption coefficient per unit length.

Table 1.2 Limit of sensitivity† of the calorimetric methods of laser spectroscopy in units of absorption coefficients per unit length (cm^{-1}). (From [1.46].)

Methods and effects used	Aggregate state of substance		
	Gas	Liquid	Solid state
Opto-acoustic			
Direct detection	10^{-9}–10^{-10}	10^{-7}–10^{-8}	10^{-5}–10^{-6}
Indirect detection		10^{-4}	10^{-4}–10^{-5}
Optothermal			
Direct detection of excited particles	10^{-5}–10^{-6}		
Heating of sample	10^{-5}–10^{-7}	10^{-5}–10^{-6}	10^{-6}–10^{-7}
Change of geometrical sizes		10^{-4}–10^{-5}	10^{-3}–10^{-4}
Change of refractive index			
Thermo-optical	10^{-6}–10^{-7}	10^{-7}	10^{-5}–10^{-6}
Interferometric	10^{-3}–10^{-4}	10^{-5}	10^{-4}–10^{-5}

† Values of sensitivity are given for the same power (or energy) of laser radiation—1 W (or 1 J).

1.4.4 *Optogalvanic spectroscopy*

The excitation of atomic or molecular electron levels by laser radiation can be detected from the variation of conductivity of the medium or the formation of charged particles (figure 1.6). There are several quite different approaches here. One of them is based on the so-called optogalvanic effect arising at optical excitation of a conducting gas (a discharge in a low-pressure gas, a flame) [1.51]. This effect consists of the conduction varying on account of photoexcitation of higher lying levels which can be more easily ionised in collisions with electrons (figure 1.11). As the excitation energy of electrons is increased, the temperature of the electrons also increases which, in its turn, causes the conductivity to increase. With the aid of tunable lasers this effect is now widely used for measuring the absorption spectrum at the transitions between excited levels, analytical applications included [1.52–1.54]. Figure 1.11(*b*) shows a typical scheme in which the optogalvanic analytical spectrometer is used to detect traces of elements in a flame. The advantages of the method are its simplicity, the absence of scattered light effect, and the possibility of applying the known technique of flame spectroscopy at a high sensitivity. For many elements the threshold sensitivity ranges from 10^{12} to 10^{14} cm^{-3}, in some cases it is better.

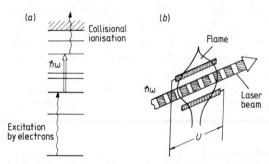

Figure 1.11   Optogalvanic laser spectrometry: (*a*) optical excitation of atoms increases the probability of collisional ionisation; (*b*) simplified scheme of optogalvanic measurement.

The sensitivity of optogalvanic spectroscopy is limited by background conductivity due to the presence of charged particles in a flame or a discharge, even without laser excitation. To overcome this limitation, one must apparently work with a nonconducting medium, i.e. get rid of the participation of electrons in the ionisation of excited particles. This can be easily achieved in *resonance photoionisation laser spectroscopy* based on the multistep excitation of atoms or molecules from the ground state only under the action of laser light. In optogalvanic spectroscopy this approach is based on nonlinear interaction of laser light with atoms and molecules, so it is considered below in §1.5.

At the end of this brief discussion of optogalvanic spectroscopy we should mention a high-sensitivity method for impurities in semiconductors based on the photoionisation of impurity levels by submillimetre radiation [1.73]. Such *photoelectric semiconductor spectroscopy* can be used successfully for direct nondestructive analysis of impurities in semiconductors, without any change of their state, from the variation of their conduction under irradiation.

1.4.5   *Fluorescence spectroscopy*

The fluorescence method is based on detecting the energy absorbed in the laser beam from the fluorescence of excited particles in the sample (figure 1.6). This method finds application in quantum transitions of atoms and molecules followed by a radiative decay of excited states. For the absorption–transmission method the future state of the particles excited by radiation is of no importance; for the optocalorimetric methods it is necessary that the excitation of most of the particles should relax to heat; the optogalvanic methods require that the excitation of particles should increase the number of charged particles in the irradiated volume; and for the fluorescence method it is desirable that the excitation relaxation channel should be considerably radiative (figure 1.6). In this respect the fluorescence method supplements the absorption–calorimetric methods, and these can both, in principle, be applied to any type of quantum transition for any atom or molecule.

The ultimate sensitivity of the fluorescence method on laser excitation can be obtained in the case when the particle being excited quickly reradiates a fluorescent photon, comes back to its initial state and, as a result, repeats this cycle many times during the observation time. This ideal situation, however, can be realised quite differently for atoms and molecules and for molecules in gas and condensed phases. Therefore it is advisable to consider the cases of atomic and molecular analysis separately.

(a)   *Detection of atoms*

The sensitivity of the fluorescence method with laser excitation of atoms can be estimated from the simplest considerations. Let the fluorescence radiation be collected by a recording system from the volume $V_{irr} = Al$ of the laser-irradiated cell (l is the length of the region, A is the cross sectional area) in a solid angle Ω and directed into a photodetector with quantum efficiency η. The quantum fluorescence yield of the particle will be $\eta_{fl} = \tau A_{21}$, where τ is the relaxation time of excited state 2 to state 1 through all channels and A_{21} is the radiative decay rate of the excited state (figure 1.12(a)). If P_{min} is the minimum fluorescence power detected by the photodetector, the least number of particles detected by the fluorescence method is given by the relation

$$\eta_{fl} \eta_{geom} V_{irr} \kappa(I) P = P_{min} \tag{1.14}$$

where $\eta_{geom} = \Omega/4\pi$ is the geometric efficiency factor of fluorescence collection and $\kappa(I)$ is the absorption coefficient per unit length determined from relation

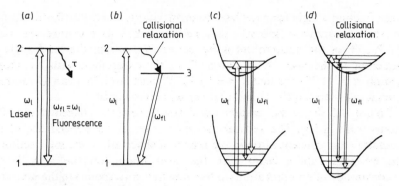

Figure 1.12 Schemes of excitation and fluorescence of atoms (*a*), (*b*) and molecules (*c*), (*d*) used for their detection by fluorescence power.

(1.2). In the optimal case when $I = I_{sat}$, the minimum number of detected particles is estimated as

$$n_{min} = N_0^{min} V_{irr} = \frac{4\tau P_{min}}{\hbar \omega_1 \eta_{fl} \eta_{geom}}. \tag{1.15}$$

Since there is no intense radiation P entering the photodetector, the minimal detectable power is determined by the photodetector noise power due to quantum fluctuations, P_{noise}, that is described by relation (1.7). As a result, n_{min} will be expressed as

$$n_{min} = 4 \frac{B}{A_{21}} \frac{\omega_{fl}}{\omega_1} (\eta \, \eta_{geom})^{-1} \tag{1.16}$$

where ω_{fl} and ω_1 are the frequencies of fluorescence and laser radiation, respectively, and B is the detection band (in Hz).

Thus, sensitivity of the fluorescence method is lower by about two or three orders than the theoretical limit ($\eta = 1$, $\eta_{geom} = 1$) because of the low quantum efficiency of the photodetector ($\eta \ll 1$) and geometrical factor ($\eta_{geom} \ll 1$). This reduction however can be fully compensated for by the factor B/A_{21}, i.e. by multiple absorption and re-emission of an atom throughout the observation time $1/B$. It is evident that if there are no other limitations to sensitivity, the fluorescence method, in principle, provides an ultimate sensitivity which corresponds to the detection of one particle in the quantum state in the volume under irradiation.

The estimations given for the ultimate sensitivity of laser fluorescence detection of atoms are confirmed by many experiments. The experiments may be divided into two types: $\omega_1 = \omega_{fl}$ (figure 1.12(*a*)) and $\omega_1 \neq \omega_{fl}$ (figure 1.12(*b*)). In the experiments with two-level cyclic excitation of atoms the fluorescence frequency is equal to the laser radiation frequency. In this case the real threshold sensitivity is determined by a low background noise because a very small fraction of light from the laser beam inevitably enters the photodetector. In such experiments special measures are taken to minimise the ratio $\beta = P_{back}/P$ of the background noise power P_{back} within the solid angle of detection to the laser beam power P. In the experiment with Na atoms described in [1.74] the factor $\beta = 3 \times 10^{-10}$, and in [1.75] with U atoms a special cell with $\beta = 10^{-13}$ and $\Omega = 1$ sr was constructed.

Figure 1.13 shows the temperature dependence of the concentration of U atoms from [1.75]. The vertical axis on the left is the dependence of the uranium atom concentration in the region of observation (the intersection of the uranium atomic beam with the laser beam), on the right is the concentration of uranium atoms in the oven that corresponds to the saturated vapour pressure. The lowest detectable concentration of uranium atoms is 3×10^3 atoms/cm^3 and the minimum number of atoms detected in the volume under irradiation is about ten.

For further increase of sensitivity, particularly when it is difficult to design a cell with a low background noise level, it is possible to use excitation and fluorescence schemes in which the fluorescence frequency differs from the excitation frequency (figure 1.12(b)). The principal shortcoming of the three-

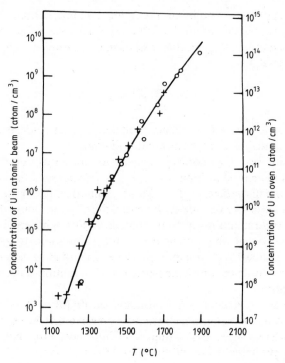

Figure 1.13 Concentration of uranium atoms from laser-excited fluorescence as a function of temperature (from reference [1.75]).

level cyclic interaction of an atom with a resonance light field is a significant increase in cycle duration. This shortcoming can be overcome, for example, in the collisional relaxation of atoms at the transition 3–2 by the presence of a buffer gas. Such an experiment was carried out in [1.76] with Na atoms where the sensitivity attained was as high as 10 atoms/cm^3 (or 0.2 atoms in the volume of detection).

The fluorescence method with laser excitation allows the detection of a single atom flying through the laser beam in real time without signal accumulation. This follows from the simplest estimations [1.77]. Let an atom in a vacuum pass through a laser beam which excites the fluorescence state of the atom. The atom–field interaction time is determined by its average time of flight through the beam, $\tau_{fl} = a/v_0$, where a is the atomic path in the laser beam (of the order of its diameter d), v_0 is the mean thermal velocity of atoms. Assume that the excited atom spontaneously returns to its initial state in the time $\tau_{sp} = A_{21}^{-1}$, where A_{21} is the spontaneous decay probability. If the laser beam intensity I exceeds the saturation intensity I_{sat} of the transition 1–2 determined by expression (1.3) the number of photons reradiated by the atom during its flight through the beam will be $N = \tau_{fl}/2\tau_{sp}$. For example, when the D$_1$ line of Na is excited ($\lambda = 5890$ Å, $\tau_{sp} = 1.6 \times 10^{-8}$ s, $\sigma_{12} = 8 \times 10^{-12}$ cm^2, $I_{sat} = 1.2$ W cm^{-2}, $v_0 = 2 \times 10^4$ cm s^{-1} at 20 °C), an atom of Na reradiates $N \simeq 10^3$ photons as it crosses the beam on a path of $a = 0.5$ cm ($\tau_{fl} = 2.5 \times 10^{-5}$ s) with power $P = I_{sat} a^2 = 0.3$ W. If the solid angle of the light-collecting system is $\Omega = 1$ sr, about 10^2 photons fall onto the photodetector. At the same time with a quite realistic factor of background noise, $\beta = 10^{-13}$, only 2 or 3 photons from the laser beam enter the photodetector, i.e. two orders less than the useful signal. Even with the real quantum efficiency of a detector of $\eta = 0.05$–0.1 it is possible to produce 5 to 10 photoelectrons as one atom flies through the laser beam. Such a 'multiple-photoelectron' signal can be easily discriminated from noise 'single-photoelectron' signals. This principle is fundamental to the successful experiments on fluorescence detection of free single atoms of Ba [1.78] and Na [1.79].

For most of the elements the resonance excitation lines lie in the uv region where it is still difficult to apply cw tunable laser radiation. So the pulsed excitation method has gained wider application for laser fluorescence analysis of atoms. A comparatively low frequency of laser pulse repetition (1–100 Hz) results in a corresponding decrease of sensitivity. At the same time, however, the pulsed method has certain advantages in discriminating the useful signal against the scattered light background. The method of laser fluorescence excitation for atomic spectral analysis is considered comprehensively in Chapter 2 (see also reviews [1.55, 1.56]).

(b) *Detection of molecules*

The fluorescence method offers excellent possibilities for laser excitation of molecules. The high sensitivity of the method permits experiments on the detection of the products of chemical reaction in molecular beams to be performed [1.80]. The sensitivity of laser fluorescence excitation here is quite comparable with the sensitivity of the best mass spectrometers but, unlike the latter, the laser fluorescence spectrometer is a highly selective detector which is able to detect molecules in a definite electronic–vibrational–rotational state resonating with monochromatic laser radiation (figure 1.12(c)). In [1.81] the fluorescence method was used to study the nonequilibrium vibrational distribution of the BaO molecule resulting from the chemical reaction $Ba + O_2 \rightarrow BaO + O$. Using a standard pulsed dye laser it was possible to detect 5×10^4 BaO molecules in a cubic centimetre on a certain vibrational–rotational (V, J) level.

The method of laser fluorescence excitation in a molecular beam is ideal for measuring microscopic (elementary) rate constants of chemical reactions not averaged over all states of an equilibrium distribution [1.80]. The laser pulse probes the presence of particles in certain quantum states before collisions induce relaxation to the equilibrium distribution. This method is particularly successful when applied to the study of the kinetics of chemical reactions with the formation or the participation of free radicals [1.82]. The detection of the OH radical illustrates the strong possibilities of the fluorescence method in detecting low concentrations of simple molecules.

The detection of the OH radical is essential for atmospheric photochemistry since its low concentrations (10^6–10^8 cm^{-3}) can control the conversion $CO \rightarrow CO_2$ in the atmosphere and are an important intermediate product in smog chemistry [1.83]. It is very difficult to detect the OH radical by a mass spectrometer due to the influence of the mass peak of OH^+ caused by the inevitable background of H_2O. But the OH radical can easily be detected by laser radiation in the region of 3000 Å. In [1.83], for example, the vibrational–rotational distribution of the OH radical in the electronic ground state was studied in the flame of a Bunsen burner with a sensitivity of 10^6 cm^{-3}. As an illustration, figure 1.14 shows the relative population distribution as a function of rotational energy at two different points in the flame. It can be clearly seen that at some points the distribution is nonequilibrium and cannot be described by rotational temperature alone.

It should be noted that in the experiment with the OH radical, as the exciting laser beam was focused onto a spot with a diameter of 0.1 mm, good spatial resolution was obtained. This is an essential feature of the laser fluorescence method that facilitates performance of microfluorescence analysis.

In exciting the molecular fluorescence by cw laser radiation it is impossible to obtain such high sensitivity as for atoms. This is due to the fact that it is difficult to realise multiple cyclic reradiation for molecules because of their radiative return to many vibrational and rotational states according to the Frank–Condon factor (figure 1.12(c)). For example, when excited by a cw Ar laser at $\lambda = 514.5$ nm the molecules of I_2 can be detected only at a

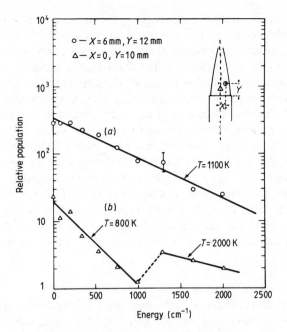

Figure 1.14 Relative populations of the rotational levels of the OH radical in the ground electronic–vibrational state as a function of the rotational energy at two different points in a Bunsen burner flame, measured by the fluorescence method (from reference [1.83]).

concentration of 5×10^8 cm^{-3} (or 2×10^6 cm^{-3} in one quantum state) [1.84]; that is much higher than the detection concentration for atoms. The addition of a buffer gas increases the rate of vibrational–rotational relaxation of molecules but is followed by quenching of electron-excited molecules.

It is far more difficult to realise fluorescent laser detection of more complex molecules, for several reasons. Most complex molecules do not have sharp electronic lines in the gas phase. For some classes of molecule (aromatic hydrocarbons) it is possible to produce narrow fluorescence lines in low-temperature matrices. Using monochromatic laser excitation it has been found [1.25] that this effect exists for many molecules but under ordinary wide-band excitation is not observable because of inhomogeneous broadening of molecular spectral lines. In the case of monochromatic laser excitation of molecular solutions at low temperatures a narrow-band fluorescence can be observed for many molecules which is used with advantage for highly sensitive selective detection of complex organic molecules at low concentration. This method of fluorescence laser spectroscopy is discussed in detail in Chapter 6 (see also review [1.57]).

When molecular solutions at room temperature have wide fluorescence bands, their selective detection is impossible. Nevertheless, a high sensitivity of the fluorescence method in the case of a high quantum yield is possible and useful for measuring the concentration of molecules of a given species. In [1.85] for example, the method of laser fluorescence excitation has made it possible to detect about 10^3 molecules of rhodamine 6G in solution (figure 1.12(d)). Another possibility for analysis is based on a combination of the high sensitivity of the fluorescence method with preliminary separation and accumulation of molecules of a certain species using, for example, a chromatograph. Such an approach has been successfully demonstrated in [1.86] where the laser fluorescence method was used to obtain a good sensitivity (about 10^{-12} g) for detection of aflatoxin molecules.

Finally, many complex molecules do not have intense excitation bands in the accessible spectral region, and the fluorescence is suppressed by quenching and photochemical transformation of excited molecules. In this case the fluorescence method does not turn out to be suitable, even with laser excitation.

1.4.6   Comparison of linear methods

None of the methods of linear laser spectroscopy is universal and therefore these methods complement rather than duplicate each other. Table 1.3 presents the fields of application of some methods and their main characteristics. Absorption spectroscopy is the most widely applied method since it covers, in principle, any atomic and molecular transitions in the UV, visible and IR regions. The disadvantage of this method, however, as well as of the two subsequent methods, is low spatial resolution (especially along the beam). The sensitivity of the method can only be high with a sufficiently long observation time that materially limits the temporal resolution. The method of intracavity absorption is applicable only in the UV, visible and near IR regions where lasers with wide amplification lines are available. Unlike the method of single-pass transition, intracavity absorbtion provides an extremely high sensitivity with a rather good temporal resolution. The opto-acoustic method can be applied for molecular transitions in the IR region with nonradiative excitation relaxation. The time constant of the opto-acoustic method is comparatively large because of slow relaxation of the cell to the initial state. The sensitivity of the opto-acoustic method is comparable with that of the absorption method at the same observation time, but the former is easier to perform. The best parameters can be compared using the fluorescence method but it is suitable only for transitions which have a radiative channel of excitation relaxation, i.e. for electronic transitions of atoms and molecules in the UV and visible spectral regions. The spatial resolution of this method is very high, up to 10 λ^3. The ultimate sensitivity, i.e. detection of a single atom or a single molecule in the given quantum state, can be attained in a comparatively short time. But at a lower sensitivity it is possible to work with a time constant no higher than 10^{-8} s. Thus, for allowed electronic transitions of atoms and

Table 1.3 Comparison of characteristics of various methods of linear laser spectroscopy.

Characteristics	Absorption method	Intracavity absorption	Opto-acoustic method	Fluorescence method
Field of applications	Absorption lines in visible and IR ranges	Absorption lines in visible and near IR ranges	Vibrational–rotational and electronic lines of molecular absorption	Electronic transitions in UV and visible range
Spatial resolution	Low	Low	Low	Up to λ^3 (practically about 10^{-6} cm^{-3})
Sensitivity	Up to 10^2 atoms or 10^8 molecules	10^5 atoms	10^8–10^9 molecules	Up to a single atom in quantum state
Temporal resolution	1 s	10^{-6} s	10^{-3} s	10^{-4} s (in principle up to 10^{-9} s)

molecules the fluorescence method is best. This, however, does not necessarily mean that it satisfies all the requirements of atomic and molecular spectral analysis. In particular, the chosen method, apart from its great sensitivity, must have high selectivity or high spectral resolution. But the electronic absorption bands of most molecules in the UV region are similar in character and, despite the high sensitivity, it is still difficult today to detect complex molecules by laser radiation.

1.5 Methods of nonlinear laser spectroscopy

Laser radiation can induce various nonlinear effects in matter. Of most interest for spectroscopy are the nonlinear resonance effects arising during excitations of atoms and molecules. They form the basis for quite new methods of nonlinear laser spectroscopy. These new methods have usually no analogue in linear spectroscopy and cannot be realised with the use of ordinary light sources because of their low spectral brightness.

1.5.1   Classification of nonlinear effects

The nonlinear effects applied in nonlinear spectroscopy can be subdivided into three types as follows.

(1) *The absorption saturation* effect means a significant change in the initial distribution of level population for the quantum transition under study. In the simplest case this effect is described by relation (1.2). This effect is useful for spectroscopy when monochromatic laser radiation interacts with an inhomogeneously broadened spectral line since it changes the line shape and makes it possible to reveal the structure of spectra screened by inhomogeneous broadening.

(2) The effect of *multistep excitation* of high-lying quantum levels takes place when a particle is excited subsequently at several quantum transitions under the action of multifrequency laser radiation. Such a method of strong excitation provides the production of highly excited particles which can be detected very easily, for example by their ionisation. Besides, it is usually impossible to populate excited high-lying states using single-quantum transitions from the ground state, and so it is necessary to apply the effect of subsequent absorption of several resonance photons by one particle. The methods of multistep laser spectroscopy are characterised by a very high sensitivity (up to single atoms and molecules) and high selectivity since it is possible to detect a multidimensional absorption spectrum at subsequent quantum transitions.

(3) The effect of *multiquantum excitation* differs from the effect of multistep excitation by the fact that the radiation frequency need not be in resonance with intermediate quantum levels. In this case not the real but the so-called virtual states are intermediate. The resonance conditions for, say, two quantum transitions in a two-frequency laser field have the form

$$\hbar\omega_1 + \hbar\omega_2 = E_{exc} - E_0 \qquad (1.17a)$$

or

$$\hbar\omega_1 - \hbar\omega_2 = E_{exc} - E_0 \qquad (1.17b)$$

where ω_1 and ω_2 ($\omega_1 > \omega_2$) are the laser radiation frequencies, E_{exc} and E_0 are the energies of final and initial quantum states, respectively. The case of two-photon resonance (1.17a) is successfully applied in two-photon Doppler-free gas spectroscopy [1.87] (see monographs [1.17, 1.18]). The case of two-photon resonance (1.17b) has found wide application in spectroscopy of Raman anti-Stokes coherent scattering [1.23] (see monographs [1.88, 1.89]). The two-frequency laser field here provides resonance excitation of molecular vibrations, coherent with respect to the excited volume of the medium. The additional probe laser radiation scattering at the excited vibrations to the anti-Stokes region makes it possible to perform highly sensitive spectroscopy of molecular vibrations.

It should be noted that the saturation effect can arise simultaneously at multistep and multiphoton excitation when the transition of particles from the initial to an excited quantum state is highly probable.

The basic methods of nonlinear laser spectroscopy are considered in more detail below.

1.5.2 Absorption saturation spectroscopy

This method of laser spectroscopy is based on the distortion of the inhomogeneously broadened absorption line shape under the action of monochromatic laser radiation. There are two quite different methods used for two materially different mechanisms of inhomogeneous broadening: Doppler broadening of any spectral lines due to thermal motion of absorbing particles; and inhomogeneous broadening of electronic absorption lines of molecules in a low-temperature matrix. Since these methods differ greatly, they will be considered separately.

(a) Doppler-free gas spectroscopy

The gas medium under study is under low pressure (no more than several torr) when the collisional broadening of spectral lines is small and then they are inhomogeneously broadened due to the Doppler effect at the thermal motion of particles by the value (1.1). The mechanism of narrowing a Doppler-broadened absorption line can easily be understood in the case shown in figure 1.15. An intense monochromatic running wave excites to the upper level those molecules whose velocity projection on the light wave direction satisfies the resonance condition:

$$kv_{res} = \omega - \omega_0 \tag{1.18}$$

where $k = 2\pi/\lambda = \omega/c$ is the wave vector and ω is the wave frequency. With the wave intensity $I \simeq I_{sat}$ a considerable fraction of 'resonance' molecules will pass to the upper level. As a result, a narrow hole with its width determined by homogeneous broadening arises at the absorption line.

Assume now that the absorption is measured by a weak counter-running wave of the same frequency as the strong counter-running wave. Since the weak wave has the same frequency, but is opposite in direction, it interacts with molecules having the same value of velocity projection but opposite to the strong wave. If the light-wave frequency ω does not coincide with the Doppler line centre frequency ω_0, the probe wave is not responsive to the strong wave. But, as the wave frequency coincides with ω_0, the weak probe wave interacts with the molecules whose absorption is reduced by the counter-running strong wave. As a result, the probe wave absorption has a resonance minimum, with its width equal to the homogeneous width $\Delta\omega_{hom}$, located exactly at the centre of the Doppler-broadened absorption line.

The nonlinear laser Doppler-free spectrometer has two advantages which are important for analytical applications. First, the information capacity of the

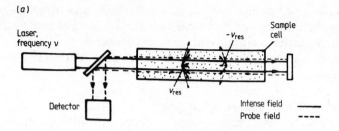

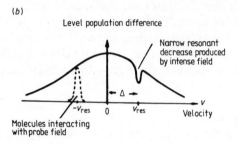

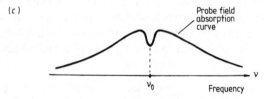

Figure 1.15 Observation of the narrow resonance using the single strong coherent travelling wave and counter-travelling weak probe wave. (*a*) Experimental arrangement. A small part of intense wave is reflected back through the cell. The attenuation of this weak wave is measured as a function of laser frequency. (*b*) Molecular velocity distribution, showing velocity groups that resonantly interact with the strong wave and the probe wave. (*c*) Probe wave absorption as a function of laser frequency.

spectral interval $\Delta\omega_{\mathrm{obs}}$ is very high for this spectrometer:

$$\rho = \rho_0 \frac{\Delta\omega_{\mathrm{obs}}}{\Delta\omega_{\mathrm{hom}}} \qquad (1.19)$$

where ρ_0 is the number of bits obtained in the resolved spectral interval $\Delta\omega_{\mathrm{hom}}$ due to the absorption value being measured at the given frequency. For example, if the resolution is standard for absorption saturation spectroscopy $R = \omega_0/\Delta\omega_{\mathrm{hom}} = 3 \times 10^8$, within the limits of the spectral interval $\Delta v_{\mathrm{obs}} = \Delta\omega_{\mathrm{obs}}/2\pi$ at $v_0 = \omega_0/2\pi = 3 \times 10^3$ cm^{-1} the number of bits will be $\rho_0 \times 10^5$, i.e.

$\rho = 10^6$ bits with $\rho_0 = 10$. At the same time for a linear IR laser spectrometer, with its resolution limited by the Doppler effect, the same spectral interval contains approximately 10^3 bits.

Secondly, in spectra of complex molecules the vibrational–rotational absorption lines are often overlapped and spectral analysis has to be performed using only vibrational bands. This greatly limits the analytical possibilities of IR molecular gas spectroscopy. For nonlinear Doppler-free spectroscopy this limitation is absent and the information on molecular gas mixtures sufficient for quantative and qualitative spectral analysis can be obtained from a comparatively narrow spectral interval even inside overlapping bands of molecular absorption in mixtures. The first successful experiments on the application of absorption saturation spectroscopy for such purposes are described in [1.90]. In this work the absorption saturation spectrum of a gas mixture of some hydrocarbons in the region of 3.4 μm corresponding to the CH bond was measured. As the He–Ne frequency was scanned with a magnetic field at the 3.39 μm wavelength, in a spectral interval of just 0.2 cm^{-1} narrow spectral resonances were observed, their width being only 0.5 MHz (about 10^3 times narrower than the Doppler width). They belonged to the vibrational–rotational lines of the molecules CH_4, C_2H_4, C_2H_6 and others. With the development of highly monochromatic tunable IR lasers this method should find wide application in analytical molecular spectroscopy.

(b) 'Hole-burning' spectroscopy

The electronic–vibrational molecular lines in a low-temperature glass-like matrix are inhomogeneously broadened which makes it more difficult to use these lines for spectral analysis. When such molecules are excited by monochromatic laser radiation tuned in resonance to the absorption line of a pure electronic transition (0–0) it is possible to excite only a definite group of molecules in exact resonance with the laser field. Such excited molecules emit narrow fluorescence lines without inhomogeneous broadening [1.25]. In some cases the electronic excitation of molecules results in their photochemical transformation and hence their escaping a narrow frequency interval within the limits of an inhomogeneously broadened absorption line. Because of this a hole is formed or, as is often said, 'hole-burning' occurs in the absorption line. This interesting effect was discovered in [1.91, 1.92] (see review [1.93]).

The laser radiation intensity needed for hole-burning is very low since the time of return or relaxation of molecules to the initial absorbing state τ (see expression (1.2)) is very long. So the real parameter of hole-burning is the laser radiation energy flux passing through the sample

$$\Phi_{burn} = \frac{\hbar\omega}{2\sigma_{el}}\eta_{burn}^{-1} \tag{1.20}$$

where σ_{el} is the electron transition cross section and η_{burn} is the quantum phototransformation yield of the electron-excited molecule.

The spectroscopy of electron transitions by the hole-burning method with the appropriate formation of narrow absorption lines gives rise to new possibilities for analysis of complex molecules which are briefly discussed in Chapter 6.

1.5.3   Spectroscopy of multistep excitation and ionisation

Multistep resonance excitation of high-lying atomic and molecular levels and their subsequent ionisation is today the most sensitive method of atomic and molecular spectroscopy. The method is based on the successive transition of a selectively excited particle from the ground state up the ladder of quantum transitions to a highly excited final state 'f' which can be easily detected, for example, from the subsequent ionisation yielding ions. Such multistep excitation of a multilevel quantum system (atom, molecule) by multifrequency laser radiation can be effective when the rate of induced transitions from each intermediate level W_{kn} is much higher than the relaxation rate of its excitation $1/\tau_{rel}^{(k)}$ (figure 1.16). The optimal case is when the multilevel system is simultaneously irradiated by a sequence of laser pulses of about the same duration $\tau_{pul} \ll \tau_{rel}^{(k)}$ ($k = 1, \ldots, f$) at the resonance frequencies ω_{kn} of successive quantum transitions.

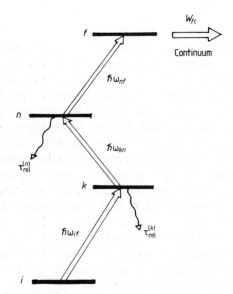

Figure 1.16   General scheme of multistep excitation of a multilevel quantum system by means of several laser pulses with properly tuned frequencies.

If the energy fluence $\Phi(\omega_{kn})$ of each such pulse is sufficient to saturate the corresponding resonance quantum transition, i.e.

$$\Phi(\omega_{kn}) = \Phi_{sat}^{kn} = \frac{\hbar\omega_{kn}}{\sigma(\omega_{kn})} \qquad (1.21)$$

where $\sigma(\omega_{kn})$ is the cross section of the transition $k \to n$, under such multifrequency pulsed laser radiation, the population of the n_k quantum levels (i = initial, k = intermediate and f = final) interacting with the laser radiation becomes equalised

$$\frac{n_i}{g_i} = \cdots = \frac{n_k}{g_k} = \cdots = \frac{n_f}{g_f} \qquad (1.22)$$

where g_k is the statistical weight of the kth state. Such behaviour of a multilevel quantum system can easily be explained. The 'temperature' of quantum states tends to be in equilibrium with the high-temperature laser radiation that resonantly interacts with them. If the statistical weights grow with increasing number of quantum state, the final quantum state 'f' will be most populated.

An atom or a molecule in a highly excited quantum state can pass to the continuum, for example, and be ionised either spontaneously or under additional laser radiation or some other perturbation. If the rate of decay to continuum W_{fc} is smaller than the excitation rate of the final state and hence ionisation mainly occurs after laser excitation, the ionisation yield η_{ion} is limited by the relative population of the final excited level

$$\eta_{ion} = \frac{g_f}{\Sigma_k g_k}. \qquad (1.23)$$

If the ionisation decay rate is higher than the excitation rate, ionisation comes about in the course of laser irradiation and its yield, η_{ion}, can reach 100%.

The method of multistep resonance excitation and ionisation has two important advantages which are essential for analytical applications. First, the multistep character of excitation increases drastically the *excitation selectivity*. To excite a certain atom, it is necessary that two or more laser pulse frequencies should coincide accurately with the resonance frequencies of successive quantum transitions. In this case the coincidence of frequencies of several successive quantum transitions for different particles is extremely low and hence the detection selectivity of particles by this method is very high. Second, it is rather easy to ionise highly excited particles, and hence it is possible to detect multistep absorption by detection of ions. Such a method of absorption detection has an ultimate *sensitivity*, i.e. it provides a detection of *single atoms and molecules*.

(a) *Photoionisation spectroscopy of atoms*
The main advantage of the photoionisation method for the detection of atoms is its extremely high sensitivity which can reach its ultimate value, i.e. detection

of each atom in resonance with the laser radiation. To achieve such an ultimate sensitivity it is necessary to realise a maximum yield of multistep photoionisation η_{ion}, i.e. to fulfil condition (1.21) at each step of excitation and ionisation. Various schemes of multistep photoionisation are considered in Chapter 3. The methods for laser detection of single atoms are considered in more detail in reviews [1.94–1.96].

In the method of multistep photoionisation the atoms are excited first by laser radiation to a certain intermediate quantum state from which only the excited atoms are ionised. Depending on the method of ionising the last intermediate state, there are two different approaches to the problem: nonresonance photoionisation of resonantly excited atoms; and resonance ionisation of such atoms.

In the first approach the excited atom is ionised by an additional laser pulse at the direct phototransition to continuum. This case is characterised by a small photoionisation cross section $\sigma_i = 10^{-17}$–10^{-19} cm^2 and, according to (1.21), by a high value of saturation energy fluence $\Phi_{ion} \simeq 0.01$–1 J cm^{-2}. For example, in the first experiment on two-step photoionisation [1.97] the yield of Rb ions was $\eta_{ion} \simeq 10^{-3}$ because of the low energy of the ionising pulse; only when its energy was substantially increased was it possible to realise the conditions with $\eta_{ion} \simeq 1$ [1.98].

In the second approach the excited atom is transferred either to an autoionisation state or to a high-lying Rydberg state with subsequent ionisation by an electric field pulse. The cross sections of resonance excitation of the autoionisation state or the high-lying Rydberg state are several orders larger than the cross section of nonresonance photoionisation. This fact reduces the requirements for ionising laser pulse energy. Besides, the use of resonance excitation at such transitions to high-lying states is of particular interest in increasing the selectivity of multistep ionisation, especially when the resonance excitation selectivity at the first stage is not high enough, for example in the case of detection of rare isotopes.

The first experiments on the detection of single atoms by the method of multistep photoionisation were successfully carried out in [1.99, 1.100]. In [1.99] the caesium atoms in a buffer gas were both excited and ionised by the same laser pulse. The resulting electron–ion pairs were detected with a proportional counter. In such an experiment the time during which the atoms to be detected stay in the irradiated region is sufficiently long since it is determined by the diffusion time of atoms in the buffer gas (figure 1.17(a)). In this case it is advisable to detect each atom under the action of single laser pulses. But in this experiment it is impossible to attain high spectral resolution since the spectral lines of the atoms to be detected are broadened by the buffer gas.

When it is necessary to have high spectral resolution, the atoms can be detected in an atomic beam (figure 1.17(b)) as has been done in [1.100] with Na atoms. It should be noted that a very high degree of detection selectivity of rare

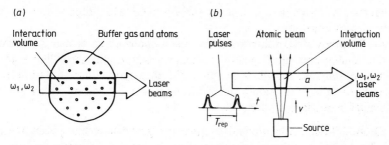

Figure 1.17 Laser photoionisation detection of single atoms. (*a*) Detection in the buffer gas with low spectral resolution. (*b*) Detection in the atomic beam with high spectral resolution.

isotopes can be achieved just in this case by using the natural or artificial isotope frequency shift at several successive excitation steps [1.101]. The atomic beam method can also be combined with a mass spectrometer which also gives an increase in the detection selectivity of single atoms and isotopes (see review [1.102]).

The analytical possibilities of laser spectroscopy of multistep atomic photoionisation were pointed out in the first work in this field [1.103]. These possibilities became manifest with the wide use of tunable pulsed dye lasers. This point has been covered in several short review papers [1.104, 1.105]. Chapter 3 gives a more detailed description of analytical photoionisation spectroscopy of atoms.

(b) *Photoionisation spectroscopy of molecules*
In one of the first experiments [1.21] on stepwise photoionisation of atoms and photodissociation of molecules the idea was conceived of creating a laser-selective detector of molecules based on selective stepwise photoionisation of molecules in combination with mass spectrometric detection of the produced photoions. In practice, however, it is more difficult to realise photoionisation detection for molecules than for atoms, particularly with regard to the requirements for the analytical method. Some of these difficulties are discussed below.

(1) The excited states of polyatomic molecules are more complex and the molecular absorption spectra do not often have a distinct structure. This impedes the photoionisation of molecules with high selectivity.

(2) As opposed to atoms, the molecules in intermediate excited states can undergo different photophysical and photochemical transformations. They can, for example, dissociate to neutral fragments at a high rate competing with the rate of the photoionisation process.

(3) The photoionisation products of polyatomic molecules, particularly when high-power laser radiation is applied, can vary greatly and this, in

principle, calls for the use of mass spectroscopic techniques of analysis and detection of photoions.

(4) The photoabsorption cross sections of molecules, are, as a rule, several orders smaller than those of atoms which necessitates the use of higher-power radiation sources. At the same time the absorption and photoionisation spectra of molecules usually lie in the UV and VUV regions which are still difficult for the laser technique.

Despite these serious difficulties, great progress has recently been made in the development of photoionisation laser spectroscopy of molecules. This has been considered in several special reviews [1.34–1.36, 1.106, 1.107]. This progress is explained, first of all, by the achievements in the development of laser sources for the UV and VUV range, excimer lasers in particular [1.108]. Secondly, the combination of resonance laser photoionisation of molecules with mass-spectrometric identification of resulting photoions in the form of the so-called 'two-dimensional mass-optical spectrometer' suggested in [1.21, 1.24] has proved to be rather effective. This idea is illustrated in figure 1.18. A laser with the tunable frequency ω_1 performs selective excitation of the vibrational or electronic molecular state. Due to this excitation the edge of the molecular photoionisation band usually lying in the VUV region shifts by a certain value towards the long-wavelength region. The second laser performs photoionisation of excited molecules, so its frequency ω_2 is chosen to be in the region of preferential photoabsorption of excited molecules. The resulting photoions are directed into the mass spectrometer which measures the mass spectrum, i.e. the photocurrent $i = f(M/e)$.

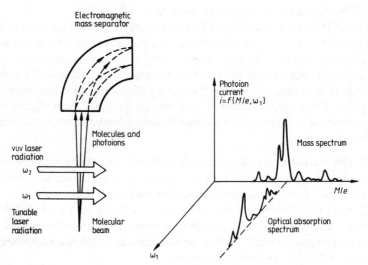

Figure 1.18 Simplified scheme of a two-dimensional optical spectrometer and mass spectrometer (from reference [1.24]).

This version of mass spectrometer can measure the photocurrent value for the given ω_2 as a function of the tunable laser frequency ω_1. In this case the molecular absorption spectrum will be measured from the value of the dependence $i = f(\omega_1)$. The laser mass spectrometer with selective laser photoionisation of molecules will produce, simultaneously, an optical absorption spectrum and a mass spectrum. This contrasts with the situation arising with conventional nonselective ionisation, for example by an electron beam or wide band VUV radiation.

After the first successful experiments [1.109, 1.110] on two-step photoionisation of molecules via an intermediate electronic state in a two-frequency (UV + VUV) laser field many works were published about multiphoton ionisation of polyatomic molecules by one-frequency UV or visible laser radiation (see review [1.106]). Such experiments proved the possibility of obtaining a high efficiency of molecular photoionisation (10 to 100%). As a result of this work a simpler version of 'two-dimensional optical mass spectrometer' was developed to detect ultrasmall amounts of molecules by photoionising them via an intermediate electronic state with one-frequency laser UV or visible radiation. Although this method of photoionisation has much lower potential selectivity, it can provide a very high sensitivity at moderate selectivity. This conclusion was supported in successful experiments on photoionisation detection of single molecules [1.111].

Laser photoionisation spectroscopy and mass spectrometry of molecules is a rapidly growing field of research at present. In 1978 there were just a few works published in this field; they now number some two hundred. Of particular interest are the analytical potentialities of this new method. Therefore Chapters 7 and 8 in the present monograph are specially concerned with this method and review the basic trends and methods of research as well as presenting the results on detection of polyatomic molecules which are most important for analytics.

1.5.4 Spectroscopy of coherent anti-Stokes Raman scattering

Experiments on stimulated Raman scattering in intense laser fields became practicable after high-power lasers were developed in 1962. But the use of stimulated Raman scattering in spectroscopy turned out to be greatly limited by such nonlinear effects as strong competition of stimulated Raman lines, the influence of other nonlinear processes (self-focusing and self-modulation) and the existence of an excitation intensity threshold. All this is due to the fact that the stimulated Raman process, in essence, results from instability in the medium in which the formation of photons at Raman frequencies is uncontrollable. In the 1970s, after tunable lasers had been developed, a new method of spectroscopy, of Raman light scattering, called the method of active spectroscopy or CARS method, was elaborated. It stands between the methods of spontaneous Raman scattering and stimulated Raman scattering. This new method was first used in [1.23] and was then developed in many experiments in many laboratories (see monographs [1.88, 1.89]. It combines the wide spectral possibilities of the spontaneous Raman scattering method and the most important feature of stimulated Raman scattering, that is, a high intensity of scattering line.

The CARS method is based on the generation of molecular vibrations in the field of two rather intensive laser beams with frequencies ω_1 and ω_2, their difference being equal to the molecular vibration frequency Ω. Two spatially coherent waves E_1 exp i$(\omega_1 t - \boldsymbol{k}_1 \boldsymbol{r})$ and E_2 exp i$(\omega_2 t - \boldsymbol{k}_2 \boldsymbol{r})$ generate molecular vibrations in the medium coherently over the volume, i.e. the amplitude of molecular vibrations in the medium is proportional to $E_1 E_2$ exp i$(\boldsymbol{k}_1 - \boldsymbol{k}_2)\boldsymbol{r}$. If the medium is exposed to a probe field with frequency ω, the field is scattered over the molecular vibrations and, as a result, at the frequencies $\omega_a = \omega + \Omega$ and $\omega_s = \omega - \Omega$ spectral lines appear (figure 1.19), their intensity being that of the corresponding spontaneous Raman scattering. But the intensities of strong waves ω_1, ω_2 and the probe wave are lower than the stimulated Raman scattering threshold, and so there cannot be uncontrollable instabilities and competition of scattering lines. By changing the laser wave frequency difference $\omega_1 - \omega_2$, it is possible to excite any active molecular vibrations of the medium, i.e. to attain the whole Raman spectrum of the medium.

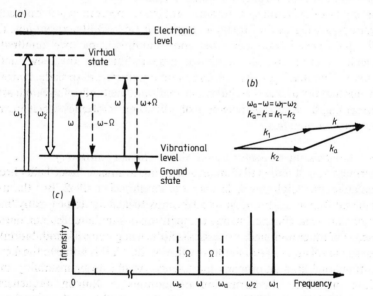

Figure 1.19 Coherent anti-Stokes Raman scattering of a probe wave (ω, k) by coherent molecular vibrations of frequency Ω, excited by two coherent waves $(\omega_1, \boldsymbol{k}_1)$ and $(\omega_2, \boldsymbol{k}_2)$: (a) energy level scheme; (b) condition for frequency and spatial phase matching at the anti-Stokes frequency; (c) field frequencies.

The probe light wave $E \exp i(\omega t - \mathbf{k}\mathbf{r})$ is scattered on the spatial grating of volume-phased molecular vibrations. Therefore, as usually happens in the case of interaction of plane waves, there is a direction of phase matching in which the scattered wave amplitude $E_{s,a} \exp i(\omega_{s,a} t - \mathbf{k}_{s,a}\mathbf{r})$ is maximum

$$\mathbf{k}_1 - \mathbf{k}_2 = \mathbf{k} - \mathbf{k}_s$$
$$\mathbf{k}_1 - \mathbf{k}_2 = \mathbf{k}_a - \mathbf{k}. \tag{1.24}$$

Under phase matching the scattered wave amplitude is proportional to the length of the medium l, and hence the scattering intensity is proportional to l^2. In spontaneous Raman scattering the intensity is proportional to l since the molecular vibrations of scattered particles are not coherent. The scattered wave intensity is related to the intensities of other waves by

$$I_{a,s} \propto I_1 I_2 I \tag{1.25}$$

i.e. it is proportional to the intensity of all three waves. In the CARS method the intensities of the Stokes and anti-Stokes components are the same in the direction of phase matching whereas in spontaneous isotopic Raman scattering the Stokes component prevails.

The first experiments on the CARS technique over coherent molecular vibrations were carried out in [1.23]. Later this effect was studied in a number of experiments using lasers at two fixed frequencies. The great possibilities of the CARS method, however, have been realised since tunable lasers have been used to excite molecular vibrations. The CARS method is discussed in more detail in monograph [1.88].

The CARS method has considerable advantages over the method of spontaneous Raman scattering in measurements which call for a high sensitivity. First, a coherently scattered signal is collimated in a definite angle of phase matching. With the use of simple schemes this enables spatial filtration of the useful signal and its discrimination from the background radiation. Secondly, the efficiency of coherent scattering is many orders higher than that of spontaneous scattering and may be as high as 1%. This allows observation of a scattered signal at normal illumination. Thirdly, a scattered signal is formed in the anti-Stokes region where there is no fluorescence from the exciting radiation. This is particularly important in the case of irradiation of biomolecules which usually fluoresce intensely in the Stokes region. In the very first experiments with biomolecules the CARS method made it possible to discriminate the anti-Stokes signal by 10^5 times against fluorescence [1.112].

The high efficiency of coherent anti-Stokes scattering provides a detectable signal at very low concentrations of molecules. For example, it is possible to attain a CARS spectrum from a molecular gas at a pressure of about 0.1 Torr. This is widely used in diagnostics of gas flows [1.113] and flames [1.114].

Further development of the analytical applications of this method is expected. This method is not discussed here as it has already been described in an excellent monograph [1.88].

1.6 Conclusions

To conclude this short introduction to various applications of laser light for analytical purposes it can be emphasised that this field has just been started. In future years rapid progress is expected in the methods of analytical laser spectroscopy and the development of various commercial laser analytical instruments.

It is obvious that the present analytical methods, particularly the separation methods, will be combined with the methods of laser analytical spectroscopy. For example, we can hope for the development of analytical instruments which will combine liquid chromatography with laser photoionisation detectors with their sensitivity being on the level of single atoms and molecules. This trend will be of particular importance for analysis of biomolecules and biosamples.

It is quite probable that new laser analytical methods will be essentially different from the numerous methods of analytical spectrochemistry. A new analytical method for very rare isotopes is expected to appear. It will not use radioactive decay effects for isotope detection and, at the same time, it will be much simpler than acceleration mass spectrometry [1.115]. The first proposal in this direction has already been published [1.116]. It will probably be possible to overcome the spatial resolution limit of local analysis on a surface preset by the light wavelength, for example by combining the Müller field-ion microscope [1.117] with selective laser photoionisation [1.118]. In other words, not only will the available laser analytical methods be developed but new methods will also have to be discovered.

This book is not, of course, a comprehensive consideration of all the problems in the progressing field of laser analytical spectrochemistry (see the recent review [1.119]). Some of the methods are briefly considered only in this chapter. But, on the whole, the reader who is also acquainted with other published monographs of this kind [1.13–1.15] should get a clear understanding of the state of the basic methods and the possible direction of their progress in the near future.

References

1.1   *Laser Spectroscopy* 1974 ed R G Brewer and A Mooradian. Proc. Int. Conf. Vail, Colorado, USA, June 25–29, 1973 (New York: Plenum)
1.2   *Laser Spectroscopy* 1975 ed S Haroche, J C Pebay-Peyroula, T W Hänsch and S E Harris. Proc. Second Int. Conf. Megeve, France, June 23–27, 1975 (Lecture Notes in Physics vol. 43) (Berlin: Springer)

1.3 *Laser Spectroscopy III* 1977 ed J L Hall and J L Carlsten. Proc. Third Int. Conf. Jackson Lake Lodge, Wyoming, USA, July 4–8, 1977 (Springer Series in Optical Sciences vol. 7) (Berlin: Springer)

1.4 *Laser Spectroscopy IV* 1979 ed H Walther and K W Rothe. Proc. Fourth Int. Conf., Rottach-Egern, FRG, June 11–15, 1979 (Springer Series in Optical Sciences vol. 21) (Berlin: Springer)

1.5 *Laser Spectroscopy V* 1981 ed A R W McKellar, T Oka and B P Stoicheff. Proc. Fifth Int. Conf., Jasper Park Lodge, Alberta, Canada, June 29–July 3, 1981 (Springer Series in Optical Sciences vol. 30) (Berlin: Springer)

1.6 *Laser Spectroscopy VI* 1983 H P Weber and W Lüthy. Proc. Sixth Int. Conf., Interlaken, Switzerland, June 27–July 1, 1983 (Springer Series in Optical Sciences vol. 40) (Berlin: Springer)

1.7 *Fundamental and Applied Laser Physics* 1973 ed M S Feld, A Javan and N A Kurnit. Proc. Esfahan Symp. Aug. 29–Sept. 5, 1971 (New York: Wiley)

1.8 *Tunable Lasers and Their Applications* 1976 ed A Mooradian, T Jaeger and P Stokseth. Proc. Int. Conf., Loen, Norway, June 6–11, 1976 (Springer Series in Optical Sciences vol. 3) (Berlin: Springer)

1.9 *Advances in Laser Chemistry* 1978 ed A H Zewail. Proc. Int. Conf., Pasadena, USA, March 20–22, 1978 (Springer Series in Chemical Physics vol. 3) (Berlin: Springer)

1.10 *Laser Spectroscopy of Atoms and Molecules* 1976 ed H Walther (Springer Series 'Topics in Applied Physics' vol. 2) (Berlin: Springer)

1.11 Letokhov V S 1977 *Laserspektroskopie* (Braunschweig: Vieweg)

1.12 Demtröder W 1981 *Laser Spectroscopy* (Springer Series in Chemical Physics vol. 5) (Berlin: Springer)

1.13 *Analytical Laser Spectroscopy* 1979 ed N Omenetto (New York: Wiley)

1.14 *Lasers in Chemical Analysis* 1981 ed G M Hieftje, J C Travis and F E Lytle (Clifton, NJ: Humana)

1.15 *Ultrasensitive Laser Spectroscopy* 1983 ed D S Kliger (New York: Academic)

1.16 Kelley P L and Hinkley E D. Tunable semiconductor lasers and their spectroscopic uses in [1.7] p. 723

1.17 *High Resolution Laser Spectroscopy* 1976 ed K Shimoda (Springer Series 'Topics in Applied Physics' vol. 13) (Berlin: Springer)

1.18 Letokhov V S and Chebotayev V P 1977 *Nonlinear Laser Spectroscopy* (Springer Series in Optical Sciences vol. 4) (Berlin: Springer)

1.19 *Ultrashort Light Pulses* 1977 ed S L Shapiro (Springer Series 'Topics in Applied Physics' vol. 18) (Berlin: Springer)

1.20 *Picosecond Phenomena III* 1982 ed K B Eisenthal, R M Hochstrasser, W Kaiser and A Laubereau. Proc. Third Int. Conf., Garmish-Partenkirchen, FRG, June 16–18, 1982 (Springer Series in Chemical Physics vol. 23) (Berlin: Springer)

1.21 Ambartzumian R V and Letokhov V S 1972 *Appl. Opt.* **11** 354

1.22 Pakhomicheva L A, Sviridenkov E A, Suchkov A V, Titova A F and Churilov S S 1970 *Pis'ma Zh. Eksp. Teor. Fiz.* **12** 60 (in Russian); Peterson N C, Kurylo M J, Braun W, Bass A M and Keller R A 1971 *J. Opt. Soc. Am.* **61** 746

1.23 Maker P D and Terhune R W 1965 *Phys. Rev.* **137** A801; Giordmaine J A and Kaiser W 1966 *Phys. Rev.* **A144** 676

1.24 Letokhov V S Future applications of selective laser photophysics and photochemistry in [1.8] p. 122

1.25   Personov R I, Al'shitz E I and Bykovskaya L A 1972 *Opt. Commun.* **6** 169
1.26   Bowman M R, Gibson A J and Sandford M C W 1969 *Nature* **221** 456
1.27   *Laser Monitoring of the Atmosphere* 1976 ed E D Hinkley (Springer Series 'Topics in Applied Physics' vol. 14) (Berlin: Springer)
1.28   *Optical and Laser Remote Sensing* 1983 ed D K Killinger and A Mooradian (Springer Series in Optical Sciences vol. 39) (Berlin: Springer)
1.29   Moenke H and Moenke L 1973 *Laser Microspectrochemical Analysis* (London: Adam Hilger)
1.30   Hillenkamp F, Unsöld E, Kaufmann R and Nitsche R N 1975 *Appl. Phys.* **8** 341; Hillenkamp F 1981 *Kvant. Elektron.* **8** 2655 (in Russian)
1.31   Brech F and Cross L 1962 *Anal. Spectrosc.* **16** 59
1.32   Laqua K Analytical spectroscopy using laser atomizers in [1.15] p. 47
1.33   Bykovskii Yu A, Dorofeev V I, Dimovich V I, Nikolaev B I, Ryzjikh S V and Sil'nov S M 1969 *Zh. Techn. Fiz.* **39** 1272 (in Russian)
1.34   Antonov V S and Letokhov V S 1981 *Appl. Phys.* **24** 89
1.35   Lichtin D A, Zandee L and Barnstein R B Potential analytical aspects of laser multiphoton ionization mass spectrometry in [1.14] p. 125
1.36   Parker D H Laser ionization spectroscopy and mass spectrometry in [1.15] p. 234
1.37   Ashkin A 1970 *Phys. Rev. Lett.* **24** 156; **25** 1321
1.38   Letokhov V S and Minogin V G 1981 *Phys. Rep.* **73** 2
1.39   Balykin V I, Letokhov V S and Minogin V G 1984 *Appl. Phys.* **B33** 247
1.40   Letokhov V S and Moore C B 1977 in *Chemical and Biochemical Applications of Lasers* vol. III ed C B Moore (New York: Academic)
1.41   Egorov S E, Letokhov V S and Shibanov A N 1983 in *Surface Studies with Lasers* ed F R Aussenegg, A Leitner and M E Lippitsch. Proc. Int. Conf., Mauterndorf, Austria, March 9–11, 1983 (Springer Series in Chemical Physics vol. 33) (Berlin: Springer) p. 156
1.42   Evenson K M, Saykally R J, Jennings D A, Curl R F Jr and Brown J M 1980 in *Chemical and Biochemical Applications of Lasers* ed C B Moore, vol. 5 (New York: Academic) p. 95
1.43   Pao Y H (ed) 1972 *Optoacoustical Spectroscopy and Detection* (New York: Academic)
1.44   Kreuzer L B and Patel C K N 1971 *Science* **173** 45
1.45   Tam A C Photoacoustics: spectroscopy and other applications in [1.15] p. 2
1.46   Zharov V P and Letokhov V S 1986 *Laser Opto-Acoustical Spectroscopy* (Springer Series in Optical Sciences vol. 37) (Berlin: Springer)
1.47   Swofford R L Analytical aspects of thermal lensing spectroscopy in [1.14] p. 143
1.48   Fang H L and Swofford R L The thermal lens in absorption spectroscopy in [1.15] p. 176
1.49   Gough T E, Miller R E and Scoles G 1977 *Appl. Phys. Lett.* **30** 338
1.50   Hartung C and Jurgeit R 1978 *Kvant. Elektron.* **5** 1820 (in Russian)
1.51   Penning F M 1928 *Physica* **8** 137
1.52   Green R B, Keller R A, Luther G G, Schenk P K and Travis J C 1976 *Appl. Phys. Lett.* **29** 727; *J. Am. Chem. Soc.* **98** 8517
1.53   *Optogalvanic Spectroscopy and Applications* 1983 Colloq. Int. CNRS, N352, June 20–24, 1983, Aussois, Savoie, France

1.54   Travis J C and De Voe J R The optogalvanic effect in [1.14] p. 93

1.55   Omenetto N and Winefordner J D Atomic fluorescence spectroscopy with laser excitation in [1.13] p. 167

1.56   Weeks S J and Winefordner J D Laser-excited atomic fluorescence spectrometry in [1.14] p. 159

1.57   Wehry E L, Gore R R and Dickinson R Laser-excited matrix-isolation molecular fluorescence spectrometry in [1.14] p. 210

1.58   Strojny N and de Silva J A F Laser-induced fluorometric analysis of drugs in biological fluids in [1.14] p. 225

1.59   McDowell R S *Adv. Infrared Raman Spectrosc.* **5** (1978) 1; *Vib. Spectra Struct.* **10** (1981) 1

1.60   Shimoda K 1973 *Appl. Phys.* **1** 77

1.61   Kingston R H 1978 *Detection of Optical and Infrared Radiation* (Springer Series in Optical Sciences vol. 10) (Berlin: Springer)

1.62   Harris T D Laser intracavity-enhanced spectroscopy in [1.15] p. 343

1.63   Belikova T P, Sviridenkov E A and Suchkov A F 1974 *Kvant. Elektron.* **1** 830 (in Russian)

1.64   Hänsch T W, Schawlow A L and Toschek P E 1972 *IEEE J. Quant. Electron.* **QE-8** 802

1.65   Antonov E N, Koloshnikov V G and Mironenko V R 1975 *Kvant. Elektron.* **2** 171

1.66   Curcio J A, Drummeter C F and Knestrick G L 1964 *Appl. Opt.* **3** 1401

1.67   Sarkisov O M, Sviridenkov E A and Suchkov A F 1982 *Chem. Phys.* **1** 1155 (in Russian)

1.68   Matiagin Yu V, Rabotnov N A, Savchenko A N and Sviridenkov E A 1983 *Kvant. Elektron.* **10** 1884 (in Russian)

1.69   Kreuzer L B 1971 *J. Appl. Phys.* **42** 2934

1.70   Hordvik A 1977 *Appl. Opt.* **16** 2827

1.71   Stone J 1972 *J. Opt. Soc. Am.* **62** 327

1.72   Friedrich D M Optical-phase-shift methods for absorption spectroscopy in [1.15] p. 311

1.73   Kogan Sh M and Lifshits T M 1977 *Phys. Status Solidi* a **39** 11

1.74   Fairbank W M Jr, Hansch T W and Schawlow A L 1975 *J. Am. Opt. Soc.* **65** 199

1.75   Balykin V I, Letokhov V S, Mishin V I and Semchishen V A 1976 *Pis'ma Zh. Eksp. Teor. Fiz.* **24** 475 (in Russian)

1.76   Gelbwachs J A, Klein C K and Wessel J E 1977 *Appl. Phys. Lett.* **30** 489

1.77   Hänsch T W 1973 in *Dye Lasers* ed F P Schafer (Springer Series 'Topics in Applied Physics' vol. 1) (Berlin: Springer) p. 194

1.78   Greenless G W, Clark D L, Kaufman S L, Lewis D A, Tonn J F and Broadhurst J H 1977 *Opt. Commun.* **23** 236

1.79   Balykin V I, Letokhov V S and Mishin V I *Pis'ma Zh. Eksp. Teor. Phys.* **26** (1977) 492 (in Russian); *Appl. Phys.* **22** (1980) 245

1.80   Zare R N and Dagdigian P J 1974 *Science* **185** 739

1.81   Schulz A, Cruse H W and Zare R N 1972 *J. Chem. Phys.* **57** 1354

1.82   Reisler H, Mangir M and Wittig C 1980 Laser Kinetic Spectroscopy of Elementary Processes in *Chemical and Biochemical Applications of Lasers* vol 5 ed C B Moore (New York: Academic) p. 139

1.83   Wang C C and Davis L I 1974 *Appl. Phys. Lett.* **25** 34
1.84   Balykin V I, Mishin V I and Semchishen V A 1977 *Kvant. Elektron.* **4** 1556 (in Russian)
1.85   Dovichi N J, Marton J C, Jett J H and Keller R A 1983 *Science* **219** 845
1.86   Diebold G J and Zare R N 1977 *Science* **196** 1439
1.87   Vasilenko L S, Chebotayev V P and Shishaev A V 1970 *Pis'ma Zh. Eksp. Teor. Fiz.* **12** 161 (in Russian)
1.88   Akhmanov S A and Koroteev N I 1981 *Methods of Nonlinear Optics in Light Scattering Spectroscopy* (Moscow: Science)
1.89   Levenson M D 1982 *Introduction to Nonlinear Laser Spectroscopy* (New York: Academic)
1.90   Radloff W, Below E and Stert V 1975 *Opt. Commun.* **13** 160
1.91   Kharlamov B M, Personov R I and Bykovskaya L A 1974 *Optic. Commun.* **12** 191
1.92   Gorokhovskii A A, Kaarli P K and Rebane L A 1974 *Pis'ma Zh. Eksp. Teor. Fiz.* **20** 474 (in Russian)
1.93   Rebane L A, Gorokhovskii A A and Kikas J V 1982 *Appl. Phys.* **B29** 235
1.94   Hurst G S, Payne M G, Kramer S D and Young J P 1979 *Rev. Mod. Phys.* **51** 767
1.95   Balykin V I, Bekov G I, Letokhov V S and Mishin V I 1980 *Usp. Fiz. Nauk* **132** 293 (in Russian) (Engl. transl. 1980 *Sov. Phys.–Usp.* **23** (10) 651)
1.96   Letokhov V S 1980 in *Chemical and Biochemical Applications of Lasers* vol. 5 ed C B Moore (New York: Academic) p. 1
1.97   Ambartzumian R V, Kalinin V P and Letokhov V S 1971 *Pis'ma Zh. Eksp. Teor. Fiz.* **13** 305 (in Russian) (Engl. transl. 1971 *JETP Lett.* **13** 217)
1.98   Ambartzumian R V, Apatin V M, Letokhov V S, Makarov A A, Mishin V I, Puretzkii A A and Furzikov N P 1976 *Zh. Eksp. Teor. Fiz.* **70** 1660 (in Russian)
1.99   Hurst G S, Nayfeh M H and Young J P 1977 *Appl. Phys. Lett.* **30** 229
1.100   Bekov G I, Letokhov V S and Mishin V I 1978 *Pis'ma Zh. Eksp. Teor. Fiz.* **27** 52 (in Russian)
1.101   Letokhov V S and Mishin V I 1979 *Opt. Commun.* **29** 168; Kudriavtsev Yu A and Letokhov V S 1982 *Appl. Phys.* **B29** 219
1.102   Karlov N V, Krynetzkii B B, Mishin V A and Prokhorov A M 1979 *Usp. Fiz. Nauk* **127** 593 (in Russian)
1.103   Letokhov V S *Soviet Patent* N784679. Appl. 30.03.1970 Published in *USSR Bulletin of Inventions* (1982) **N18** 308
1.104   Bekov G I and Letokhov V S 1983 *Appl. Phys.* **B30** 161
1.105   Bekov G I and Letokhov V S 1983 *Trends in Analytical Chemistry* **2** 252
1.106   Johnson P M 1980 *Acc. Chem. Res.* **13** 20
1.107   Antonov V S, Letokhov V S and Shibanov A N 1984 *Usp. Fiz. Nauk* **142** N2 177 (in Russian)
1.108   *Excimer Lasers* 1979 ed Ch K Rhodes (Springer Series 'Topics in Applied Physics' vol. 30) (Berlin: Springer)
1.109   Antonov V S, Knyazev I N, Letokhov V S, Matiuk V M, Movshev V G and Potapov V K 1978 *Opt. Lett.* **3** 37
1.110   Antonov V S, Letokhov V S and Shibanov A N 1980 *Zh. Eksp. Teor. Fiz.* **78** 2222 (in Russian)
1.111   Antonov V S, Letokhov V S and Shibanov A N 1981 *Opt. Commun.* **38** 182
1.112   Begley R F, Harvey A B and Byer R L 1974 *Appl. Phys. Lett.* **25** 3.87

1.113 Bunkin A F and Koroteev N I 1981 *Usp. Fiz. Nauk* **134** 93 (in Russian)
1.114 Moya F, Driet S A J and Taran J P E 1975 *Opt. Commun.* **13** 169
1.115 Suter M, Balzer R, Bonani G, Stoller Ch, Wölfli W, Beer J, Oeschger H and Stuffer B 1981 *IEEE Trans. Nucl. Sci.* **NS-28** 1475
1.116 Kudriavtzev Yu A, Letokhov V S and Petrunin V V 1986 *Pis'ma Zh. Eksp. Teor. Fiz.* **42** 23 (in Russian)
1.117 Muller E W and Tsong T T 1973 Field ion microscopy, field ionization and field evaporation in *Progress in Surface Science* vol. 4 (Oxford: Pergamon)
1.118 Letokhov V S 1975 *Kvant. Elektron.* **2** 930 (in Russian)
1.119 Zare R N 1984 *Science* **226** 298

2 Laser Atomic Fluorescence Analysis

M A Bol'shov

2.1 Introduction

The use of tunable lasers as sources of intense resonance radiation has given rise to rapid progress in the conventional fluorescence spectroscopy of atoms and molecules. For the last 10–15 years the laser atomic (or molecular) fluorescence (LAF or LMF) technique has taken up a sound position among the most widespread sensitive methods of spectroscopic and analytical investigation. The list of fields where LAF is successfully applied is quite extensive: determination of trace concentrations of elements in gases, liquid and solid samples of various chemical composition; measurement of the lifetime of excited states of atoms and molecules; measurement of the rates of inelastic collisions in atom–atom, atom–molecule and molecule–molecule systems; kinetics of chemical reactions; diagnostics of plasma, flames and hydrodynamic fluxes; spectroscopy of molecules in supersonic jets; fluorometry of condensed media, etc. The main advantages of LAF (and LMF) which have permitted its wide application include: high sensitivity of detection of atoms and molecules in the gaseous phase, the potential to investigate distribution of fluorescent objects with high spatial (up to 10^{-3} mm^3) and time (up to 10 ns) resolution, absence of perturbation in objects being investigated (this is especially important for applications in diagnostics) and high selectivity of the analysis.

The above mentioned fields of LAF applications have become independent directions which, being ideologically close, nevertheless have their specific features in objects and methods of investigation. It is not possible to review in detail all the above fields within the framework of this chapter.

We shall consider here briefly the following applications of LAF: measurements of trace concentrations of elements, and of the rates of inelastic collisions of atoms with molecules in the gaseous phase and in flames. It should be noted that the author has no claim to a comprehensive survey of the literature dealing with the above range of problems.

Recently several monographs and reviews have been published devoted to the physical fundamentals of the LAF method [2.1, 2.2] as well as different aspects of its analytical applications [2.3–2.7]. In the present chapter the main emphasis will be on the results of original work published after 1976, though for completeness and the convenience of readers the results already discussed in [2.1–2.7] will also be used. The review in this chapter opens with a short theoretical section required for the understanding of the main characteristics of the technique and a reasonable choice of parameters for fluorescent spectrometers being developed.

In §2.3 the principle of the construction of a laser fluorescence spectrometer is discussed in a general form and its main components are briefly described. In this section a table of chemical elements which can be analysed by the LAF method is presented. Analytical applications of the LAF method and its advantages for the analysis of real objects are considered in §2.4.

Examples of the application of LAF for measurement of fundamental spectroscopic characteristics of the rates of radiative and collisional deactivation of atomic excited states are considered in §2.5.

2.2 Physical fundamentals of laser atomic fluorescence

In a general case the behaviour of atoms in a resonance field of a light wave is described by a nonstationary equation for the density matrix [2.8]. Alternatively use of the rate equation for the population densities is possible in the case when the equation for the off-diagonal density matrix elements may be considered to be steady state [2.8, 2.9]. This approximation implies certain limitations for the resonance field characteristics. The amplitude of the electrical field E of the light wave must satisfy the condition $\mu E/\hbar \ll T_2^{-1}$, where μ is the dipole moment of the transition and T_2 is the transverse or phase relaxation time in the atomic system. Physically this means that the intensity of the resonance field is insufficient for coherent excitation of high-frequency polarisation in an ensemble of atoms (or molecules).

In addition, it is necessary that the rate of the field intensity variation is slow compared with the time T_2, i.e. that the condition $\tau_l \gg T_2$ is fulfilled (where τ_l is the duration of a laser pulse). Thus, if we are not interested in coherent effects [2.10, 2.11], the rate equation formulation is quite permissible for the laser fields commonly used in the LAF method.

Both the steady state regime of radiation interaction with two- and three-level systems [2.12, 2.13] and the temporal behaviour of two- and three-level atomic systems in pulse fields have recently been the subjects of detailed [2.14–2.21] study. Investigations of both regimes are of practical interest since the LAF method is realised in widely varying experimental conditions depending on the specific problems. A large group of problems which can be solved by use of the LAF method is diagnostics of flames, gas-dynamic flows, etc. In this case local temperatures are measured in the steady-state and in the transient modes. The main goal of the theoretical analysis for diagnostic applications is to find the relation between experimentally measurable parameters (saturation parameters, the ratio of fluorescence intensities at different transitions, the amplitude and the integral brightness of fluorescence at a fixed transition) and the characteristics being investigated (the local temperature, quenching constants for an atomic or a molecular system, concentration of fluorescent particles or total concentration of nonfluorescent particles in flames or flows).

Determination of optimum conditions for excitation and recording of fluorescence is important for analytical application of LAF as a method of measuring trace concentrations of elements. For this purpose the analysis of the nonsteady-state regime of LAF is of particular interest since the experiments are carried out mainly with pulse lasers whose pulse duration is about several nanoseconds.

2.2.1   Fluorescence in a three-level scheme with regard to saturation

(a)   General equations
If the spectral width of laser radiation is far in excess of the width of the atomic absorption line, then the rate equations for a three-level system are:

$$\frac{dn_1}{dt} = -n_1(R_{12} + R_{13}) + n_2 R_{21} + n_3 R_{31}$$

$$\frac{dn_2}{dt} = n_1 R_{12} - n_2(R_{21} + R_{23}) + n_3 R_{32} \tag{2.1}$$

$$\frac{dn_3}{dt} = n_1 R_{13} + n_2 R_{23} - n_3(R_{31} + R_{32})$$

where R_{ij} are the total rates of excitation and deactivation of the appropriate levels including radiative and collisional processes. The system (2.1) may be solved analytically in a general form for the light-wave field in the form of a step function or a rectangular pulse. In the work [2.17] a solution for the general case is given for a light pulse in the form of a step. Such an approach describes most adequately the LAF method for flames where the temperatures in the analytical region are within the limits 1500–2500 °C and the rates of collisional transitions may considerably exceed the radiative rates.

The solutions in the general form are, naturally, too cumbersome and their clear physical interpretation is difficult. In [2.17] three typical experimental situations were investigated where real atomic systems (of the Tl and Ca type) were excited by resonance radiation at transitions 1–3 or 2–3. Numerical integration of the system (2.1) was carried out in which the known experimental values of the radiative and collisional rates were explicitly taken into account, whereas unknown collisional rates were varied over wide limits as a parameter. In this way the temporal behaviour of populations $n(t)$ was analysed alongside their stationary values and the influence of the laser intensity on the temporal behaviour of the curves $n_i(t)$ was also investigated.

Below we shall demonstrate the possibilities of investigating specific problems arising both while detecting trace concentrations of elements and when the LAF method is used for diagnostic purposes. It will be exemplified by an analytical solution of the system (2.1) with some simplifying assumptions. When trace concentrations of elements are measured in nonflame atomisers the atom, as a rule, interacts with the resonance field in an inert atmosphere. In this case the rates of collisional processes may be of the same order or less than radiative rates. If, in addition, such an atomic system is investigated in which the distances between the levels $\Delta E_{ij} \gg kT$ (in nonflame atomisers the temperarure in the analytical zone is of the order of several hundreds of °C), one can deliberately neglect the processes of collisional excitation of the higher levels.

Consider a three-level model of an atom shown in figure 2.1. If the laser radiation frequency is set for transition 1–3, then with due regard to the above assumptions the constants R_{ij} in the system (2.1) become:

$$R_{12} = R_{23} = 0 \qquad R_{31} = \rho B_{31} + A_{31} + Z_{31}$$

$$R_{13} = \rho B_{13} \qquad R_{21} = Z_2 \qquad R_{32} = A_{32} + Z_{32} \qquad (2.2)$$

where ρ (J m^{-3} Hz^{-1}) is the spectral density of the laser radiation energy, B_{13}

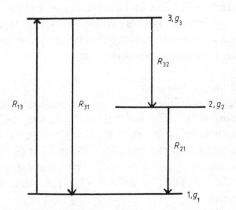

Figure 2.1 A three-level scheme of an atom. Statistical weights of the levels are g_1, g_2, g_3; R_{ij} are the rates of excitation of the levels.

and B_{31} $(J^{-1} m^3 Hz s^{-1})$ are the Einstein coefficients of induced absorption and emission, A_{31}, A_{32} (s^{-1}) are the Einstein coefficients for spontaneous radiation, Z_{21}, Z_{31}, Z_{32} (s^{-1}) are the rates of collisional deactivation of the appropriate levels. The solution of system (2.1) for ρ independent of time and with the initial conditions $n_2(0) = n_3(0) = 0$, $n_1(0) = n_0$ is [2.14]:

$$n_1(t) = n_0 \frac{R_{13}}{\xi_2 - \xi_1} \left[-\left(1 - \frac{R_{21} + R_{32}}{\xi_1} \right) \exp(-\xi_1 t) \right.$$

$$\left. + \left(1 - \frac{R_{21} + R_{32}}{\xi_2} \right) \exp(-\xi_2 t) \right] + n_0 \frac{R_{21}(R_{31} + R_{32})}{\xi_1 \xi_2}$$

$$n_2(t) = n_0 \frac{R_{13} R_{32}}{\xi_2 - \xi_1} \left(-\frac{1}{\xi_1} \exp(-\xi_1 t) + \frac{1}{\xi_2} \exp(-\xi_2 t) \right) + n_0 \frac{R_{13} R_{32}}{\xi_1 \xi_2}$$

$$n_3(t) = n_0 \frac{R_{13}}{\xi_2 - \xi_1} \left[\left(1 - \frac{R_{21}}{\xi_1} \right) \exp(-\xi_1 t) - \left(1 - \frac{R_{21}}{\xi_2} \right) \exp(-\xi_2 t) \right]$$

$$+ n_0 \frac{R_{13} R_{21}}{\xi_1 \xi_2} \tag{2.3}$$

where R_{ij} are determined in (2.2) and values ξ_1 and ξ_2 are determined by

$$\xi_{1,2} = \frac{X + R_{21}}{2} \mp \left[\left(\frac{X - R_{21}}{2} \right)^2 - R_{13} R_{32} \right]^{1/2}$$

$$X = R_{13} + R_{31} + R_{32}. \tag{2.4}$$

In the system being considered the optical dipole transition between the levels 1 and 2 is parity-forbidden and, therefore, the level 2 may decay due either to nonradiative collisional processes or to magneto-dipole and quadrupole radiation. We shall analyse the basic dependences which determine the magnitude of the fluorescent signal for two limiting cases: $R_{21} = 0$ and $R_{21} \to \infty$. Physically these cases correspond to a system with a metastable level and a system in which a strong coupling bond between the level 1 and 2 makes it in fact a two-level one.

(b) *The case of a metastable level*
Such an approximation is fulfilled rather well, e.g. for lead when Pb atoms are excited at the transition $6p^2$ $^3P_0$–$7s$ $^3P_1^\circ$ $(\lambda = 283.31$ nm) and fluorescence is recorded at the transition $7s$ $^3P_1^\circ$–$6p^2$ $^3P_2$ $(\lambda = 405.78$ nm). At atmospheric pressure of Ar the lifetime of the level $6p^2$ $^3P_2$ is 1.8×10^{-5} s [2.22] which corresponds to the value $R_{21} - 5 \times 10^4$ s^{-1}. The values A_{31} and A_{32} equal, respectively, 0.6×10^8 s^{-1} and 3×10^8 s^{-1} [2.23] so that the relationship $R_{21} \ll A_{31}$, A_{32} is valid.

In the case $R_{21} = 0$ the solution (2.3) is simplified. Steady-state values of populations are $n_1(\infty) = n_3(\infty) = n_0$ and this corresponds to a physically quite obvious fact: upon the switching on of laser radiation at the moment $t=0$ all the atoms from the ground state will follow the cycle 1–3–2 and accumulate at the metastable level. The temporal behaviour of the upper level population is determined from (2.3)

$$n_3(t) = n_0 \frac{R_{13}}{\xi_2 - \xi_1} [\exp(-\xi_1 t) - \exp(-\xi_2 t)]. \tag{2.5}$$

For ease of further calculation we assume that a laser pulse is rectangular and has an amplitude ρ and duration τ: $\rho(t) = \rho$ $(0 \leqslant t \leqslant \tau)$; $\rho(t) = 0$ $(t > \tau)$. It is well known that recording of direct-line fluorescence considerably simplifies the problem of suppression of the Rayleigh and Mie scattering; therefore in the system being considered we shall record photons at a frequency v_{32} which corresponds to the transition 3–2. We shall assume also that the recording system integrates flourescence radiation over the time interval from t_1 to t_2 and therefore we take as the signal value the total number of photons with the frequency v_{32} radiated by a unit volume of the atomic cloud within the time $t_2 - t_1$:

$$N_f = \int_{t_1}^{t_2} A_{32} n_3(t) \, dt. \tag{2.6}$$

The choice of the beginning and the duration of the integration interval $(t_2 - t_1)$ may have a considerable influence on the experimentally determined dependence of the fluorescence signal on the value ρ. This problem was investigated in detail in [2.18] for a two-level scheme as an example; we shall discuss it in more detail later on. Now for ease of calculation we assume that t_1 coincides with the beginning of the laser pulse and t_2 with its end, so that the fluorescence signal is being integrated during the laser pulse.

In this case the value N_f is determined from (2.5):

$$N_f = \int_0^\tau A_{32} n_3(t) \, dt = n_0 \frac{A_{32}}{A_{32} + Z_{32}} \left(1 - \frac{\xi_2 \exp(-\xi_1 \tau) - \xi_1 \exp(-\xi_2 \tau)}{\xi_2 - \xi_1}\right)$$

$$\tag{2.7}$$

For further calculations we express the values $\xi_{1,2}$ in an explicit form in terms of the rates of radiative and collisional transitions. For the case being considered ($R_{21} = 0$) from (2.4) it is possible to obtain approximate expressions for $\xi_{1,2}$ [2.13]:

$$\xi_1 \simeq \frac{R_{13} R_{32}}{X} = \frac{\rho B_{13}(A_{32} + Z_{32})}{A_{31} + Z_{31} + A_{32} + Z_{32} + \rho B_{13}(1 + g_1/g_3)}$$

$$\xi_2 \simeq X = A_{31} + Z_{31} + A_{32} + Z_{32} + \rho B_{13}(1 + g_1/g_3) \tag{2.8}$$

(note that $\xi_2 > \xi_1$). Now we go back to the expression (2.7) for N_f. It can easily be shown that when the condition

$$\xi_1 \tau \gtrsim 5 \tag{2.9}$$

is fulfilled, the expression for N_f is transformed with an accuracy of $\lesssim 3\%$ and becomes

$$N_f = n_0 \frac{A_{32}}{A_{32} + Z_{32}}. \tag{2.10}$$

The relation (2.10) shows that during the time τ determined from (2.9) all atoms follow the cycle 1–3–2 and accumulate at the metastable level and in this case the total number of fluorescent photons N_f is determined by the quantum efficiency of fluorescence in the channel 3–2–$[A_{32}/A_{32} + Z_{32}]$. It is interesting to point out that the rates A_{31} and Z_{31} do not influence the value N_f since all the atoms which have returned from the upper level to the ground state will be excited again by the field into the level 3 and finally they will be at the metastable level 2.

In the case under consideration a saturation regime of the total fluorescence signal is realised since N_f in (2.10) is independent of ρ. With regard to (2.8) the condition (2.9) may be rewritten as follows

$$\frac{\tau \rho B_{13}(A_{32} + Z_{32})}{A + Z + \rho B_{13}g} \gtrsim 5 \tag{2.11}$$

where the notations $A = A_{31} + A_{32}, Z = Z_{31} + Z_{32}$ and $g = 1 + g_1/g_3$ have been introduced. The condition (2.11) may be fulfilled in various experimental situations. If the field energy density is large so that the condition

$$\rho B_{13}g \gg A + Z \tag{2.12}$$

is realised, the condition (2.10) is fulfilled with the duration of the laser pulse

$$\tau \gtrsim \frac{5g}{A_{32} + Z_{32}}. \tag{2.13}$$

The condition (2.13) has a distinct physical sense. In a field with high energy density (2.12) saturation of the transition 1–3 is obtained so that populations of these levels are maintained in the saturated state $(n_3/n_1 \simeq g_3/g_1)$. In this situation the rate accumulation of the particles at the metastable level is independent of the radiation power and is determined by the total rate of deactivation of the level 3 in the channel 3–2. In absence of quenching $(Z_{32} \ll A_{32})$ the condition (2.13) is fulfilled for Pb at $\tau \gtrsim 20$ ns.

In a weak field, when the relation

$$\rho B_{13}g \ll A + Z \tag{2.14}$$

is realised, the condition (2.11) is fulfilled for τ which satisfies the relation

$$\tau \gtrsim \frac{5}{\rho B_{13}} \left(1 + \frac{A_{31} + Z_{31}}{A_{32} + Z_{32}}\right). \tag{2.15}$$

This result is also quite vivid. In a weak field the accumulation rate of particles at level 3 is determined by the rate of the induced absorption 1–3 and by the ratio of decay probabilities of level 3 in the channels 3–1 and 3–2. Note a considerable difference in the character of fluorescence saturation. In a strong field (2.12) the signal N_f would be saturated, as it would also be in the case where level 2 had a finite lifetime. The finite value of R_{21} would result in the growth of the saturation value ρ. In contrast to this, saturation in a weak field is possible only if a metastable is available.

Calculations made in [2.14] show that for Pb-type atoms in an inert gas atmosphere and at a laser pulse duration $\tau = 5 \times 10^{-9}$ s the approximation $\xi_1 \tau \lesssim 1$ and $\xi_2 \tau \gg 1$ is valid. In this case, neglecting $\exp(-\xi_2 \tau)$ in (2.7) and retaining the first two terms in the expansion $\exp(-\xi_1 \tau)$ we obtain

$$N_f = n_0 \frac{A_{32}}{A_{32} + Z_{32}} \left(1 - \frac{\xi_2(1 - \xi_1 \tau)}{\xi_2 - \xi_1}\right) \simeq n_0 \frac{A_{32}}{A_{32} + Z_{32}} \xi_1 \tau$$

$$= n_0 \frac{\rho B_{13} A_{32} \tau}{A + Z + \rho B_{13} g}. \tag{2.16}$$

The expression (2.16) illustrates a well known dependence of the fluorescence signal: the linear growth at small levels of ρ which satisfy (2.14), slowing down of the growth in the region $\rho B_{13} g \sim A + Z$ and saturation in the region of ρ values satisfying (2.2). The asymptotic value N_f is determined by the relationship

$$N_f^{\max} \simeq n_0 \frac{A_{32} \tau}{g}. \tag{2.17}$$

It should be noted that in (2.16) the coefficient before n_0 is less than 1, i.e. $A_{31}/A_{31} + Z \lesssim 1$ and $\xi_1 \tau \lesssim 1$ (only in this case the approximation (2.16) is valid). This result has a clear physical sense. When a metastable level is available atoms follow the cycle 1–3–2 only once within time τ and when quenching occurs ($Z_{32} \neq 0$) only a part of these atoms will radiate photons of the frequency ν_{32}.

(c) *The case of rapid decay of level 2*

In this limiting case a three-level system degenerates into a two-level one in which the upper level may decay with the probabilities A_{32} and A_{31} and emit photons of the frequencies ν_{31} and ν_{32}.

Provided that $R_{21} \gg R_{13}$, R_{31}, R_{32} the expression (2.4) for $\xi_{1,2}$ is transformed to

$$\xi_{1,2} \simeq \frac{X + R_{21}}{2} \mp \frac{R_{21} - X}{2}$$

where

$$\xi_1 \simeq X \qquad \xi_2 \simeq R_2 \qquad \text{and} \qquad \xi_2 \gg \xi_1. \qquad (2.18)$$

In this case the solution of the system (2.3) is transformed in such a way that $n_2(t)=0$ and for $n_1(t)$ and $n_3(t)$, with regard to (2.4),

$$n_1(t)=\frac{n_0}{X}\,[R_{13}\exp(-Xt)+R_{31}+R_{32}]$$

$$n_3(t)=\frac{n_0 R_{13}}{X}\,[1-\exp(-Xt)] \qquad (2.19)$$

is valid.

For a rectangular laser pulse we obtain a fluorescence signal as follows

$$N_f = n_0 \frac{A_{32}R_{13}}{X}\left(\tau - \frac{1}{X}\,[1-\exp(-X\tau)]\right). \qquad (2.20)$$

In the case when the condition $X\tau \gg 1$ is fulfilled the relationship (2.20) becomes

$$N_f \simeq n_0 \frac{A_{32}R_{13}}{X}\left(\tau - \frac{1}{X}\right) \simeq n_0 \frac{A_{32}R_{13}}{X}\,\tau. \qquad (2.21)$$

Physically fulfilment of the condition $X\tau \gg 1$ means that the system reaches a steady-state regime at which the fluorescence signal becomes $N_f = n_3^{ss} A_{32}\tau$, where n_3^{ss} is the steady-state value of the population of the upper level. A small correction $(\tau - 1/X)$ just includes the initial moment when the system is reaching the stationary level at the rate X^{-1}.

The expression (2.21) for N_f coincides exactly with (2.16). At small levels of ρ a linear mode of fluorescence is observed which transits to saturation at $\rho B_{13}g \gg A + Z$. In the saturation mode, however, in contrast to the case of a metastable level, the parameter $A_{32}\tau/g$ may be far in excess of 1, so that $N_f \gg n_0$. From the physical viewpoint it means the presence of a cyclic interaction of the two-level system with the resonance field in the saturation regime; as a result of this interaction one atom repeatedly radiates photons during the interaction with the resonance field of radiation.

2.2.2   Influence of spatial–time effects

The LAF method is widely used for diagnostics of flames and, in particular, for measurements of unknown quenching rates. One of the ways to determine the values Z_{ij} is the construction of an experimental saturation curve $N_f(\rho)$ and determination of the saturation parameter ρ^s

$$N_f(\rho^s)=0.5N_f^{max}.$$

The value ρ^s is connected explicitly with radiative and collisional rates so that when the values A_{ij} are known from independent measurements, the constants

Z_{ij} can be determined. It is rather important therefore that experimental conditions of excitation and fluorescence recording were adequate for a theoretical model. The influence of experimental conditions on the correctness of ρ^s determination is illustrated by two examples.

(a) *A two-level model of an atom*
In this case the values of X and R_{ij} determined before in (2.4) become

$$R_{13} \rightarrow \rho B_{12} \qquad R_{31} \rightarrow \rho B_{21} + A_{21} + Z_{21} \qquad X \rightarrow \rho B_{12}g + A_{21} + Z_{21}.$$

$$(2.22)$$

Substituting (2.22) into (2.19) we obtain for the population of the upper level $n_2(t)$

$$n_2(t) = n_0 \frac{\rho B_{12}}{A_{21} + Z_{21} + \rho B_{12}g} \{1 - \exp[-(A_{21} + Z_{21} + \rho B_{12}g)t]\}$$

$$\equiv n_0 \frac{\rho_0}{g\rho_0 + A} \{1 - \exp[-(g\rho_0 + A)t]\} \qquad (2.23)$$

where the notations $\rho B_{12} = \rho_0$ and $A_{21} + Z_{21} = A$ are introduced [2.18].

Now we assume that a gate of the integrating system has an arbitrary position at the time scale, i.e. by a fluorescence signal we mean the value N_f determined in (2.6). Then, with regard to (2.23), we obtain

$$N_f = A_{21} \int_{t_1}^{t_2} n_2(t)\, dt$$

$$= n_0 \frac{\rho_0 A_{21}}{g\rho_0 + A} \left\{ (t_2 - t_1) - \frac{1}{g\rho_0 + A} [\exp - (g\rho_0 + A)t_1 - \exp - (g\rho_0 + A)t_2] \right\}$$

$$(2.24)$$

The complicated dependence $N_f(\rho_0)$ in this case may lead to an appreciable deviation of the experimentally determined parameter ρ_0^s from the parameter ρ_{stat}^s determined in the stationary regime and connected explicitly with Z_{ij}. In figure 2.2 taken from [2.18] saturation curves for different values t_2 are shown. These curves were obtained by numerical integration of (2.24). In this case all the values have been normalised in such a way that $n_0 = 1$, $A_{21} + Z_{21} = 1$ and $g = 2(g_1 = g_2 = 1)$, the initial moment coincided with 0 and not the value $N_f(\rho)$ determined in (2.24), but $N(\rho_0) = N_f/A_{21}$ is calculated. The broken curve in figure 2.2 shows the trajectory of the apparent saturation parameter ρ_0^s depending on the gate duration (the value t_2). At $t_2 \rightarrow \infty$ the expression (2.24) tends to its 'stationary' value (2.21). At the value of the parameters chosen for the numerical calculation ρ_{stat}^s determined from (2.21) equals 0.5 ($g = 2$, $A = 1$, $g\rho_{stat}^s = A$). The value ρ_0^s varies from 2 for $t_2 = 0.5$ to 0.5 for $t_2 \rightarrow \infty$.

The numerical calculation made in [2.18] for a laser pulse as $\rho_0(t) =$

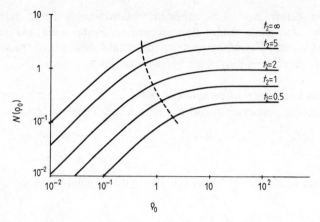

Figure 2.2   Saturation curves according to equation (2.24); $n_T = 1$, $A = 1$, $g = 2$, $t_1 = 0$, t_2 ranges from 0.5 (lower curve) to a value much in excess of $(g_0 + A)^{-1}$ (upper curve). The latter curve yields to the true curve of the saturation parameters, i.e. 0.5. The broken line shows the trajectory of the apparent saturation parameter.

$P\tau^{-1}t \exp(-t/\tau)$ showed that, depending on the duration and position on the time scale of the rectangular gate of the recording system, the value ρ_0^s being measured may differ from the 'stationary' one by more than an order. In this case, in order to obtain 'diagnostically' correct values of ρ_0^s, it is recommended to shorten the gate duration and bring it into coincidence with the maximum of the fluorescence pulse, where $dn_2/dt \simeq 0$. An analogous consideration for the case of a three-level system is given in [2.19].

(b)   *Transverse inhomogeneity of intensity*
Up to now inhomogeneity of the intensity of laser radiation in the cross section of the beam has been neglected. At the same time the transverse structure of the beams, at best, has the Gaussian shape (TEM_{00} mode of the resonator). It is physically clear that if transverse dimensions of the analytical region are comparable or exceed the cross section of the laser beam, then as the laser intensity increases, the total fluorescence signal of the atomic cloud will increase due to the wings of the transverse distribution even at total saturation in the centre of the beam. The influence of the Gaussian structure of the beam on the shape of the saturation curve was analysed theoretically in [2.24]. The equation for the stationary population of the upper level has its general form as follows

$$n_k = \beta n_0 \frac{\rho}{\rho + \rho^s} \tag{2.25}$$

where the coefficient β takes into account a specific model of the atom (two-, three- or multi-level). Let the profile of the laser beam be

$$\rho(r) = \rho \exp(-r^2/\omega^2).\tag{2.26}$$

The intersection region of the atomic beam of thickness δ and a transverse dimension h with the laser beam may have the shape of a cylinder (if the laser beam is perpendicular to the atomic beam) or of a slab (if $\delta \ll \omega$ and the axes of the laser and the atomic beams lie in the same plane).

For these cases total intensity of fluorescence is determined by

$$I_f = h\nu \frac{A_{ki}}{4\pi} \Omega\delta \int_0^\infty n_k(r)2\pi r\,dr \tag{2.27a}$$

for the cylindrical shape of the region and

$$I_f = h\nu \frac{A_{ki}}{4\pi} \Omega\,\delta h\,2 \int_0^\infty n_k(r)\,dr \tag{2.27b}$$

for the slab, where $h\nu$ is the energy of the fluorescent photon and Ω is the solid angle of the fluorescence radiation collection. Calculations of (2.27a,b) with regard to (2.25) and (2.26) yield the following result

$$I_f = h\nu \frac{A_{ki}}{4\pi} \Omega\,\delta\beta n_0 \pi\omega^2 \ln\left(1 + \frac{\rho}{\rho^s}\right) \tag{2.28a}$$

for the cylindrical case and

$$I_f = h\nu \frac{A_{ki}}{4\pi} \Omega\,\delta\beta n_0 2\omega h \int_0^\infty \frac{\exp(-x^2)\,dx}{\rho^s/\rho + \exp(-x^2)} \tag{2.28b}$$

for the slab. Thus, if the transverse dimension of the beam is less than the dimension of the analytical zone, the linear part of the saturation curve at small level ρ turns to a logarithmic dependence at $\rho \gg \rho^s$.

An experimental saturation curve with a shape close to the predicted one was observed in [2.25, 2.26].

2.2.3 Optimisation of laser radiation parameters

Here we consider briefly optimisation problems concerned with the parameters of laser radiation in analytical applications of the LAF method. Detection of trace concentrations of elements requires that the maximum possible signal-to-noise ratio (S/N) is realised. The methods for increasing and optimising S/N depend considerably on experimental conditions and on the nature of the background signal in a specific analytical technique. The problem was analysed in detail in [2.3, 2.27, 2.28].

We shall now discuss the problem of the choice of optimum laser power. In the case when the background, which restricts the detection limits of the analysis, is not connected with the scattered laser radiation but is determined by the thermal radiation of the atomiser, noise in the recording system, etc, it is advantageous to operate in the region of fluorescence saturation. In this case the maximum level of the fluorescence signal N_f is attained and, moreover, the precision of the analysis increased due to decreasing contribution of the source fluctuations to the signal ones.

However, in the case when the background level is determined by the scattered radiation (Rayleigh, and Mie scattering in the analytical volume, luminescence of the atomiser windows, etc), there exists an optimum level of the value ρ at which the maximum S/N ratio is realised. The presence of the optimum value of ρ is explained by the fact that the background increases linearly with increasing ρ, whereas the rate of the growth of the signal $N_f(\rho)$ slows down in the region of a transition from the linear regime of fluorescence to saturation.

We write the general expression for the fluorescence signal in the form

$$N_f = C_1 n_0 \frac{\rho}{\rho + \rho^s} \qquad (2.29)$$

where C_1 is the coefficient accounting for the conditions of collection and recording of fluorescence and also for the specific model of the atom. In the case of the Poisson distribution of noise photoelectrons for the background fluctuations, which is determined by the scattering of the exciting radiation, one may write $N = C_2\sqrt{\rho}$ where C_2 is the coefficient analogous to C_1. The signal-to-noise ratio in this case is

$$\frac{S}{N} = \frac{C_1}{C_2} \frac{n_0 \sqrt{\rho}}{\rho + \rho^s}. \qquad (2.30)$$

It can easily be shown that the maximum of this expression determined from the condition $\partial(S/N)\, \partial\rho = 0$ is attained at the optimum energy density

$$\rho^{opt} = \rho^s. \qquad (2.31)$$

If one takes into account the Gaussian distribution of the intensity across the section of the beam, then for the fluorescence signal the expression (2.28a) is valid and the maximum S/N ratio will be attained at $\rho^{opt} \simeq 4\rho^s$.

It should be noted that the condition (2.31) is rather 'soft'. Thus, at a five-fold variation of ρ with respect to ρ^{opt} the ratio S/N in (2.30) varies by $\lesssim 35\%$. Therefore, in analytical practice the saturation curve may be constructed, with sufficient approximation, by 6–7 points in order to evaluate ρ^{opt} near a transition from the linear part of the curve to saturation.

One should note that when the laser radiation is sufficiently intense it is advantageous to broaden the beam cross section and increase the analytical volume, maintaining the density of the excitation energy at the level ρ^{opt}.

When trace concentrations of Pb-type elements are analysed in nonflame atomisers, certain requirements arise with respect to the duration of a laser pulse. As has already been noted above, Pb atoms in an inert gas atmosphere represent a typical example of a three-level system with a metastable intermediate level. In this situation, when the power of laser radiation is close to the saturating one, the majority of atoms accumulate at the metastable level during a laser pulse. In fact, in experiments with Pb atoms in an Ar atmosphere it was found that at an intensity of laser radiation of about 10 kW cm^{-2} during a 5 ns pulse approximately 40% of the atoms are transferred into the metastable level [2.15]. Increasing the duration of the pulse at a fixed energy density ρ would lead to a decrease in the ratio S/N, i.e. the background grows in proportion to the pulse duration and the growth rate of the fluorescence signal decreases due to depletion of the ground state.

If we use the expression (2.7) we may write for the number of signal photoelectrons recorded per single laser pulse

$$s = C_1 \left(1 - \frac{\xi_2 \exp(-\xi_1 \tau) - \xi_1 \exp(-\xi_2 \tau)}{\xi_2 - \xi_1} \right) \tag{2.32}$$

where C_1 is the coefficient accounting for the parameters of the atomic system and the conditions of registration (the solid angle, the volume, the efficiency of the photoreceiver, etc). During the total time of evaporation of the sample, Δt, the total number of signal photoelectrons is

$$S = s \, \Delta t f \tag{2.33}$$

where f is the laser repetition frequency. The total number of background photoelectrons within the same time is

$$N = (C_1 n_b + n_d) \tau \, \Delta t f \tag{2.34}$$

where n_d is the number of dark photoelectrons per second and n_b is the number of background photoelectrons per second due to scattering of laser radiation and to the thermal radiation of the atomiser. If we again assume Poisson distribution of the background photoelectrons we may write an expression for the signal-to-noise ratio as follows

$$\frac{S}{N} = k \sqrt{\Delta t f} \sqrt{\tau} \left(1 - \frac{\xi_2 \exp(-\xi_1 \tau) - \xi_1 \exp(-\xi_2 \tau)}{\xi_2 - \xi_1} \right)^{-1}. \tag{2.35}$$

It can be shown [2.15] that in experiments with Pb the maximum value of S/N determined from the condition is attained at the duration of the laser pulse

$$\tau_{opt} \simeq \frac{1}{\xi_1} + \frac{2}{\xi_2}. \tag{2.36}$$

Numerical evaluations for Pb at an intensity of laser radiation of approximately 10 kW cm^{-2} give the value $\tau_{opt} \simeq 7$ ns.

Thus, realisation of the maximum sensitivity in the analysis of trace concentrations in inert gas atmospheres requires a thorough analysis of spectroscopic characteristics of the element being investigated. In experiments with flames the collisional rates are, as a rule, large enough and the existence of metastable traps is hardly probable. At the same time other channels may exist for decreasing the total number of atoms interacting with radiation (e.g. chemical reactions of excited atoms with the flame components). Such processes may also require optimum conditions of excitation and LAF recording.

2.2.4   Laser-induced fluorescence in molecules

Recently the techniques and results of theoretical investigations of LAF have been widely introduced into analytical practice for laser-induced fluorescence of molecular systems [2.29–2.36]. On the one hand the analysis of molecular systems provoked great interest from investigators from the viewpoint of the application of molecular fluorescence to measuring the distributions of important molecular components in flames and gas-dynamic flows. It also seems promising for investigation of the relaxation processes in molecules. On the other hand, investigations have been undertaken to analyse the fluorescence of atoms interacting with molecular systems, using it as a tool to study the nature and properties of collisional processes in atom–molecule complexes and as a method to measure low concentrations of molecules.

A theoretical analysis of laser molecular fluorescence (LMF) is based on the results obtained in the course of investigations of transient and steady-state LAF processes. When approaching the problem from this angle one has to consider carefully the specificity of the molecular system by putting into the calculations additional radiation channels for the decay of excited states. Vibrational bath levels bonded to upper and lower laser-coupled levels by collisional transitions should also be considered.

The stationary conditions of the LMF excitation for the two- and three-level models have been considered in [2.30]. The stationary regime of LMF at excitation of the transition $1e \rightarrow 2e$ and recording of fluorescence in the channel $2i \rightarrow 1k$ (the index 1 refers to the ground electronic state, the index 2 to the excited one) has been discussed in [2.31]. The analysis was carried out with an assumption of a high energy density of laser radiation, at which the transition $1e \rightarrow 2e$ is close to saturation. The problem discussed there was the influence of relaxation inside electronic states (Z_{11}, Z_{22}) and quenching of the upper state (Z_{21}) on the distribution of populations of the ground and excited states about the vibrational grid and on the intensity of fluorescence. It was shown by numerical calculations that at the ratio Z_{22}/Z_{21} and $Z_{11}/Z_{21} \leqslant 100$ an appreciable deviation from Boltzmann distribution is observed in both electronic states. The behaviour of an OH molecule in a H_2–O_2–N_2 flame was analysed in [2.32, 2.33], also on the basis of a steady-state approximation. Instead, four-level model solutions for the upper level populations and fluorescence intensities were attained by assuming the approximate equality of the total rates for direct and inverse transitions among laser-coupled levels and vibrational bath levels. The analysis of rate equations for OH in an air–

methane flame [2.34a,b] revealed appreciable discrepancies of the results with those obtained on the basis of rate equations for a two-level model. The results of this work show that in order to obtain an accuracy of calculation better than an order of magnitude, precise values of the rates Z_{22}, Z_{21} are required and also a detailed experimental investigation of the time behaviour of OH fluorescence with nanosecond resolution. In [2.34b] the system of rate equations for OH in a flame was integrated numerically with regard to the processes of vibrational–translational and rotational–translational relaxation and quenching in the ground and excited electronic states. The investigation was aimed at finding the conditions for correct measurement of OH concentration in a flame by the LMF technique. Regions of experimental parameters were found, at which the calculated results [2.34b] differ greatly from those given by a simplified two-level model. The results of the studies show that at pressures in the combustion region below 1 atmosphere OH concentrations may be measured with accuracies of 10–20% at saturating laser powers and the resolution of the fluorescence pulse of the order of several nanoseconds. The possibility of correctly determining OH concentrations at high pressures in the combustion zone was also examined.

An LMF analysis on the basis of rate equations for two- and three-level models was made in [2.35]. Both temporal behaviour of populations of the upper levels $n_2(t)$ and $n_3(t)$ and their stationary values were subject to analysis. It was shown that at a high laser energy density the temporal behaviour of the system is totally determined by the rates of induced transitions. On the basis of the results attained a numerical calculation was made of the dynamics of $n_2(t)$ and $n_3(t)$ for benzophenol, fluorene and rhodamine 6G (with due regard to literature data for the values A_{ij}, Z_{ij}).

In [2.36] a system Na–N_2 was analysed in detail. In this system, laser excitation of the transition $3^2S_{1/2}$–$3^2P_{1/2}$ in a Na atom was accompanied by an efficient collisional bond with the vibrational bath levels of N_2. Collisional population of vibrational levels of N_2 and the associated deviation from the Boltzmann distribution were considered. For the Na–N_2 complex numerical calculations within the three-level model for both populations $n_2(t)$, $n_3(t)$ and fluorescence intensities of Na D_1 and D_2 lines (I_3/I_2) were carried out with regard to the known energy transfer rates. It was shown that with Na concentrations of 10^{10}–10^{13} cm^{-3} the total concentration of N_2 molecules (N_T) may be measured up to 10^{17} cm^{-3} (at $N_T > 10^{17}$ cm^{-3} the decay of the levels 2 and 3 is totally determined by quenching Z_{21} and Z_{31} and the ratio I_3/I_2 no longer depends on N_T). An advantage of this method is the low sensitivity of the value I_3/I_2 to the temperature of the system. As the temperature changes from 300 to 2000 K there is less than a factor of two shift in the ratio I_3/I_2. In figure 2.3 taken from [2.36] results for the I_3/I_2 ratio calculated from the total concentration N_T are presented. These results show the prospects for using fluorescent marks (of the Na-type atom) for diagnostics of profiles of nonfluorescent molecular components in gas-dynamic flows.

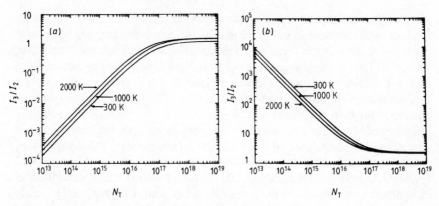

Figure 2.3 (*a*) $3^2 P_{3/2}$–$3\,^2P_{1/2}$ population ratio for the $3\,^2P_{1/2}$ excited level. (*b*) $3\,^2P_{3/2}$–$3\,^2P_{1/2}$ population ratio for the $3\,^2P_{3/2}$ excited level.

2.3   Experimental procedure and apparatus

The structure of laser atomic fluorescence spectrometers (LAFS) is relatively simple and is used with slight variations in practically all analytical applications. A block diagram of an LAFS in its most general form is shown in figure 2.4. The spectrometer consists of the following main units: a source of resonance radiation (a laser), an atomisation system and a recording system. We shall now describe briefly the main units of an LAFS.

2.3.1   Lasers for excitation
The main device which provides smooth tuning of the wavelength of

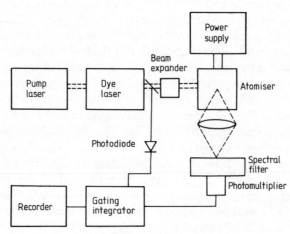

Figure 2.4   General scheme for experiments on laser atomic fluorescent analysis. For explanation see the text.

resonance radiation over a broad spectral range is a dye laser (DL). Both coherent or incoherent DL-pumping are used. When coherent pumping is used a DL is excited by laser radiation with a fixed frequency and in the case of incoherent pumping a DL is excited by xenon flashlamps of different types. Solid state Nd–YAG, nitrogen and excimer lasers are most frequently used for DL excitation. With any kind of excitation direct generation in dye solutions can be obtained only in the spectral range above 360 nm. To obtain tunable generation with wavelengths below 360 nm nonlinear optics techniques are used: generation of harmonics and sum frequencies of DL radiation in nonlinear crystals of the types KDP, ADP, LiF, KPB and also generation of Stokes and anti-Stokes components of the stimulated Raman scattering (SRS) in compressed gases. Specifications of the dye lasers which have developed up to now are summarised in table 2.1.

Peak intensities of radiation given in the third column of the table correspond to the maximum of the tuning curve of the most intense dye (as a rule it is from the rhodamine group). When evaluating the powers attainable in the UV region of the tuning range (below 360 nm) and in the IR region (above 850–900 nm) one should bear in mind that the efficiency of nonlinear

Table 2.1 Specifications of some commercial dye lasers.

Excitation source	Tuning (μm)	Peak power (W)	Pulse duration (ns)	Repetition frequency (Hz)	Line-width (nm)	Manu-facturer	Model
Nd–YAG	0.19–3	10^7	8	1, 5, 10	0.002	Quantel Internat.	YG581–10
	0.19–4.5	10^7	2–6	2–40	0.001	Quanta-Ray	PDL–1
	0.2–0.76	10^7		10.20	0.001	Molectron	DL18P
	0.27–0.44	10^5	10	12.5, 25, 50	0.02–0.033	SIU	LZhI501
	0.55–0.76	4×10^5	10	12.5, 25, 50	0.02–0.033	'Poljus'	
N_2 laser	0.217–0.95	1.3×10^5	6	1–100	0.001	Molectron	DL14P
	0.217–0.84	1.8×10^5	3–4	2–200	0.003	Lambda Physik	FL200E
Excimer	0.21–0.96	5×10^6	5–15	1–200	0.001	Molectron	DL19D
	0.217–0.97	1.5×10^6	6–30	1–100	0.001	Lambda Physik	FL2002E
	0.217–1.175	2×10^5	4–10	1–100	0.002	Instrum. SAJ.Y	Spectrolas F1T
Flashlamps	0.265–2.6	10^5	1000	30	0.2	Chromatix	CMX-4
	0.22–0.96	10^7	500	0.3	0.4–0.001	Phase-R	DL2100B
	0.34–0.75	1.5×10^6	300	1	0.2	Candela	CDL-250
Cu laser	0.53–0.71	3×10^2	20	10^4	0.01	SIU 'Zenit'	LZhI-504

processes, conversion into harmonics, summed and differential frequencies and generation of SRS components is at best 20% at each cascade. As seen from the table the systems based on Nd–YAG lasers provide greater radiation power and this is especially important when one has to use generation of harmonics and sum frequencies of DL radiation in order to obtain the required wavelength. Dye lasers based on N_2 lasers have a lower peak power, but their design is simpler, they can operate with higher repetition frequencies and they are cheaper. Lasers pumped by excimer lasers and by flashlamps are also sufficiently convenient and reliable systems, though flashlamps are not always optimal in analytical practice because of large pulse durations (see §2.2.3).

Table 2.2 gives the elements which have suitable analytical lines in the spectral region covered by current DLs [2.37]. The transition probabilities columns give data from [2.23]. In the last column a method for obtaining the required wavelength is stated. For this table a DL excited by the second harmonic of the Nd–YAG laser has been chosen. The range I–550–670 nm is covered with the fundamental frequency of a DL with two dyes: rhodamine 6G and 6-aminophenolene. The range II–275–335 nm is covered by the DL second harmonic. The range III–360–410 nm is covered by radiation of the sum frequency of the DL and Nd–YAG laser. The range IV–220–260 nm is covered by radiation of the sum frequency of the Nd–YAG laser and the DL second harmonic.

2.3.2   Atomisers
In modern analytical practice of LAF the atomisers most widely used are flame and electrothermal ones.

(a)   Carbon electrothermal atomisers (CETA)
These may be of two types: open and closed. In both versions atomisation is performed in an inert gas atmosphere. A CETA includes a carbon cup or a profiled carbon rod or boat with a recess for sample insertion. The rod, boat or cup is fixed between massive water-cooled metal electrodes. The dimensions of the cup or the recess in the rod allow variation of the sample volume within the limits 2–100 μl. The maximum heating temperature of the cup is 3000 °C. In the closed version of the CETA a metal box is used which covers the platform with the electrodes.

The box is provided with flanges for the input and output windows for laser radiation and the output window for fluorescence radiation. When operating with the elements having an atomisation temperature of more than 1500–1700 °C the CETA box should be water cooled because of its strong heating caused by radiation of the carbon rod or cup. In the open version of the CETA no box with windows is provided. The corrugated shims within the chimney assembly mounted under the rod or cup provide uniform flow of inert gas around the cup or the rod. The CETA is operated automatically by means of a control unit. A set programme controls the rate of heating, maintains the cup

Table 2.2 List of elements which can be analysed by the LAF technique.

| | Wavelength (nm) | | Transition probabilities | | |
| | | | Excitation | Fluorescence | Method of |
Element	Excitation	Fluorescence	$(10^8 \, s^{-1})$	$(10^8 \, s^{-1})$	excitation
Ag	328.1	338.3	3.3	1.3	II
Al	308.2	308.2	2.7	2.7	II
		309.2		5.5	
Au	242.8	312.3	1.8	1.9	IV
	267.6	627.8	1.1	0.05	
B	249.68	249.77	7.0	14	IV
Ba	307.2	472.6	1.8	1.2	II
	553.5	553.5	2.0	2.0	I
Be	234.8	234.8	2.9	2.9	IV
Bi	306.8	472.2	7.0	0.18	II
Ca	272.1	671.7	0.074	0.24	
	422.6	422.6	1.0	1.0	
Cd	326.1	326.1	0.01	0.01	II
Co	304.4	340.5	3.1	15	II
Cr	335.2	464.6	9.8	0.99	II
Cu	324.7	510.5	4.1	0.05	II
Eu	287.9	536.1	2.2	3.2	II
Fe	296.7	373.5	3.9	20	II
Ga	287.4	294.4	5.9	1.8	II
Gd	308.7	313.8	1.7	2.2	II
Ge	265.2	275.4	8.0	9.8	
Hf	288.9	356.7	2.0	0.83	II
		343.8		0.4	
Hg	253.6	253.6	3.5	3.5	IV
Ho	405.3	405.3	24.0	24.0	III
In	303.9	325.9	7.1	2.9	II
Ir	295.1	322.1	1.0	6.6	II
K	404.4	404.4	0.95	0.95	III
La	550.1	583.9	0.32	0.01	I
Li	323.3	323.3	0.34	0.34	II
Lu	308.0	328.1	0.18	3.2	II
	298.9	298.9	1.1	1.1	
Mg	285.2	285.2	9.4	9.4	II
Mn	279.5	279.5	8.3	8.3	II
Mo	294.4	444.3	0.16	0.14	II
Na	589.0	589.0	1.8	1.8	I
Nb	374.2	374.2	1.0	0.1	III
Nd	562.0	562.0	0.67	0.67	I
Ni	301.9	310.1	0.53	4.6	II
Os	280.7	304.1	4.5	5.6	II
Pb	283.3	405.8	1.8	9.2	II

continued

Table 2.2 *continued*

Element	Wavelength (nm) Excitation	Wavelength (nm) Fluorescence	Transition probabilities Excitation (10^8 s^{-1})	Transition probabilities Fluorescence (10^8 s^{-1})	Method of excitation
Pd	276.3	361.7	0.62	6.4	II
Pt	293.0	299.8	2.1	2.2	II
Rb	420.1	420.1	0.96	0.96	III
Re	289.6	439.1	0.22	0.11	II
Rh	312.4	328.1	0.24	2.5	II
Ru	287.5	297.7	6.7	2.1	II
Sb	231.1	231.1	1.5	1.5	IV
Se	297.4	534.9	4.9	3.5	II
Si	251.4	298.8	5.7	0.63	IV
Sm	380.3	392.5	1.5	2.6	III
	562.6	572.0	0.17	0.07	I
Sn	286.3	300.9	5.3	3.9	II
Sr	293.1	716.7	0.05	0.17	II
Ta	287.3	304.9	0.44	1.0	II
Tb	432.6	432.6	7.2	7.2	
Te	225.9	253.0	0.33	0.74	IV
Th	306.0	347.1	0.66	1.5	II
Ti	293.3	296.7	1.4	1.5	II
Tl	276.8	352.9	4.1	2.8	II
Tu	291.4	291.6	1.4	2.8	II
U	311.5	311.5	1.7	1.7	II
V	283.8	286.4	0.48	0.47	II
W	291.1	402.8	0.29	0.24	II
Y	297.4	302.2	2.4	0.44	II
Yb	246.4	246.4	2.7	2.7	IV
Zn	307.6	307.6	0.01	0.01	II
Zr	296.1	301.1	3.0	7.9	II

temperature at a set level during a set period and provides the required supply of inert gas at various stages of sample evaporation and atomisation.

(b) *Flame atomisers*

Flame atomisers used for LAF have some differences in their design compared with burners developed for atomic-absorption analysis. They have, as a rule, a round and not a knife-like jet since the size of the analytical zone for LAF does not exceed more than a few millimetres. Designs with separated flames are often used in which analyte aerosol is introduced with the combustible material into the central part and a pure combustion mixture is fed to the outer zone of the flame. In order to reduce the concentration of nitrogen and oxygen entering the flame from the ambient atmosphere three-section burners provided with an inert gas sheathing are used. Correct choice of burner design and of the composition of the mixture has a considerable influence on the analytical sensitivity and attainable limits of detection. The flame temperature is determined by the mixture composition and the feed rate of components to the combustion zone. Table 2.3 taken from [2.5] presents average temperatures of some flames.

Table 2.3 Composition and temperature of flames.

Combustible	Oxidant	Temperature (K)
Acetylene	Air	2500
	Oxygen	3160
Methane	Air	2150
	Oxygen	2950
Propane	Air	2200
	Oxygen	3125
Hydrogen	Oxygen	2930
	Nitrogen monoxide	3100

When flames are used as atomisers for LAF there is the problem of quenching of excited states by molecular components of the flame. Oxygen is one of the most active quenching agents and therefore one has either to give up completely the idea of using oxygen flames or to dilute the mixture with inert gases (e.g. an oxygen–argon–hydrogen flame).

Compared to CETAS, in flames a higher temperature of the analytical zone is realised and this leads not only to the increase of the analytical signal due to a higher degree of the element atomisation, but also to an increase in the background signal caused by the flame emission. For each particular analytical technique (the type of flame used, the element being analysed, the parameters of the recording system) there exists an optimal position of the analytical zone on the flame axis where the signal-to-noise ratio is a maximum.

As a rule in flames, as compared with CETAS, scattering of the exciting radiation by dust particles (Mie scattering) and by fluctuations of the optical density in the analytical zone (the Rayleigh scattering) is a more serious problem. There is also the problem of molecular luminescence of the organic flame components. All these factors are responsible for a higher level of the background in experiments with flames and, consequently, higher limits of detection compared with CETAS. At the same time such factors as the simplicity of design, operating convenience, and stationary regime of analysis which permits efficient time averaging of the signals make flames of different compositions a rather widely used type of atomiser for LAF.

(c)   *ICP atomiser*

Recently, research into the application of argon inductively-coupled plasma (ICP) as an atom and ion reservoir for the LAF technique has been very active. The major advantage of ICP which gives it wide application in analytical practice is the high temperature of the analytical zone (6000 K). The high temperature of the ICP provides complete dissociation of practically all molecular compounds of the sample being investigated and thus, on the one hand, increases the efficiency of element atomisation, and, on the other hand, diminishes appreciably the influence of the matrix on the analytical signal.

The main components of the ICP are: a radio-frequency generator, a torch and a nebuliser. Systems of generators operating at fixed frequencies in the range 10–50 MHz are most widely used. The power input to the plasma is usually varied within the limits 0.5–2 kW. Systems have been developed which provide automatic stabilisation of the power fed into the plasma and thus ensure high stability of temperature.

ICP torches are a system of three concentric tubes of different diameters made of fused silica or quartz. Aerosol of the analyte being investigated is transported into the plasma by the carrier gas flow along the central capillary. The rate of this flow is usually varied within 0.5–5 l min^{-1}. The other two tubes serve for plasma gas and the stabilising and coolant gas flows of argon. A regime with two flows (a plasma and a carrier flow), is quite common. The rate of the plasma gas flow is varied within the limits 10–20 l min^{-1}.

The nebuliser system provides suction of the sample solution from the reservoir tank, formation of an aerosol with drops of diameter 5–25 μm and transportation of the aerosol formed to the torch. Various designs of nebuliser are used. The most efficient nebulisers so far developed are pneumatic and ultrasonic ones. The optimal design of the nebuliser and its fine tuning have a considerable influence on the ICP analytical sensitivity. A detailed description of the operational principle, main characteristics and designs of separate blocks of the ICP system may be found in [2.38–2.42] and in references cited in these works.

2.3.3   *Recording of fluorescence*

The recording part of the LAF system includes: a system for collecting fluorescence radiation, a system for spectral filtration of fluorescence radiation from nonselectively scattered laser radiation and thermal radiation of atomisers, a photoelectric receiver, an attenuator, a gate integrator, and a system for processing experimental data.

Realisation of the maximum sensitivity of the analysis requires the collection of the fluorescent radiation from the maximum volume V in the maximum solid angle Ω. The signal-to-noise ratio is proportional to $V\Omega$ in the case when the background is determined by noise from the electronics and is proportional to $\sqrt{V\Omega}$ when the background is determined by the atomiser radiation and the scattered exciting radiation. The dimensions of the analytical volume are set by the atomiser design, and by the optimal intensity of laser radiation (see equation (2.23)); in fact they are determined by the parameters of the optical system forming the laser beam before the atomiser. The solid angle is determined by the system of spectral filtration of fluorescent radiation. In the case when interference filters are used, the value Ω may be of the order of 1 steradian. In this case elliptical or spherical reflectors and condensers with a large relative aperture are employed. If a monochromator is used as a spectral filter, the value Ω is set by the monochromator aperture ratio. The aperture ratio of the highest-aperture monochromators reaches about 1:2.

In the case of interference filters as well as in the case of monochromators it is necessary to use systems of screens and diaphragms to prevent direct illumination of the photoelectric receiver by thermal radiation from the atomiser. As a rule it is possible to use interference filters at relatively low atomisation temperatures ($\lesssim 1500$ °C); at high temperatures it is preferable to use monochromators because of the low level of light scattered by them and a sharply outlined spectral transmission band.

In the great majority of analytical applications of the LAF technique photomultipliers (PM) are used as photoreceivers. The best way of measuring the fluorescent signal in this case is by measuring the total charge passing in the anode circuit of the PM during a laser pulse. In LAF applications for diagnostics, as has already been noted above, situations can arise in which it is adequate to measure the instantaneous value of the fluorescence signal in the vicinity of its maximum (at $\partial N_f/\partial t \sim 0$). To avoid overloading the recording system when measuring intense fluorescence signals (in the case of large concentrations of the element being analysed) either neutral light filters before the monochromator are used or the feeding voltage of the PM is decreased.

Lasers developed up to now have a very high off–on time ratio. In such a situation the recording system must be gated for suppression of the thermal radiation of hot parts of the atomiser and noise from the electronics during pauses between pulses. A common way of achieving this is the use of gate voltmeters or gate integrators. The duration of the gate pulse is determined by the specific experimental task and the design of the gate integrators employed and may be either of the same order as the fluorescence pulse duration or appreciably less. The final result of measurements is the sum of the parameter being measured (the amplitude value or the integral of the fluorescence pulse) during the fixed time of recording. The recording time is determined by the duration of the cycle of the sample atomisation when working with a CETA and by the optimal signal-to-noise ratio when working in the stationary regime of the sample atomisation in flames and ICP.

A recently proposed method of determining two-dimensional distributions of fluorescent objects (atoms or molecules) being investigated in flames, flows, discharges, etc seems to be rather promising. In this method laser radiation illuminates the cloud being studied and the optical system focuses the definite cross section of the illuminated region on a Vidicon screen. Thus, during one laser pulse it is possible to obtain the profile of distribution of fluorescent particles. If the sensitivity of the system is high enough and the radiative decay time of the excited state of an atom or a molecule exceeds the duration of the laser pulse, then gating of the recording system also permits determination of the dynamics of spatial distribution of fluorescent particles. Such a technique was exemplified in [2.43] by the recording of the OH distribution in a flame and in [2.44] by the recording of Al in high-voltage discharge initiated by focused laser radiation.

2.4   Analytical applications

2.4.1   *Detection of small concentrations of elements in a gaseous phase*
Analytical applications of the LAF technique are based on the extremely high sensitivity with which atoms can be detected in the gaseous phase. In some experiments, under the conditions when cyclic interaction of an atom with the resonance field was realised, the theoretical limit was attained, i.e. a single atom could be detected at an average concentration of atoms in the interaction volume of 10^{-2}–10^{-4} atoms/mm^3 [2.45–2.50]. A detailed and complete review of work devoted to the single-atom detection problem can be found in [2.51].

It should be noted, however, that the LAF sensitivity at the level of single atoms was demonstrated in specific conditions with Na-type atoms and in cells of a special design with the minimum level of scattered radiation. When working with elements having a more complicated structure of energy levels and in real analytical conditions with a high level of thermal radiation of atomisers the sensitivity of the analysis, naturally, decreases, but it is still much more sensitive than classical spectral techniques.

The possibility of detecting small concentrations of atoms with a strong channel of decay of the upper excited state to the metastable level was exemplified in [2.52, 2.53] by recording lead atoms. The experiments were carried out with saturated vapours of the element in a quartz cell. Fluorescence was excited by radiation of the DL second harmonic with wavelength 283.3 nm. The DL was excited by the radiation of the second harmonic of the Nd–YAG laser. A system of lenses and diaphragms provided a homogeneous distribution of the intensity about the beam cross section in the cell with lead vapour. The intensity of the beam inside the cell was 2 kW cm^{-2}, the linewidth 0.03 nm and the pulse duration FWHM 5 ns. The quartz cell was placed in an electrically heated oven evacuated to the pressure $\leqslant 10^{-7}$ Torr and degassed at a temperature of 600 °C under evacuation.

The fluorescence radiation of wavelength 405.8 nm was collected by a lens from the volume $5 \times 6 \times 4$ mm^3 and over the solid angle 0.16 sr onto the entrance slit of the monochromator with aperture ratio 1:2.5 and recorded by the PM, the amplifier and the gating integrator. At the laser repetition frequency 50 Hz the registration time was 15 s. Each experimental point was determined by averaging the results of five to six measurements for a given temperature. Figure 2.5 shows the density of lead vapour versus temperature. The experimental results were calibrated in absolute units of density with the aid of reference data on the pressure of saturated Pb vapours over the temperature range 600–800 °C. The RSD of the experimental values in each series was 7%. The limiting concentration was determined from the crossing of the experimental curve with the level of tripled RSD of the background fluctuation. In the present experiments it was 250 atoms/cm^3 or 30 atoms in the analytical volume. Such high sensitivity was reached in conditions where each atom was radiating on average about 0.4 photons during the time of interaction with the resonance radiation [2.15] (compare with 10^4–10^5 photons in conditions of cyclic interaction).

High sensitivity of the recording of metastable (4 $^3P_{0,1,2}$) atoms of Ca in a molecular beam was demonstrated in [2.54]. In this work metastable Ca levels were populated by means of electron impact; fluorescence was excited by radiation of a narrow band CW dye laser ($\Delta\nu_L = 1$ MHz). The laser radiation

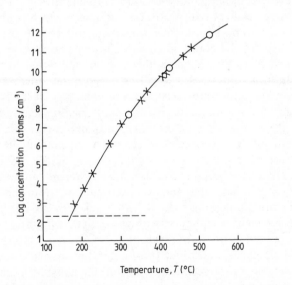

Figure 2.5 Experimental temperature dependence of the Pb saturated vapour concentration (open circles are reference data, crosses are the results of the present work); the broken line shows the level of triple mean-square fluctuations of the background (3σ).

was tuned alternately to the components of the triplet $4\,^3P\!-\!5\,^3S_1$, and frequency-shifted fluorescence was recorded at the 616.2 nm line (a transition $5\,^3S_1\!-\!4\,^3P_2$). The calibration in absolute units of Ca atom concentration was carried out using Rayleigh scattering by Ar with the use of literature data for the Rayleigh cross section. The total accuracy of the measurement of absolute concentrations with regard to the accuracy of spectroscopic data for Ar in the literature did not exceed 21%. The limit of detection in the fluorescence saturation regime was 10^2 atoms/cm³ for an S/N ratio of 2.

Determination of neon microconcentrations in a helium HF discharge by means of the LAF method was performed in [2.55]. In this work a pulsed dye laser excited by the second harmonic of a Nd–YAG laser was used. Excitation of Ne was carried out at the 588.2 nm line, fluorescence was recorded at the 616.4 nm line. Concentrations of Ne were measured in the fluorescence saturation regime. The value of the saturating power $60\,\text{kW}\,\text{cm}^{-2}\,\text{nm}$ determined experimentally turned out to be in a reasonable agreement with the value $43\,\text{kW}\,\text{cm}^{-2}\,\text{nm}$ calculated by the model of a three-level scheme with a metastable level (quenching was not taken into account in calculations: $Z_{ij}=0$). The minimal concentration determined by extrapolation of the calibration curve to the level of triple RSD of the background fluctuations was 10^{-6} per cent of the volume, i.e. 500 times below the level attainable by conventional emission analysis.

Recently some works have appeared in which the LAF technique is used in a version of the two-photon excitation of atoms or molecules [2.56–2.60]. Such an approach considerably broadens the class of objects which can be analysed since it allows excitation of resonance transitions in atoms over the spectral region below 220 nm. To obtain intense, frequency-tunable radiation in the two-photon version of the LAF method (TPLAF) an additional cascade of the frequency transformation is used. For this purpose the radiation of the DL second harmonic is focused into a cell with a molecular gas (usually H_2 or D_2) and excites the effective stimulated Raman scattering of the molecule. At a sufficiently high gas pressure the efficiency of the pump energy transformation into Stokes and anti-Stokes SRS components may reach about 10%. Such an excitation scheme for the TPLAF was suggested in [2.56]. In [2.57] this technique was used to determine concentration of the monatomic oxygen in an acetylene–oxygen flame. Two-photon excitation of a transition $2p^4\,^3P_2\!-\!2p^3\,P_{0,2}$ was carried out using radiation of wavelength 226 nm (the second anti-Stokes component of radiation with wavelength 277 nm). Fluorescence was recorded at the 845 nm line (a transition $2p^3\,3p\,^3P_{0,2}\!-\!2p^3\,3s\,^3S_1$). With regard to the theoretical value of the two-photon transition cross section and quenching constants the concentration of O in the flame being investigated was evaluated as 0.05 mol %. The linear dependence of the observed fluorescence signal on the square intensity of the exciting radiation confirms unambiguously the two-photon character of excitation.

In [2.58] the TPLAF method was used to investigate the distribution of atomic hydrogen in a H_2-O_2-Ar flame. The transition $1s^2S-3d^2D$ was excited by radiation of wavelength 205.1 nm (the first anti-Stokes component of the radiation of wavelength 224 nm). The distribution of H atoms in flames of various stoichiometric composition was examined as was the quenching rate for the $3d^2D$ level, which was evaluated by the tail of a fluorescence pulse (at atmospheric pressure $Z \sim 10^{10}$ s^{-1}). The attained limit of detection of H atoms was evaluated as 10^{16} cm^{-3} (for a flame at atmospheric pressure) and 10^{15} cm^{-3} (for a flame at the pressure of 0.1 atm). The real possibility of decreasing this limit by three to four orders of magnitude was discussed.

In [2.59] the TPLAF method was used to investigate excitation of atomic oxygen (the $3p^3P$ level) and nitrogen (the $3p^4D$ level). In experiments with cells the radiative lifetimes of these levels were determined (39 ns for O and 27 ns for N) as were quenching rates for collisions with N_2 molecules $(2.5 \times 10^{-10}$ cm^{-3} s^{-1}). The detection limit attained for these elements was 10^{14} atoms/cm^3. Increasing the intensity of the exciting radiation by approximately 30 times will make it possible to improve the limits of detection up to 1 PPM.

2.4.2 Determination of LAF analytical sensitivity by aqueous standards

One of the first applications of the LAF method was measurement of trace concentrations of elements in liquids with the use of flame and nonflame atomisers. This analytical application of LAF has maintained a separate direction and its successful development is continuing [2.61–2.73].

A conventional method of determining the potentialities of new analytical methods and to compare sensitivities of various methods is the construction of calibration curves by aqueous standards and the determination, using these curves, of limits of detection (LOD) of elements. At present according to the IUPAC nomenclature the 3σ criterion is accepted for LOD determination, i.e. the limiting concentration of an element is determined from the condition of equality of its analytical signal to triple-RSD background fluctuations. The background is measured in the regime of a blank measurement of a pure solvent which was the basis of synthetic standards. One should differentiate the LOD determined by the purity of the solvent used (the analytical limit) and the limit determined by physical conditions of the experiment: the radiation scattering, thermal radiation of the atomiser, electronic noise, etc (the physical LOD). The analytical limit is determined by a blank experiment with the laser radiation line tuned to the resonance transition of the element being determined. The physical LOD is determined by a blank experiment with the line detuned from the resonance one.

At present the amount of analytical work carried out with flames considerably exceeds that of work with electrothermal atomisers. Reference [2.69] gives a good example of the experimental procedure with a flame atomiser. In this work a flashlamp-pumped DL was used (the CMX-4 model). The iron content of aqueous standards and reference samples was analysed. Fe atoms were excited at the 296.7 nm line and fluorescence was recorded at the 373.5 nm line. The LAF signal was increased by multiple passing of the laser radiation through the analytical zone. For this purpose the burner was placed between the mirrors—plane and concave—and the laser beam was directed through a 0.3 cm aperture in a plane mirror to the multipass cell and was reflected back across the burner head by a concave mirror. Such a multipass regime allowed a threefold decrease in the LOD compared with a single-pass mode.

The calibration curves shown in figure 2.6 were constructed in two different zones of the flame: in its centre and on the periphery. Due to self-absorption effects the slope of the curve for the centre of the flame decreases at a concentration of 10^2 μg ml^{-1}. In the peripheral zone absolute values of signals are less (due to the lower concentration of Fe); however, linearity remains up to concentrations of 4×10^3 μg ml^{-1}. As can be seen in the figure the LOD is 0.06 ng ml^{-1} and in this case it is the physical limit (a measured point on the calibration curve).

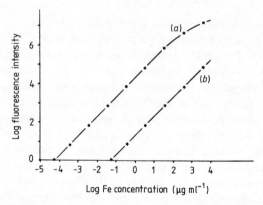

Figure 2.6 Calibration curves for: (a) multipass configuration, viewing centre of flame, with an aspiration rate of 7 ml min^{-1} and (b) single-pass configuration, viewing edge of flame, with an aspiration rate of 1 ml min^{-1}.

The analytical procedure with CETAS may be exemplified by [2.64]. The closed version of the atomiser was used in this work. In order to diminish the level of scattered radiation and of thermal radiation of the carbon boat special orifices and a conic trap were placed under the case. A relatively low atomisation temperature (1400 °C) and high quality of thermal radiation screening permitted the use of an interference filter to isolate fluorescence radiation. The filter bandwidth was 10 nm and transmission at a maximum was 26%.

The content of Tl in aqueous standard solutions was analysed. The

276.8 nm line was used to excite Tl atoms and the total fluorescence intensity at the 352.9 nm and 351.9 nm lines was recorded. Radiation of the second harmonic of the DL, excited by an N_2 laser had the following parameters: pulse energy $0.15 \mu J$, duration 5 ns, spectrum width 0.03 nm, pulse repetition frequency 50 Hz and an energy stability during two hours of operation of 7%.

Standard solutions were prepared by successive dilution by deionised water of the stock solution $(1 \, mg \, ml^{-1})$. The water purity (specific resistance $18 \, M\Omega \, cm$) permitted preparation of standards up to the concentration $0.5 \, pg \, ml^{-1}$. Atomisation of the solution was carried out in three stages: evaporation at 100 °C for 120 s; drying at 400 °C for 90 s; and atomisation of the dry residue at 1400 °C for 7–15 s. At the third stage the Ar flow to the chamber was stopped. In figure 2.7 taken from [2.64] one can see the calibration curve $(2.7(a))$ and part of the recorder trace with ten successive determinations of $50 \, \mu l$ thallium samples at $0.5 \, pg \, ml^{-1}$ $(2.7(b))$. In the concentration region close to the LOD a deviation of the calibration curve from the linear dependence is clearly noticeable. Such typical behaviour in the region of low concentrations is caused by contamination of the deionised water and by nonsterility of the sample preparation.

The authors of [2.64] evaluate the attained LOD as $0.5 \, pg \, ml^{-1}$ which corresponds to the absolute Tl content in the $50 \, \mu l$ sample, $2.5 \times 10^{-14} \, g$ or 7.4×10^7 atoms. In figure 2.7 one can see that the fluorescence peak amplitude which corresponds to the concentration $0.5 \, pg \, ml^{-1}$ is three times more than the amplitude of the maximum spike of the 'zero' line in a blank experiment. In these experiments high (for CETA) analytical accuracy was also achieved. The precision of the measurement was 9% for a concentration of $10 \, pg \, ml^{-1}$; 6% for $0.1 \, ng \, ml^{-1}$ and 4% for $1 \, ng \, ml^{-1}$.

Summing up one may note that, as in analytical practice with classical sources, in the LAF method with graphite atomisers a much lower LOD can be achieved compared with flames, though the precision of the analysis in an electrothermal atomiser is, as a rule, worse. The detection limits currently attained with flame atomisers are given in [2.3]. As a rule they are inferior to the results obtained in emission spectroscopy with an inductively coupled plasma; however the 'LAFS-flame' method has a considerable advantage which will be considered below. The LOD obtained by the 'LAFS–graphite atomiser' method are presented in table 2.4 (from data in [2.63, 2.64, 2.71, 2.72]).

The first results to be obtained were concerned with the investigation of the analytical potentialities of an ICP atomiser in the LAF method. The high temperature of the ICP analytical zone provides not only efficient atomisation, but also ionisation of elements. In this situation it is particularly important to search the optimum plasma region having the maximum concentration of atoms of the element being investigated. As a rule this region is situated some 20–30 cm above the torch in a relatively cooler plasma zone. Distribution profiles of the concentration of Ba atoms and ions were investigated in [2.74]. Distribution of Ca and Y atoms and ions was determined in [2.75]. The

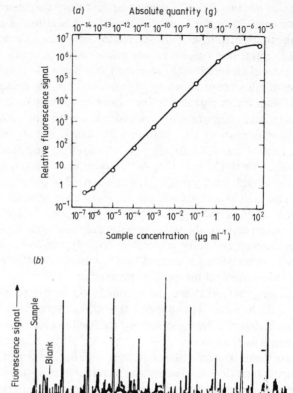

Figure 2.7 (*a*) Calibration curve for thallium. (Error bars are smaller than the symbols used to indicate data points.) (*b*) Ten successive determinations of 50 μl thallium samples at 0.5 pg ml^{-1}. Time scale: (A) sample atomisation, 7 s; (B) chart recorder continues to run after atomisation completed (about 30 s); (C) chart recorder off during evaporation cool-down period (about 60 s), blank insertion, dry and ash cycles.

character of the distribution, naturally, depends on the power introduced into the plasma and on the flow rate of the carrier gas. In these investigations it was shown that at the parameters of the ICP used in the experiment (power 1 kW, the flow rate of the carrier gas 0.9–1.2 l min^{-1}) the plasma temperature in the region 10–25 mm above the torch was too high for fluorescence recording since the large thermal population of the upper levels lead to a strong emission background and deterioration of the signal-to-noise ratio. One of the ways to improve this ratio is fluorescence recording within the zone 50–80 mm above

Table 2.4 Limits of detection of the LAFS–graphite atomiser method.

Element	Limits of detection				
	LAFS		AAS†		AEICP‡
	(ng ml^{-1})	(pg)§	(ng ml^{-1})	(pg)‖	(ng ml^{-1})
Ag	3×10^{-3}	0.1	5×10^{-3}	0.4	0.2(4)
Co	2×10^{-3}	0.06	3×10^{-2}	3	0.1(2)
Cs	2×10^{-2}	1.2			–
Cu	2×10^{-3}	0.06	2×10^{-2}	2	0.06(2)
Ir	0.2	6	20¶	2×10^{3}	–
Fe	1×10^{-3}	0.03	5×10^{-2}	5	0.09(2)
Na	0.02	0.6	0.5¶	50	0.02(10)
Pb	2.5×10^{-5}	7.5×10^{-4}	2×10^{-2}	2	2(20)
Tl	5×10^{-4}	2.5×10^{-2}			–(75)

† Data taken from Bulletin for Perkin-Elmer, model HGA–500.
‡ Data taken from [2.41]. The figures given in the column correspond to the best results available in the literature; the figures in brackets have been taken from a Bulletin for the commercial instrument by Jarrell-Ash.
§ Sample volume 30 μl.
‖ Sample volume 100 μl.
¶ Data from Bulletin for GHA–2100.

the torch where the temperature is considerably lower; however, this requires a special torch design with minimised diffusion of the analyte atoms from the flow at such long distances.

The first results of the determination of the LOD for elements by aqueous standard solution by the LAF–ICP method [2.75–2.77] were not as impressive as could have been expected. They were inferior to the results of the emission analysis with the ICP. The authors [2.75–2.77] related the comparatively high LOD obtained by the LAF–ICP method to the high level of the plasma emission background on the analytical lines of the element being determined and to the anomalously high degree of element ionisation in the analytical zone, which is a sign of the distortion of the local thermodynamic equilibrium (LTE). Thus, in [2.76], even at the minimum power of 0.6 kW introduced into the plasma, the ratio of line intensities BaII/BaI exceeded by 90 times that corresponding to the LTE at 5850 K. In this situation the sensitivity of LAF analysis by ion lines turns out to be higher than that by atomic lines and this is in agreement with the results obtained in [2.41] by emission analysis with the ICP.

An appreciably higher sensitivity of analysis with the ICP was reached in [2.78]. The LOD of the elements investigated are shown in table 2.5, which is an abridged version of a table given in [2.78]. In this work a generator at a frequency of 27 MHz was used, the power was varied within the limits 0.75–1.0 kW and the rates of argon flow were 12 l min^{-1} (plasma gas) and 0.2–0.8 l min^{-1} (carrier gas). A dye laser was excited by the radiation of a 10 MW

Table 2.5 Limits of detection (ng ml^{-1}) obtained by the LAF–ICP method.

Element	Wavelength (nm)		LOD		
	Excitation	Fluorescence	LAF–ICP	AES–ICP†	LAF–ICP‡
Al I	394.403	396.053	0.4	22	7.5
B I	249.678	249.773	4	4	
Ba II	455.404	614.172	0.7	1	1.5(RF)
Ga I	287.424	294.418	1	46	6
		294.364			
Ma I	313.259	317.035	5	8	150(RF)
Pb I	283.307	405.782	1	42	150
Si I	288.158	251.433	1	12	
		251.612			
		251.921			
Sn I	300.915	317.502	3	25	
Ti II	307.865	316.257	1	4	
		316.852			
Tl I	276.787	352.943	7	40	12(RF)
V II	268.796	290.882	3	5	600(RF)
Y II	508.743	371.029	0.6	3	105
Zr II	310.658	256.887	3	7	9000
		257.139			

† Data obtained by emission analysis with the ICP in [2.79].
‡ Data obtained by LAF in [2.80].

excimer laser with pulse repetition rate 100 Hz. The range 240–510 nm was covered by seven dyes and their harmonics. The radiation power in this spectral region was varied from 1 to 200 kW at the spectral linewidth 0.015 nm.

The improvement in the detection limits attained in [2.78] points to the considerable promise of the LAF–ICP method which is now making only its first steps. The high potential of the ICP justifies the large amount of work which will have to be done in order to investigate the mechanisms of fluorescence excitation and of optimisation of the analysis conditions.

2.4.3 Analysis of real objects
Recently the analytical potentialities of the LAF technique have been demonstrated not only by analysis of aqueous standards, but also by analyses of real samples of complicated composition.

Complicated process solutions were analysed for Ir in [2.81]. When the procedures for determination of platinum metals are developed it usually becomes necessary to first isolate and pre-concentrate the elements being determined since their content in the solutions being analysed is on or below the thresholds of their detection by classical methods (emission and absorption analysis). Such solutions also contain a wide spectrum of macrocomponents, acids and organics and this makes direct analysis of platinum elements in process solutions still more difficult. The experiments were carried out in a set-up described in detail in [2.71, 2.72].

The stock solution with a concentration of 1000 PPM was prepared on the basis of iridium chloride in 6N hydrochloric acid. The reference solutions were prepared by successive dilution of the stock standard by 1N hydrochloric acid. The solutions were atomised in three stages: a temperature of 100 °C for 45 s; a temperature of 2100 °C for 60 s; and a temperature of 3000 °C for 12 s. At all stages intensive smoking of the sample occurred. At the first and the third stages the rate of Ar was 11 min^{-1}, at the second stage it was 101 min^{-1}.

Atoms of Ir were excited at the 295.12 nm line and fluorescence was recorded at the 322.28 nm line. The solutions investigated 'poison' the carbon cup and this results in a monotonic decrease of the analytical signal from a single sample. Therefore, in the course of the analysis a permanent comparison with the signal from an aqueous standard with a similar Ir content was made. The analytical signal for the given sample was assumed to be the ratio of the measured signal from the sample to the arithmetic average of two signals from standards analysed before and after the sample measurement. In the investigated samples the accompanying elements Cu, Ni, Fe, Co, Se, Te were in concentrations of 1–10%; the concentration of microcomponents of the platinum group was $10^{-7}\%$ or less.

A strong matrix effect was observed for two samples. The signal for one of them was 10 times lower and for the other one 3 times lower than the signal from the same samples with the modified matrix (when modification of the matrix was carried out the matrix was chemically eliminated without any change in the concentration of the element being determined). Ir determination was performed for the same samples by the method of additions. In other samples this effect was not observed and Ir content was determined by reference solutions.

The results of measurements for some investigated samples are presented in table 2.6. Parallel analyses were carried out by atomic absorption (AAS) with a graphite furnace and with preliminary concentration. Good agreement between the results of the analysis by these two methods should be noted.

A possibility for the direct analysis of solid agricultural products for Co content and soil extracts of different compositions for Co, Fe and Cu contents was demonstrated in [2.71, 2.72]. In the case of direct analysis of a solid matrix in a graphite atomiser calibration against a primary standard is required since aqueous standards are not adequate for solid samples. In the abovementioned experiments a discrepancy of more than an order of magnitude was observed in the analytical signals when a solid sample and an aqueous standard with the same Co content were analysed.

The concentration of Fe in standard reference water samples (SRM–1643),

Table 2.6 Ir content in process solutions.

Solution being analysed	LAFS (μg ml^{-1})	AAS with preconcentration
Sulphuric acid solution 1 (300 g l^{-1} H$_2$SO$_4$)	0.59 $\pm$ 0.07	0.60 $\pm$ 0.15
Sulphuric acid solution 2 (100 g l^{-1} H$_2$SO$_4$)	0.06 $\pm$ 0.02	0.08 $\pm$ 0.02
Sulphuric acid solution 3 (60 g l^{-1} Ni, 70 g l^{-1} Cu, 5 g l^{-1} Fe)	1.1 $\pm$ 0.13	1.0 $\pm$ 0.25
Hydrochloric solution (3M HCl)	1.9 $\pm$ 0.15	2.1 $\pm$ 0.54

pure copper (SRM–394) and fly ash (SRM–1633) was determined in [2.69] by the LAF method with atomisation in the flame. The measured concentrations are in excellent agreement with the certified ones.

The great analytical potential of the LAF with a flame atomiser was demonstrated in [2.70] where standard samples were analysed for Ni and Sn contents. Ni atoms were excited at the 300.249 nm, 299.260 nm and 299.446 nm lines. The total intensity of fluorescence lines over the range 330–360 nm was recorded. In this case the monochromator, with a spectral slitwidth set at 1.6 nm, was adjusted approximately for the central region of the fluorescence spectrum at 342 nm. Sn atoms were excited at the 300.914 nm line and fluorescence was recorded at the 317.505 nm and 380.102 nm lines.

In the analysis of standard samples strong spectral interference with the Fe lines arises (as Fe has a rich spectrum in the region 300 nm). This interference was eliminated in [2.70] by narrowing the spectrum of the exciting radiation.

The results of measuring Ni and Sn concentrations in standard water samples (SRM–1643), pure copper (SRM–394, 396) and fly ash (SRM–1633) are presented in table 2.7 taken from [2.70]. In this table the column Fe/Ni gives the ratio of Fe and Ni concentrations in the samples being analysed and the last column gives the lines of excitation (for Ni) and fluorescence (for Sn) and the spectral width of the exciting radiation.

The results presented in the table demonstrate, firstly, the excellent analytical precision of LAF. In addition, in experiments with SRM-396 and, especially, with SRM-1633, a strong influence of the neighbouring Fe lines (299.4429 nm and 299.4507 nm) on the result of the analysis is observed. Narrowing of the laser line and the choice of an optimum excitation line make it practicable to eliminate this influence.

Similar methods of elimination of strong spectral interference were exemplified in [2.66] by the analysis of Mn in the presence of Ga. It should be noted that at the small spectral interval between the lines of interfering

Table 2.7 Ni and Sn contents in standard samples.

Element	Sample	Certified concentration	Fe/Ni	LAFS	λ (nm)	$\Delta\lambda$ (nm)
Ni	1643	$49\pm1\,\mathrm{ng\,g^{-1}}$	2	$50\pm2\,\mathrm{ng\,g^{-1}}$	300.249	0.002
	396	$4.2\pm0.1\,\mu\mathrm{g\,g^{-1}}$	34	$4.1\pm0.1\,\mu\mathrm{g\,g^{-1}}$	300.249	0.002
				$5.4\pm0.4\,\mu\mathrm{g\,g^{-1}}$	299.446	0.002
	1633	$98\pm3\,\mu\mathrm{g\,g^{-1}}$	633	$99\pm3\,\mu\mathrm{g\,g^{-1}}$	300.249	0.002
				$99\pm4\,\mu\mathrm{g\,g^{-1}}$	290.260	0.002
				$150\pm12\,\mu\mathrm{g\,g^{-1}}$	299.260	0.05
				$623\,\mu\mathrm{g\,g^{-1}}$	299.446	0.002
Sn	394	$65\pm5\,\mu\mathrm{g\,g^{-1}}$		$66\pm3\,\mu\mathrm{g\,g^{-1}}$	317.505	0.002
	396	$0.8\pm0.3\,\mu\mathrm{g\,g^{-1}}$		$0.7\pm0.2\,\mu\mathrm{g\,g^{-1}}$	317.505	0.002

elements the attempts to attenuate spectral interference by isolating—with a monochromator—a narrow region in the fluorescence spectrum turned out either to be completely useless or they resulted in a drastic decrease in sensitivity due to a decrease in the fluorescence signal. In [2.70] the effective elimination of Ni–Fe spectral interference was achieved by narrowing of the laser linewidth up to 0.002 nm while the spectral slitwidth of the monochromator was set at 1.6 nm.

Thus the great analytical potential of LAF for the analysis of real samples is determined not only by the high sensitivity of the method, but also by its high selectivity.

2.4.4 Combination of the LAF method with laser evaporation of a sample

A combination method, LAF–laser ablation (TABLASER), may become a promising direction for the local analysis of pure materials. That this method was of possible analytical use was first demonstrated in [2.82, 2.83]. The idea of this method is evaporation by focused laser radiation of the sample being investigated in a vacuum chamber and the analysis of the laser plume by means of LAF. The direction of the evaporating laser beam is perpendicular to the target, and the axes of the dye laser and the fluorescence recording system are parallel to the target and form three mutually orthogonal vectors.

This method of analysis has some undoubted advantages.

(1) Heavy particles of the laser plume, which are responsible for strong Mie scattering, expand away from the target much more slowly than neutral atoms and ions. Therefore, by the proper choice of the appropriate distance from the target and of the pulse delay of the dye laser one can analyse an atomic cloud in the absence of Mie scattering (see figure 2.8).

(2) Rapid electron–ion recombination and collisional deactivation of atomic

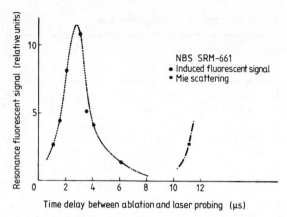

Figure 2.8 Laser-induced resonance fluorescent signal versus delay time at a fixed location 0.55 cm out from target along ruby laser axis.

levels at an early stage of the plume expansion provide predominant concentration of neutral atoms in the ground state in the analytical volume.

(3) Plasma background emission in a cold analytical volume is minimum.

(4) Radial expansion in a vacuum diminishes the probability of particle collisions in the cold zone and the loss of the atoms being analysed due to the formation of molecules and complexes.

(5) Analysis in conditions of saturation provides high concentration sensitivity of the method.

The potentialities of the method were investigated by analysing microcontents of Cr in NBS standard steel samples and specially manufactured pellets of two synthetic samples. For the analysis resonance Cr fluorescence at the 428.9 nm line was used. The sample was evaporated by ruby laser radiation with a pulse energy of 10–20 mJ and pulse duration 30 ns. In the LAF scheme the dye laser was excited by N_2 laser radiation. The radiation power of the dye laser was 1 kW at a pulse duration of 6 ns and the spectrum width was 0.01 nm. The detection limit for Cr attained in the investigated samples was around 1 PPM. It seems quite likely that a well designed TABLASER could achieve sub-PPB sensitivity. A more complicated and improved system and its first analytical tests for Li analysis in a film emulsion are described in [2.84].

2.5 Measurement of spectroscopic parameters

LAF and LMF are used intensively for determination of fundamental spectroscopic parameters of atoms and molecules, e.g. radiative lifetimes and collisional deactivation constant of excited levels. Such investigations are of great interest for atomic and molecular physics and their results make a basis for the improvement of calculation techniques for energy states of systems and elementary interaction processes—atom–atom and atom–molecule collisions. However, as has been noted in §2.2.4, precise data on spectroscopic parameters are also absolutely necessary for analytical and diagnostic applications of LAF and LMF methods. It is not possible to review the whole literature on the problems concerned and here, therefore, we shall only discuss briefly the methods of measuring, by means of the LAF method, the constants of inelastic collisions of atoms with molecules.

In [2.85] saturation of Na fluorescence in an H_2–O_2–Ar flame ($T = 1700$ K) was investigated. In this work a flashlamp-pumped dye laser with a pulse duration of 460 ns was used. At such a pulse duration excitation conditions may be considered stationary and the connection of the saturation parameters ρ^s with the spectroscopic constants of the problem is determined from balance equations. It may easily be shown that the ratio of populations of the upper 3P levels and the saturation parameters ρ^s are determined by the relations:

$$\frac{n_3}{n_2} = \frac{g_3}{g_2} + \frac{A}{YZ_{32}} \tag{2.37a}$$

$$\rho^s = \frac{g_1}{g_3} \frac{8\pi h}{\lambda^3} \frac{g_2+g_3}{g_1+g_2+g_3} \frac{1}{Y} \tag{2.37b}$$

where all the parameters were determined in §2.2 and the quantum efficiency of fluorescence Y is determined by the relation $Y = A/A + Z$, Z being the total quenching rate of the upper level into the ground state and Z_{32} the rate of the nonradiative transition 3–2. By measuring the ratio of intensities of D lines (proportional to n_3/n_2) and knowing the constants A and Z for the flame being investigated it is possible to determine the unknown constant Z_{32}. The authors of the work [2.85] investigated thoroughly the influence of the transverse distribution of the intensity in the laser beam and of the method of fluorescence recording (integral or amplitude) on the saturation curve shape. The influence of these effects on the correctness of the results obtained has already been discussed in §2.2.2. In figure 2.9 the shape of the saturation curves for integral and amplitude regime of fluorescence recording is shown. One can see that for the integral measuring regime the fluorescence intensity continues to grow in the region $\rho > \rho^s$ due to the exponential tails of the pumping pulses. In experimental conditions adequate for the theoretical model (homogeneous distribution of the intensity across the section in the analytical zone, amplitude recording) a saturation curve was constructed and ρ^s was determined (see figure 2.10). Excellent agreement between experimental and theoretical values of ρ^s was obtained and the quenching constant was determined to be $Z_{32} = 3 \times 10^{-8}$ s^{-1}.

In reference [2.86] the cross section of the transition $^3P_1^\circ$–$^3P_0^\circ$ in lead was determined under collisions with atoms and molecules of buffer gases. In these

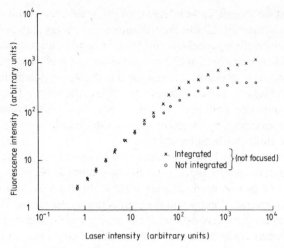

Figure 2.9 Experimental demonstration of the effect of time integration of the fluorescence pulse on the saturation curve.

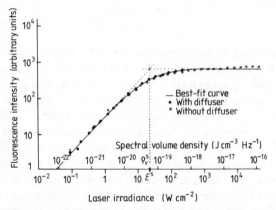

Figure 2.10 An experimental saturation curve of the Na D_1 line with the laser tuned at the D_2 line. The curve represents the best fit of the theoretical curve to the experimental points. $\rho_v^s = 5.6 \times 10^{-20}$ J cm^{-3} Hz^{-1}; $E^s = 22$ W cm^{-2}.

experiments laser radiation of wavelength 283.3 nm excited a level $^3P_1^o$ and the fluorescence from the collisionally populated level $^3P_0^o$ was recorded. Part of the energy scheme of lead levels is shown in figure 2.11 where the above levels are indicated by 7 and 6. It can be shown that the ratio of integral fluorescence intensities at the 405.8 nm (transition 7–3) and 368.3 nm line (transition 6–2) is

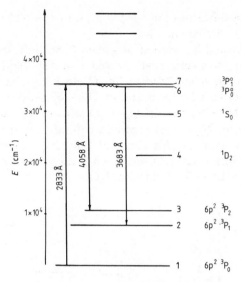

Figure 2.11   A scheme of lead levels.

determined by the relation

$$\alpha = \frac{I_{62}}{I_{73}} = B\,\frac{A_{62}}{A_{73}}\,\frac{Z_{76}}{A_{62}+Z_6} \tag{2.38}$$

where B is the coefficient which takes into account the sensitivity of the recording system, Z_{76} is the rate of excitation transmission between the levels 7 and 6, and Z_6 is the total rate of nonradiative decay of the level 6. Using the relation $Z = N\langle\sigma v\rangle$ (where N is the concentration of quenching particles and σ is the cross section of the appropriate process) and $P = Nk\,T$ (where P is the buffer gas pressure and T is the temperature) we obtain for the cross section

$$\langle v\sigma_{76}\rangle = \frac{A_{73}}{BA_{62}}\,(A_{62}+Z_6)kT\,\frac{\alpha}{P}. \tag{2.39}$$

In the region of low concentrations of the buffer gas $Z_6 \ll A_{62}$ and the value $\langle v\sigma_{76}\rangle$ depends only on A_{73}. The value $\langle v\sigma_{76}\rangle$ is determined by the slope of the graph for the dependence of α on P in the region of low pressures of the buffer gas. At large pressures P a reverse situation may be realised, when $Z_6 \gg A_{62}$. In this case the population of level 6 reaches saturation and the same is true for the value α. Thus, having determined $\langle v\sigma_{76}\rangle$ for the linear part, one can obtain the cross section $\langle v\sigma_6\rangle$ from the relation

$$\langle v\sigma_6\rangle = \frac{\langle v\sigma_{76}\rangle}{\alpha^{max}}\,\frac{A_{62}B}{A_{73}}. \tag{2.40}$$

From the found values $\langle v\sigma\rangle$ the cross sections of the appropriate processes may be defined by the expression $\sigma = \langle v\sigma/\bar{v}\rangle$, where $\bar{v} = (8kT/\pi M)^{1/2}$ (M is the reduced mass of colliding particles).

The values Z_{76} and Z_6 were measured in [2.86] in a specially designed quartz cell placed into an electrically heated oven. The temperature in the cell was varied from room temperature to $700\,°C$, He, Kr, Ar, Ne, NH_3, H_2, CH_4, CO_2, N_2 and H_2O were used as quenchers. The dependence of α on the buffer gas pressure for N_2, CH_4 and an $Ar + N_2O$ mixture is shown in figure 2.12. One can see that for N_2 the curve reaches saturation in the pressure region around 400 Torr, and for this case not only Z_{76} but also Z_6 may be determined. The measured cross sections σ_{76} vary from $(0.6-3) \times 10^{-20}$ cm^2 for noble gases to 1.5×10^{-16} cm^2 for H_2O.

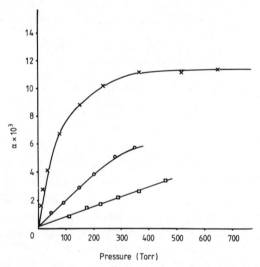

Figure 2.12   The ratio of fluorescence intensities α versus the pressure of the buffer gas: $\times$, N_2; $\bigcirc$, CH_4; $\square$, $N_2O + Ar$.

Recently the LAF method has been used intensively for spectroscopic investigations of the ICP. Spatial profiles of the distribution of atoms and ions of the elements under investigation required for the choice of optimum conditions of analysis and the fundamental characteristic lifetime of excited states have been determined [2.87–2.90]. The values of τ were determined in [2.91] for Na, Li, Al, Ga and Tl and in [2.92] for a number of transitions in Cr and Fe atoms and in Sc and Y ions. If the values of radiative lifetimes of the same levels τ_0 are known from independent experiments the quantum efficiency of fluorescence of the given level in the ICP is determined as the ratio $Y = \tau/\tau_0$. The values Y allow evaluation of the role of collisional deactivation of excited levels. In particular the high quantum efficiency of fluorescence of Na D lines determined in [2.92] permits one to make a conclusion about the minor role of collisional deactivation processes (including the electron impact $M^* + e \rightarrow M + e$) in the ICP systems that have been investigated. The same conclusion is valid for elements of the III group investigated in [2.92].

The rates of deactivation and mixing of Cu levels ($^2P_{1/2,3/2}$ and $^2D_{3/2,5/2}$) in various gases were determined in [2.93]. These levels are used as operational ones in Cu lasers. Quenching of In levels ($5p^2\,^4P_{1/2}$) and ($6p^2\,P$) in Ar, H_2, N_2, D_2 and CH_4 was investigated in [2.94]. The quenching cross sections of the level $^4P_{1/2}$ vary from 1.6×10^{-16} cm^2 for Ar to 37×10^{-16} cm^2 for H_2. For the level $6p^2P$ these cross sections are about half those values. Other examples of the use of the LAF method for determination of radiative and collisional decay rates of excited levels can be found in [2.95–2.102] and in references therein.

2.6 Conclusions and prospects

The LAF method (and the allied LMF method) is currently widely used in various physical and physico-chemical investigations. In analytical applications for detection of trace concentrations of elements the LAF method has demonstrated its high sensitivity which permits direct analysis of real samples in cases when conventional analytical methods require the stages of preliminary chemical preparation and concentration. The LAF method is also characterised by high selectivity when used for analysis of real samples and this permits elimination of spectral interference caused by close spacing of lines of the element being determined and by the lines of macrocomponents of the complex matrix. Selectivity of the LAF method may play the same important role in analytical practice as high sensitivity does.

Further progress in analytical applications of the LAF method is connected with improvement of the characteristics of the lasers used. This is primarily a question of increasing the stability of spectral and energy parameters of the radiation, and the reliability and longevity of laser systems. An important source of increasing the sensitivity of the analysis is an increase in spectral brightness of the radiation and a decrease of the off–on time ratio. From this viewpoint a copper-vapour laser operating with a repetition frequency of 10^4 Hz seems to be a promising system (see table 2.1).

A dye laser with acousto-optical tuning of the radiation wavelength [2.103–2.106] may also turn out to be a promising radiation source for the LAF method. In such a laser tuning of the wavelength is performed by varying the frequencies of electric voltage fed to the acousto-optical crystal. Stability and reproducibility of frequencies of electric voltage generators may turn out to exceed appreciably the stability of mechanical systems of mirrors or grating and étalon rotation used to tune wavelengths in conventional schemes of dye lasers. In addition to the abovementioned advantage, lasers with an acousto-optical crystal may pave the way to multi-element analysis since they may operate simultaneously at several wavelengths. On-line control of such a laser by means of a minicomputer may provide various modes of operation: generation at several wavelengths in each pulse or alternation of laser operational wavelengths in various pulses in accordance with a preset law. With such modes of laser operation it becomes possible not only to analyse several elements simultaneously, but also to measure the background signal during the same cycle of the sample evaporation.

At present in the LAF method photoelectric recording of fluorescent radiation by means of photomultipliers (in some cases a photodiode array or a Vidicon) is widely used. Another new method was recently suggested in reference [2.106]. It is suggested in that reference that fluorescent photons can be detected in a special cell by means of two- or three-stage resonance ionisation of the gas, which fills the cell, with subsequent detection of the ions formed. A gain, compared to the conventional scheme monochromator–PM, may be attained owing to the large solid angle of accumulation of fluorescent photons into a cell of the resonance ionisation detector and to the high quantum efficiency of multiphoton ionisation. The method proposed now looks somewhat exotic, but its experimental realisation may turn out to be promising.

The LAF and LMF techniques are promising for diagnostics of flames, flows, burning processes, etc, since they permit, in principle, determination of concentrations of particles and radicals, local temperatures and also the rates of chemical reactions and excitation transmission. However, a large amount of work needs to be done in this direction in order to accumulate spectroscopic information required for unambiguous association of the fluorescence intensities being measured with the parameters being diagnosed (absolute concentration of atoms or molecules and the local temperature). The problems arising in connection with the use of the LAF and LMF techniques for diagnostics of combustion processes are analysed in [2.107, 2.108]. In obtaining such information the LAF and LMF methods play a leading role alongside the supplementary methods of spontaneous and coherent Raman scattering and absorption spectroscopy.

References

2.1   Omenetto N and Wineford ner J D 1979 Laser-excited atomic fluorescence spectroscopy in *Analytical Laser Spectroscopy* ed N Omenetto (New York: Wiley)

2.2   Zaidel A N 1980 *Atomic Fluorescence Analysis: Physical foundations* (Moscow: Nauka) (in Russian)

2.3   Omenetto N and Wineford ner J D 1981 Laser in analytical spectroscopy in *CRC Critical Reviews in Analytical Chemistry* vol. 13, p. 59

2.4   Weeks S J and Winefordner J D 1981 Laser-excited atomic fluorescence spectrometry in *Lasers in Chemical Analysis* ed G N Hieftje, J C Travis and F E Lytle (Clifton, NJ: Humana)

2.5   Zaidel A N 1983 *Atomic Fluorescence Analysis* (Leningrad: Khimiya) (in Russian)

2.6   Grishko V I and Judelevich I G 1982 *Zavod. Lab.* **48**(4) 1 (in Russian)

2.7   Harvis T D and Lytle F E 1983 Analytical applications of laser absorption and emission spectroscoy in *Ultrasensitive Laser Spectroscopy* ed Kliger (New York: Academic)

2.8   Allen L and Eberly J H 1975 *Optical Resonance and Two-level Atoms* (New York: Wiley)

2.9   Daily J W 1977 *Appl. Opt.* **16** 2322

2.10   Daily J W 1979 *Appl. Opt.* **18** 360

2.11   *Laser and Coherence Spectroscopy* 1978 ed J I Steinfeld (New York: Plenum)

2.12   Daily J W 1977 *Appl. Opt.* **16** 568

2.13   Boutilier G D *et al* 1978 *Appl. Opt.* **17** 2291

2.14   Bolshov M A, Zybin A V, Zybina L A and Koloshnikov V G 1976 *Preprint* 2/27 Inst. Spectrosc. Acad. Sci. (in Russian)

2.15   Bolshov M A, Zybin A V, Koloshnikov V G and Koshelev K N 1977 *Spectrochim. Acta. B* **32** 279

2.16   Olivares D R and Hieftje G M 1978 *Spectrochim. Acta B* **33** 78

2.17   Zizak G, Bradshaw J D and Winefordner J D 1980 *Appl. Opt.* **19** 3631

2.18   Omenetto N, Van Dijk C A and Winefordner J D 1981 *Appl. Spectrosc.* **35** 389

2.19   Omenetto N, Van Dijk C and Winefordner J D 1982 *Spectrochim. Acta B* **37** 703

2.20   Muller K G and Stania M 1978 *Appl. Phys.* **49** 5801

2.21   Mailander M 1978 *J. Appl. Phys.* **49** 1256

2.22   Deakin J J and Husain D 1966 *J. Chem. Soc. Faraday Trans.* II **68** 1603

2.23   Corliss C H and Bozman W R 1962 *Experimental Transition Probabilities* NBS Monograph 53 (Washington, DC: NBS)

2.24   Daily J W 1978 *Appl. Opt.* **17** 225

2.25   Blackburn M B, Mermet J M, Boutilier G B and Winefordner J D 1979 *Appl. Opt.* **18** 1804

2.26   Sharp B L and Goldwasser A 1976 *Spectrochim. Acta B* **31** 431

2.27   Boutilier G D, Bradshaw J D, Weeks S J and Winefordner J D 1977 *Appl. Spectrosc.* **31** 307

2.28   Matveyev O O 1983 *Zh. Prikl. Spektrosk.* **39** 709 (in Russian)

2.29   Baronavski A P and McDonald J R 1977 *Appl. Opt.* **16** 1897

2.30   Boutilier G D, Winefordner J D and Omenetto N 1978 *Appl. Opt.* **17** 3482

2.31   Lucht R P and Laurendeau N M 1979 *Appl. Opt.* **18** 856

2.32   Lucht R P, Sweeney D W and Laurendeau N M 1980 *Appl. Opt.* **19** 3295

2.33   Lucht R P, Sweeney D W and Laurendeau N M 1983 *Combust. Flame*

2.34   (a) Campbell D H 1982 *Appl. Opt.* **21** 2913; (b) 1984 *Appl. Opt.* **23** 689

2.35   Boutilier G D, Omenetto N and Winefordner J D 1980 *Appl. Opt.* **19** 1838

2.36   Campbell D H and Lewis J W L 1981 *Appl. Opt.* **20** 4103

2.37   Zybin A V 1981 *PhD Thesis* Institute of Spectroscopy Acad. Sci. Troitsk

2.38   Boumans P W J M and de Boer F J 1972 *Spectrochim. Acta B* **27** 391

2.39   Fassel V A and Kniseley R N 1974 *Anal. Chem.* **46** 1110A

2.40   Nixon R N, Fassel V A and Kniseley R N 1974 *Anal. Chem.*

2.41 Boumans P W J M and de Boer F J 1975 *Spectrochim. Acta* B **30** 309
2.42 Boumans P W J M 1982 *Recent Advances in Analytical Spectroscopy* ed K Fuma (Oxford: Pergamon)
2.43 Deyer M J and Crosley D R 1982 *Opt. Lett.* **7** 382
2.44 Dougal R A, Williams P F and Pease D C 1983 *Rev. Sci. Instrum.* **54** 572
2.45 Balykin V I, Letokhov V S, Mishin V I and Semchishen V A 1977 *Pis'ma Zh. Eksp. Teor. Fiz.* **26** 492 (in Russian)
2.46 Gelbwachs J A, Klein C F and Wessel J E 1977 *IEEE J. Quantum Electron.* **13** 11D
2.47 Gelbwachs J A, Klein C F and Wessel J E 1978 *IEEE J. Quantum Electron.* **14** 121
2.48 Gelbwachs J A, Klein C F and Wessel J E 1977 *Appl. Phys. Lett.* **30** 489
2.49 Lewis D A, Tonn J F, Kaufman S L and Greenless G W 1979 *Phys. Rev.* **A19** 1580
2.50 Pan C L, Prodan J V, Fairbank W M and She C Y 1980 *Opt. Lett.* **5** 459
2.51 Alkemade C Th J 1981 *Appl. Spectrosc.* **35** 1
2.52 Bolshov M A, Zybin A V and Koloshnikov V G 1980 *Sov. Quantum Electron.* **7** 1808 (in Russian)
2.53 Bolshov M A, Zybin A V, Koloshnikov V G and Vasnetsov M V 1981 *Spectrochim. Acta* **36B** 345
2.54 Dobryshin V E, Rachovsky B I and Shustriakov V M 1983 *Optika i Spectroskopiya* **54** 68 (in Russian)
2.55 Bolshakov A A, Oshemkov S V and Petrov A A 1983 *Zh. Priklad. Spektrosk.* **39** 757 (in Russian)
2.56 Bishel W K, Perry B E and Crosley D R 1981 *Chem. Phys. Lett.* **82** 85
2.57 Alden M, Edner H, Grafstrom P and Svanberg S 1982 *Opt. Commun.* **42** 244
2.58 Lucht R P *et al* 1983 *Opt. Lett.* **8** 365
2.59 Bishel W K, Perry B E and Crosley D R 1982 *Appl. Opt.* **21** 1419
2.60 Tiee J J, Ferris M J, Loge G W and Wampler F B 1983 *Chem. Phys. Lett.* **96** 422
2.61 Mayo S, Keller R A and Travis R B 1976 *J. Appl. Phys.* **47** 4012
2.62 Bolshov M A, Zybin A V, Zybina L A and Koloshnikov V G 1976 *Spectrochim. Acta* **31B** 493
2.62 Hohimer J F and Hargis P J Jnr 1977 *Appl. Phys. Lett.* **30** 344
2.64 Hohimer J P and Hargis P J Jnr 1978 *Anal. Chim. Acta*
2.65 Smith B W, Blackburn M B and Winefordner J D 1977 *Can. J. Spectrosc.* **22** 57
2.66 Weeks S J, Haraguchi H and Winefordner J D 1978 *Anal. Chem.* **50** 360
2.67 Blackburn M B, Mermet J M and Winefordner J D 1978 *Spectrochim. Acta* **33B** 847
2.68 Haraguchi H, Weeks S J and Winefordner J D 1979 *Spectrochim. Acta* **34B** 391
2.69 Epstein M S *et al* 1980 *Spectrochim. Acta* **35B** 233
2.70 Epstein M S *et al* 1980 *Appl. Spectrosc.* **34** 372
2.71 Zybin A V and Smirenkina I I 1981 *Preprint* Institute of Spectroscopy Acad. Sci. No 1, Troitsk (in Russian)
2.72 Bolshov M A, Zybin A V and Smirenkina I I 1981 *Spectrochim. Acta* **36B** 1143
2.73 Fujiwara K *et al* 1980 *Appl. Spectrosc.* **34** 85
2.74 Omenetto N *et al* 1980 *Spectrochim. Acta* **35B** 507
2.75 Uchida H, Kosinski N A and Winefordner J D 1983 *Spectrochim. Acta* **38B** 5
2.76 Pollard B D *et al* 1979 *Appl. Spectrosc.* **33** 5

2.77   Epstein M S *et al* 1980 *Anal. Chim. Acta* **113** 221

2.78   Omenetto N, Haman H G C, Cavalli P and Rossi G 1984 *Spectrochim. Acta* **39B** 115

2.79   Boumans F W J M 1980 *Line Coincidence Tables for Inductively Coupled Plasma Atomic Emission Spectrometry* (Oxford: Pergamon)

2.80   Kosinski M A, Uchida H and Winefordner J D 1983 *Talanta* **30** 339

2.81   Bolshov M A *et al* 1984 *Zh. Analitich. Khim.* **39** 320 (in Russian)

2.82   Measures R M and Kwong H S 1979 *Appl. Opt.* **18** 281

2.83   Kwong H S and Measures R M 1979 *Anal. Chem.* **51** 428

2.84   Lewis A L, Beenen G J, Hosch J W and Piepmeier E H 1983 *Appl. Spectrosc.* **37** 263

2.85   Van Calcar R A *et al* 1979 *J. Quantum Spectrosc. Radiat. Transfer* **21** 11

2.86   Bolshov M A, Zybin A V and Koloshnikov V G 1979 *Optika i Spektroskopiya* **46** 417 (in Russian)

2.87   Havey M D, Balling L C and Wright J J 1977 *J. Opt. Soc. Am.* **67** 49

2.88   Klose J Z 1979 *Phys. Rev.* **19A** 678

2.89   Rudolph J and Helbig V 1982 *J. Phys. B: At. Mol. Phys.* **15** 21

2.90   Kwiatkowski M, Micali G, Werner K and Zimmermann J 1982 *J. Phys. B: At. Mol. Phys.* **15** 4357

2.91   Uchida H, Kosinski M A, Omenetto N and Winefordner J D 1983 *Spectrochim. Acta* **38B** 529

2.92   Uchida H, Kosinski M A, Omenetto N and Winefordner J D 1984 *Spectrochim. Acta* **39B** 63

2.93   Hao-Lin Chen and Ebert G 1983 *J. Chem. Phys.* **78** 4985

2.94   Hovis F E and Gelbwachs J A 1983 *J. Chem. Phys.* **78** 6680

2.95   Himmel G and Sowa L 1983 *J. Quantum Spectrosc. Radiat. Transfer* **30** 357

2.96   Harvis M and Lewis E L 1982 *J. Phys. B: At. Mol. Phys.* **15** 2613

2.97   Husain D and Schifino J 1983 *J. Chem. Soc. Faraday Trans. II* **79** 919 and 1265

2.98   Brewer P, Das P, Ondrey G and Bersohn R 1983 *J. Chem. Phys.* **79** 720 and 724

2.99   Hannaford P and Lower R M 1983 *J. Phys. B: At. Mol. Phys.* **16** 243

2.100   Duquetto D W, Salih S and Lawler J F 1982 *Phys. Rev.* A **26** 2623

2.101   Duquetto D W and Lawler J E 1982 *Phys. Rev.* A **26** 330

2.102   Salih S, Duquetto D W and Lawler J E 1983 *Phys. Rev.* A **27** 1193

2.103   Taylor D J, Harris S E, Nich S T K and Hansch T V 1971 *Appl. Phys. Lett.* **19** 269

2.104   Dmitriev V G and Cherednichenko O B 1980 *Izv. Akad. Nauk* **44** 1720 (in Russian)

2.105   Stelmach M F *et al* 1984 *Zh. Priklad. Spektrosk.* **40** 181 (in Russian)

2.106   Matveyev O I 1983 *Zh. Anal. Khim.* **38** 736 (in Russian)

2.107   Schofield K and Steinberg M 1981 *Opt. Eng.* **20** 501

2.108   Crosly D R and Smith G P 1983 *Opt. Eng.* **22** 545

3 Laser Atomic Photoionisation Spectral Analysis

G I Bekov and V S Letokhov

3.1 Introduction—statement of the problem

The development of new analytical methods for determining ultra-low contents of elements in different substances is of prime importance for science and engineering at the present time. This is because a wide range of problems currently faced in the processing of high-purity materials, in geology and geochemistry, in toxicology, in environmental protection, etc could be solved by controlling the trace elements which may be present in amounts of 10^{-8}– $10^{-10}\%$. In some cases such sensitivity of analysis can be provided by conventional analytical methods [3.1] or their modifications: atomic absorption and atomic fluorescence spectrometry, neutron activation analysis, spark mass spectrometry and others. But in most cases their sensitivity is limited to the level of $10^{-7}\%$.

Recently developed laser methods of single-atom detection are of particular interest for analytical applications [3.2–3.4]. They are based on the methods of laser excitation of atomic fluorescence and on laser stepwise resonant photoionisation of atoms. However, it is necessary to solve a number of additional problems before these methods can be directly used in analytical tasks.

The process of determination of ultra-low element traces in the substance consists of three successive stages:

(i) production of free atoms of the element;

(ii) transport of these atoms into the region of the laser beam;
(iii) detection of the atoms by the laser radiation.

All these stages of analysis can be solved, as a whole, by combining the methods of atomisation used in conventional analytics with the laser methods of detection. The use of laser radiation in atomic fluorescence analysis has made it possible to reduce the detection limits of some elements by two or three orders of magnitude (see Chapter 2). In practice, however, it is very difficult to improve detection limits and to reach a high sensitivity in the analysis of real objects because of some restrictions typical also of conventional analytical methods. The need to atomise the substance in an inert gas is a disadvantage of the method. The atoms formed can be lost as a result of their interaction with the atoms and molecules formed by atomisation of the base material and with the microimpurities in the gas.

The use of a preliminary chemical separation of the matrix by means of different chemical reagents may lead to both an uncontrollable addition of the element with the reagents and a loss of the element in the process of chemical extraction of the matrix. And finally, the detection limit of the chosen element in the substance will be defined by the content of this element in the inert gas. These shortcomings can be eliminated by preventing interaction between the atoms of the element and the environment, for example by atomisation of the substance in a vacuum when direct analysis of the substance in its natural form is possible. Evaporation and atomisation can be realised by thermal heating up to 3000 °C, a laser radiation pulse, an electron beam, an electric discharge, etc. The choice of one or other method of atomisation is defined by the specific task. But the simplest and most universal method, in our opinion, is thermal atomisation in vacuum.

The method of laser stepwise resonant photoionisation of atoms suggested and realised more than ten years ago [3.5] is the most suitable for the detection of the resulting atoms. The high sensitivity of this method is due to the fact that the ionisation products (ions or electrons) can be detected with an efficiency of nearly unity. Thus, the detection efficiency of a neutral atom will be defined mainly by the photoionisation efficiency which depends on the laser radiation and the method of ionisation, and may reach unity.

The basic consideration of this chapter will be the method of laser stepwise photoionisation of atoms combined with thermal atomisation of the substance in a vacuum. The specific features and the advantages of such a combination will be discussed here, too. Some experiments on the determination of trace elements in real substances—high-purity materials, natural and biological objects—will be considered.

The method of stepwise resonant photoionisation of atoms has been considered in detail in reviews [3.2–3.4, 3.6] and monograph [3.7]. It is useful, however, to consider the basic features and characteristics of this method so that readers may familiarise themselves.

3.2   The method of laser multistep photoionisation of atoms

In §1.5.3 the principle of laser spectroscopy of multistep atomic and molecular excitation was briefly described. The present section will be concerned with the properties of this method as applied to atoms. The most essential characteristics for analytical applications are an ultimate detection sensitivity, up to the level of single atoms, and a high selectivity that allows detection of the analysed atoms against the background of a large number of foreign particles. Below we shall consider the requirements relating the parameters of the laser radiation and atomic system which allow us to attain the ultimate results.

3.2.1   Stepwise photoionisation schemes

In the method of laser multistep photoionisation the atoms are excited by laser radiation to an intermediate high-lying state through one or several steps and then only the excited atoms are subjected to photoionisation. Three approaches in the method of stepwise photoionisation may be singled out, differing from each other in the way of ionisation of the atom from an intermediate state (figure 3.1):

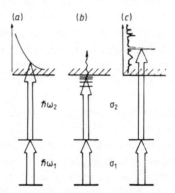

Figure 3.1 The schemes of stepwise photoionisation of atoms: (*a*) nonresonant ionisation to a continuum; (*b*) atomic ionisation by the electric field through Rydberg state; (*c*) photoionisation of an atom via the autoionisation state.

(i) nonresonant photoionisation of excited atoms to the continuum;

(ii) atom ionisation from Rydberg states by an electric field, as a result of the collisions with the buffer gas particles, etc;

(iii) resonant photoionisation of excited atoms by their excitation to a narrow autoionisation state.

Each of these approaches is considered in more detail below.

(a) Ionisation by transition to the continuum

In this case the excited atom is ionised by additional laser radiation or by the radiation used at one of the resonant excitation steps (figure 3.1(a)). Such nonresonant photoionisation is characterised by comparatively low cross sections ($\sigma_{ion} = 10^{-17} - 10^{-19}$ cm^2) as compared with the cross section of resonant excitation of intermediate levels ($\sigma_{exc} = 10^{-11} - 10^{-14}$ cm^2). For effective excitation and subsequent photoionisation of excited atoms the pulsed laser energy fluxes must comply with the following conditions:

$$\phi_{exc} \gtrsim \phi_{exc}^{sat} = \frac{\hbar\omega_{exc}}{2\sigma_{exc}} \qquad \phi_{ion} \gtrsim \phi_{ion}^{sat} = \frac{\hbar\omega_{ion}}{\sigma_{ion}}. \tag{3.1}$$

For an ionising pulse the saturation energy flux ϕ_{ion}^{sat} lies in the range from 0.01 to 1 J cm^{-2} (for exciting pulses the corresponding values of ϕ_{exc}^{sat} are smaller by $2\sigma_{exc}/\sigma_{ion}$ times). Such parameters of laser radiation are attained by available pulsed lasers if the required pulse repetition rate does not exceed several tens of hertz.

Several instances of such photoionisation may be cited. In [3.8] helium atoms in the excited state 2^1S were detected using two-step photoionisation via the intermediate state 3^1P by a laser radiation pulse with $\lambda = 5015$ Å. With the photoionisation cross section $\sigma_{3^1P} = 4.4 \times 10^{-18}$ cm^2 every He (2^1S) atom was ionised at a photon flux equal to 1.8×10^{18} cm^{-2} or 0.7 J cm^{-2}. In [3.9] the two-step photoionisation of Rb atoms through the intermediate states $6^2P_{1/2,3/2}$ was studied experimentally. When the atoms were ionised by ruby laser radiation ($v = 14\,403$ cm^{-1}) the ionisation yield was saturated with the pulse energy flux $\phi_{ion} \gtrsim 0.03$ J cm^{-2} corresponding to $\sigma_{ion} \simeq 1.7 \times 10^{-17}$ cm^2. In experiment [3.10] on the detection of Cs atoms the photoionisation was accomplished via the intermediate state $7^2P_{3/2}$ by the laser pulse ($\lambda = 4555$ Å) used to excite the atoms. To saturate the transition to the continuum with $\sigma_{ion} = 8.8 \times 10^{-18}$ cm^2 and to reach 90% ionisation the laser pulse energy flux required was 0.1 J cm^{-2}.

(b) Ionisation through Rydberg states

In this method, suggested in [3.11], the atom from an intermediate state is excited to a Rydberg state lying near the ionisation limit and then ionised by an electric field pulse. The studies [3.12–3.16] have demonstrated that Rydberg atoms have a unique ability to be ionised comparatively easily in an electric field, no matter what the type of the atom. And each Rydberg state is characterised by its value of critical electric field near which the ion yield has threshold character. The dependence of the critical electric field strength on the effective quantum number n^* of the state for a major part of the elements is well described by the classical formula

$$\mathscr{E}_{crit} = \frac{1}{16(n^*)^4} \text{ atomic units (1 atomic unit} \simeq 5 \times 10^9 \text{ V cm}^{-1}). \tag{3.2}$$

If the electric field strength is higher than the critical one for this Rydberg state, the atoms excited to this state are ionised with an ion yield of nearly unity. The cross section of atomic ionisation from an intermediate state is defined in this case by the cross section of resonant excitation of the Rydberg state. This cross section is several orders higher than the cross section of nonresonant ionisation to the continuum.

The atom can be excited to a Rydberg state in two or three steps by the radiation of synchronised pulsed dye lasers. The choice of excitation scheme depends on the particular atom. Two-step excitation, for example, is convenient for alkali metal atoms. For heavy elements with a complex spectrum of atomic states and with an ionisation potential of over 6 eV the best excitation scheme is the excitation of an atom to a Rydberg state in three steps. Since the processes of stepwise atomic excitation to Rydberg states are resonant, the saturation of transitions requires rather low laser pulse energies which can be attained with the use of available dye lasers. Typical values of saturation energy flux lie in the range $\phi_{exc}^{sat} = 10^{-6}-10^{-4}$ J cm^{-2}. So even at a high pulse repetition rate essential for detecting thermal-velocity atoms the required average laser radiation power is rather moderate (of the order of 1 W).

The choice of the principal quantum number n of the Rydberg state is very important for attaining a high efficiency of atomic ionisation via Rydberg states. The excitation cross section of such states decreases drastically as the principal quantum number of a level increases (figure 3.2(a), $\sigma_{exc} \sim n^{-3}$). On the other hand, the critical field, i.e. a field in which the ion yield approximates to unity, increases materially with a decrease of n (figure 3.2(b), $\mathscr{E}_{crit} \sim (n^*)^{-4}$).

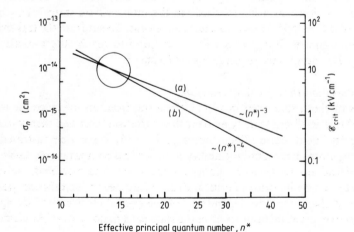

Figure 3.2 The dependences of the cross section of Rydberg state excitation $\sigma_n(a)$ and of critical electric field strength $\mathscr{E}_{crit}(b)$ on the effective principal quantum number n^*.

The case is optimal when the Rydberg state excitation cross section is as high as possible, and the field strength necessary for effective ionisation can be attained rather easily under experimental conditions. In [3.17, 3.18], for example, where Na and Yb atoms were detected, the saturation of the transitions $3p \rightarrow 13d$ (Na) and $7s \rightarrow 17p$ (Yb) was realised. The values of ϕ_{exc}^{sat} were equal respectively to 10^{-3} J cm^{-2} and 3×10^{-3} J cm^{-2}, i.e. two orders smaller than the values of ϕ_{ion}^{sat} given in the previous section. It should be noted that the values of ϕ_{exc}^{sat} obtained are not optimal because the spectral width of the laser radiation in use was $\Delta v_{las} \simeq 1$ cm^{-1} with the absorption linewidth in the atomic beam $\Delta v_{abs} \simeq 10^{-2}$ cm^{-1}. Thus, the values of ϕ_{exc}^{sat} can be reduced by almost two orders of magnitude through laser line narrowing. In [3.19] the dependence of the Rydberg state excitation cross section on the level principal quantum number in the range $n = 14$–35 was studied for the Ga atom. The numerical values $\sigma_{np} = 5 \times 10^{-14}$–$3 \times 10^{-15}$ cm^2 obtained for the transitions $5^2S_{1/2}$–$n^2P_{3/2}$ agree well with the law $\sigma_{np} \sim n^{-3}$.

It is most convenient to act on the atom by electric pulse after the laser pulsed excitation. In this case the atomic spectrum is not complicated by the Stark shift and splitting of Rydberg states. But the absolute photoionisation yield in this case is limited by the relative population of the last excited discrete state defined by expression (1.23). It would seem to be very promising to use autoionisation states in order to reach an ultimate ($\eta_{ion} = 1$) yield of stepwise photoionisation with moderate requirements on the average power of photoionising laser radiation.

(c) Ionisation via autoionisation states

Another possibility for increasing the atomic photoionisation cross section is the excitation of the atom at the last step to an autoionisation state. Autoionisation states (AS) are states of the discrete spectrum which are conditioned by the excitation of inner electrons of an atom and lie above the ionisation limit of atom, i.e. in the continuum. For multi-electron atoms such states can be rather narrow and the cross section of such an autoionisation transition can be several orders higher than that of nonresonant photoionisation [3.20–3.22]. On the other hand, even with a very small autoionisation state width, for example $\Delta v_{AS} \simeq 0.01$ cm^{-1}, its lifetime relative to the decay to the continuum is several nanoseconds. Consequently, when such a state is excited by a laser pulse with a typical duration of 10^{-8} s, it will be effectively depleted during a single laser pulse. This provides an ultimate absolute ionisation yield $\eta_{ion} = 1$.

Systematic studies of autoionisation states became possible with the development of the methods of stepwise excitation of high-lying atomic states by tunable laser radiation. Such states are of practical interest for increasing the cross section of laser atomic photoionisation, and of scientific interest for understanding and identifying spectra of multi-electron atoms. Since autoionisation states have been studied only for a few elements it is a matter of importance to perform experiments on their investigation. In [3.22] the AS of the gadolinium atom were studied. In the region of 300 cm^{-1} above the ionisation limit an AS was found with its lifetime of 0.5 ns ($\Delta v_{AS} \simeq 0.05$ cm^{-1}) and excitation cross section of about 10^{-15} cm^2 comparable with that of Rydberg states. The studies of the lutetium Rydberg and autoionisation states [3.23] have shown that the optimal scheme of photoionisation for Lu is the photoexcitation of the transition 5d 6s 6p $^2D^o_{3/2} \xrightarrow{460.7 \, nm}$ AS with its cross section of 1.5×10^{-15} cm^2. The width of the AS was found to be 0.6 cm^{-1}, its energy difference from the ground state equals 43 828 cm^{-1}. The experimental and theoretical studies of the ytterbium narrow AS [3.24] have shown that the method of multistep laser excitation is the most promising for such studies, and the theoretical approach applied can be used to calculate the AS of other multi-electron systems with a good accuracy.

(d) Comparison of different ionisation schemes

Each of the schemes of stepwise photoionisation considered above has both its pros and cons when used for atom detection including laser photoionisation spectral analysis.

The method of nonresonant ionisation enables almost all the atoms interacting with the laser radiation to be ionised during a laser pulse. But this method cannot be applied if a high pulse repetition rate is needed. Indeed, the condition for detection of each atom during its stay in the region of interaction with the laser radiation dictates a necessary pulse repetition period, T_{rep}

$$T_{rep} \leqslant \tau_{fl} \qquad (3.3)$$

where τ_{fl} is the time of atom flight through the laser beam. For effective detection of atoms in an atomic beam, when the atoms cross the laser beam with a thermal velocity and $\tau_{fl} \simeq 2 \times 10^{-5}$ s, the laser pulses must be delivered at a rate $f_{rep} \geqslant 50$ kHz. Under these conditions the average intensity of laser radiation necessary for effective ionisation must be

$$I_{av} = (10^4 - 10^5) \text{ Hz} \times (0.01 - 1) \text{ J cm}^{-2} = (10^2 - 10^5) \text{ W cm}^{-2}.$$

If the atoms are photoionised in a dense vapour or in a buffer gas, the time of atom flight through the beam is defined by atomic diffusion in the gas, values of τ_{dif} lying in the range $10^{-2} - 1$ s. Despite a considerable value of ionising pulse energy flux, the average intensity of laser radiation required for effective ionisation is moderate, i.e.

$$I_{av} = (1 - 10^2) \text{ Hz} \times (0.01 - 1) \text{ J cm}^{-2} = (10^{-2} - 10^2) \text{ W cm}^{-2}.$$

In this experiment, however, it is impossible to reach a high spectral resolution since the presence of a buffer gas brings about a considerable collisional broadening of absorption lines. In cases when it is necessary to realise a maximum spectral resolution photoionisation should be carried out in a rarefied gas or in a vacuum in an atomic beam. Under such conditions the atoms move with thermal velocities almost without collisions and, as has been shown, the application of the method of nonresonant photoionisation here requires a very high average intensity of ionising laser radiation.

The method of ionisation of Rydberg atoms by an electric field reduces these requirements by several orders

$$I_{av} = (10^4 - 10^5)\, \text{Hz} \times (10^{-6} - 10^{-4})\, \text{J cm}^{-2} = (10^{-2} - 10)\, \text{W cm}^{-2}.$$

This permits effective ionisation of atoms in an atomic beam, when a high pulse repetition rate is required. When working with a beam it is possible to reach a maximum spectral resolution, and the use of a multistep excitation scheme allows an extremely high ionisation selectivity to be obtained. This can be explained by the fact that the processes of excitation at each step are independent, and so the selectivity of the entire process of excitation and ionisation of an atom will be equal to the product of excitation selectivities at all the steps [3.25]. However, this method can allow excitation to a Rydberg state for only a fraction of the atoms interacting with the laser radiation. As a result, every second atom at best will be ionised. But this factor is not always essential.

The method of atomic ionisation through narrow autoionisation states is very promising because it is a kind of synthesis of earlier methods, i.e. the excitation remains resonant and a permanent channel of ionisation decay is retained. A disadvantage of this method is the absence of such states near the ionisation limit for a considerable number of elements.

Thus, the most suitable method of ionisation of atoms for their detection is selective excitation by laser radiation of Rydberg states and their subsequent ionisation by a pulsed electric field. As far as the use of this method for analytical purposes is concerned, we think it important to stress its advantages. As has been said, the method of multistep excitation has a very high selectivity, and so even with nearby absorption lines of an element and a matrix at one step the selectivity of detection of this element can be as high as $10^{10} - 10^{12}$. As the excitation of transitions at all the steps is resonant, relatively low laser radiation energy fluxes are required for transition saturation (3–4 orders lower than for ionisation to the continuum). In this case the nonselective ion background conditioned by the multiphoton ionisation of base material will be many orders lower than in the case of nonresonant ionisation.

3.2.2 Sensitivity of photoionisation detection of element traces in a sample

In §3.2.1 consideration was given to different methods of atomic photoionisation and the conditions under which the ultimate ionisation yield $\eta_{ion} = 1$ can be obtained. In this case only the atoms interacting with the laser radiation are taken into account. However, when it is necessary to detect the atoms in a sample, besides the process of their detection, the production of free atoms and their transportation to the laser beam region must be realised.

These tasks have been mentioned in §3.1. These stages precede detection and sometimes govern the choice of one or other method of detection. As for reaching a maximum sensitivity of atom detection in a sample, we shall discuss each of the stages in more detail.

(a)　*Production of atoms in an irradiation zone—geometrical factors*
There are three different means of transportation of atoms to the region of the laser beam which are most characteristic of analytical applications (figure 3.3). One of them is the introduction of atoms into a gas cell in the form of low-pressure vapour. The atoms in low-pressure vapour cross at random the region of interaction with the laser radiation (figure 3.3(a)). In this experiment the absorption line is broadened by the Doppler effect, which limits the spectral resolution. The exception to this is the case of two-photon excitation of atoms in a standing light wave field when all the atoms, whatever their velocity, interact with the field. The time of atom flight τ_{fl} with an average thermal velocity through the beam is small, and so it is necessary to apply laser pulses with repetition period $T_{\text{rep}} < \tau_{\text{fl}}$—otherwise a considerable fraction of atoms will cross the detection region without interaction with the laser radiation. With sufficiently long irradiation it is possible to detect the majority of atoms introduced into the cell. Under these conditions it is necessary, of course, that the atoms should not interact with the walls. This restricts significantly the field of application of the method.

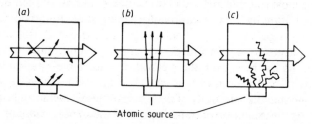

Figure 3.3　Various means of atomic motion through a laser beam: (a) free random crossing of the laser beam by vapour atoms; (b) free flight of atoms in the collimated atomic beam; (c) diffusion motion of an atom in a buffer gas.

Another possibility is the formation from free atoms of a collimated atomic beam in a vacuum (figure 3.3(b)). In this case the spectral resolution increases and at the same time the requirement on laser radiation intensity is reduced by a factor equal to the ratio of absorption line widths in vapour and in an atomic beam. All that was said for the case (a) about the conditions of irradiation holds true for the case of the atomic beam. When such a method of transportation is realised, losses of atoms are inevitable in the process of atomic beam formation. This geometrical factor may be rather significant and therefore we shall consider it a little later.

The detection of atoms can be carried out more easily if the atom is able to remain in the region of the laser beam for a long time. In this case the average laser radiation power required for detection is reduced and the total number of atoms necessary for stable detection of the element is also decreased. If the atom to be detected is in an inert buffer gas of sufficiently high pressure (hundreds of torr), the time of its stay in the region of irradiation is defined by the diffusion time and may be rather long (10^{-2}–1 s). The diffusion confinement of an atom in the region of irradiation (figure 3.3(c)) is accompanied by a significant loss of resolution because of the pressure broadening of spectral lines. It should be noted that under these conditions the pulsed operation can be simplified materially since the laser pulse repetition period T_{rep} may be increased to the value τ_{dif}.

By changing the relative geometrical position of the atomic source and the laser beam it is possible to investigate the physico-chemical processes which occur during atomic diffusion in a complex buffer gas. This approach is described in [3.26, 3.27] where the diffusion of alkali atoms (Cs and Li) in noble gases with an admixture of oxygen was studied.

(b) *Optimal conditions for multistep photoionisation*

Experiments on stepwise photoionisation of atoms aimed at obtaining a maximum photoionisation yield have been carried out for all three ionisation schemes (figure 3.1). For illustration, we shall consider below a universal regime of detection of atoms in an atomic beam through Rydberg states.

For effective excitation of atoms to a Rydberg state it is necessary to provide saturation of all the quantum transitions in use. The pulsed electric field strength must be sufficient to ionise Rydberg atoms with a probability of near unity. Figure 3.4 shows the dependences of the Yb ion yield [3.18] on the laser pulse energy fluxes of the first, second and third excitation steps (3.4(a)–(c)) and pulsed electric field strength (3.4(d)) as the Rydberg state 17 $^3P_2^o$ is excited. It follows from expression (1.23) that at simultaneous saturation of all three transitions in the state 17 $^3P_2^o$ $\eta_{exc} = 5/12$ of all the atoms in the interaction region will be excited. Ion signal saturation (figure 3.4(d)) arises when all the atoms excited to a Rydberg state are ionised during an electric pulse. Thus, under optimal experimental conditions about one half of the atoms in the registration zone are excited to a Rydberg state and each highly excited atom is ionised.

The laser pulse energy fluxes necessary to saturate the chosen transitions were determined from the dependences (3.4(a)–(c)). They are respectively equal to $\phi_1 = 8 \times 10^{-5}$ J cm^{-2}, $\phi_2 = 10^{-6}$ J cm^{-2}, $\phi_3 = 3 \times 10^{-3}$ J cm^{-2}. The value ϕ_{exc}^{sat} of the transition can be estimated by the formula (3.1)

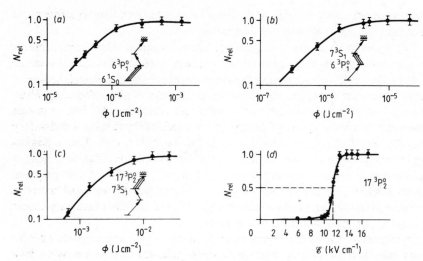

Figure 3.4 The dependence of the Yb ion yield on laser energy fluxes of the first, second and third excitation steps (a)–(c) and on pulsed electric field strength (d) (from reference [3.18]). (a) $\lambda_1 = 555.6$ nm. (b) $\lambda_2 = 680$ nm. (c) $\lambda_3 = 584.3$ nm. (d) $\mathscr{E}_{crit} = 11.5$ kV cm^{-1}.

$$\phi_{exc}^{sat} = \frac{h\nu_{exc}}{2\sigma_{exc}} \qquad \sigma_{exc} = \frac{g_b}{g_a}\frac{\lambda^2 A_{ba}}{2\pi\,\Delta\omega} \qquad (3.4)$$

where σ_{exc} is the cross section of the radiative transition a → b, $h\nu$ and λ are the quantum energy and the transition wavelength, g_b and g_a are the degeneracy factors of upper and lower levels, A_{ba} is the Einstein coefficient of the transition, $\Delta\omega$ is the transition absorption linewidth. The Einstein coefficients of the transitions $6\,^3P_1 \to 6\,^1S_0$ and $7\,^3S_1 \to 6\,^3P_1$ are known: $A_{6p\to6s} = 1.2 \times 10^6$ s^{-1}, $A_{7s\to6p} = 10^8$ s^{-1}. The absorption linewidth $\Delta\nu_{abs}$ in the atomic beam used in the experiment is equal to 0.01 cm^{-1}, and the corresponding transition cross sections are equal to $\sigma_{6s\to6p} = 1.5 \times 10^{-12}$ cm^2 and $\sigma_{6p\to7s} = 6 \times 10^{-11}$ cm^2. The laser pulse energy fluxes necessary for transition saturation are determined by the expression

$$\phi_i = \phi_i^{sat}\frac{\Delta\nu_{las}}{\Delta\nu_{abs}} \qquad (3.5)$$

and for laser radiation with $\Delta\nu_{las} \simeq 1$ cm^{-1} they are $\phi_1 = 2 \times 10^{-5}$ J cm^{-2} and $\phi_2 = 5 \times 10^{-7}$ J cm^{-2} respectively. The difference between the experimental and theoretical results is conditioned by the influence of transitions on one another and probably by an incorrect determination of the transverse dimensions of laser beams. Nevertheless, such an approach is quite suitable for estimating the optimal parameters of laser radiation.

(c) *Absolute sensitivity*.

In all the experiments described above the sources of free atoms were samples with a macroscopic (weight) amount of an element. To use the method of laser stepwise photoionisation in analytical tasks it is necessary, besides obtaining an ultimate relative sensitivity (detection of a single atom in a laser beam), to solve the problem of a minimal absolute quantity of an element which still permits us to obtain a selective signal from its atoms. An example of successful search in this direction is [3.28]. In this paper the isotopic and hyperfine structure of the transition $6\,^1S_0 \to 6\,^3P_1$ of the Yb atom was investigated using samples with about 10^{10} atoms of any Yb isotope.

The samples were pieces of tantalum foil in which 10^{10} ions of the definite Yb isotope had been implanted. The ions with an energy of about 30 keV penetrated the foil to a depth of 100 Å and were neutralised. Thus, the atoms turned out to be isolated from the outside and under normal conditions could be kept in the foil for a long time. When such foil is heated up to over 1200 °C, one can observe the diffusion of the Yb atoms from the tantalum. If the heating is done in a tantalum crucible with a narrow cylindrical outlet channel, the free atoms are collimated by such a channel into an atomic beam.

The described technique for detection of single Yb atoms in a beam was used for diagnostics of this monoisotopic Yb beam. The laser radiation and electric field parameters provided ionisation of every third atom in the excitation zone. Under optimal conditions of heating the selective ion signal from the Yb isotope was detected during 20 minutes with a signal-to-noise ratio of no less than three. This time allows the study of the structure of one of the excited Yb transitions by scanning the frequency of the corresponding laser. In studying the transition $6\,^1S_0 \to 6\,^3P_1$ a pressure-tunable dye laser with the linewidth $\Delta\nu_{las} \simeq 0.04\,\mathrm{cm}^{-1}$ was used at the first excitation step. The pulse repetition rate was $f_{rep} = 12$ Hz. Since the isotope atomic beam density was very low, the signal was detected in an ion counting regime as the laser frequency was scanned. The spectra produced allowed identification of the Yb isotopes ^{173}Yb, ^{174}Yb and ^{176}Yb.

The results of the experiment made it possible to estimate the minimum number of atoms necessary for registration of the total spectrum as well as the coefficient of atomic yield from the crucible and foil. Figure 3.5 illustrates the estimation of the absolute sensitivity of the method under the conditions of the experiment described. Three ions were sufficient for stable detection of one resolved spectral interval. The ionisation probability being 0.3, ten atoms are required in the detection zone for three ions to be produced. This zone covered just one seventh of the atomic beam cross section. If the atomic velocity in the beam is $5 \times 10^4\,\mathrm{cm\,s}^{-1}$ and the detection zone diameter is 1 mm, the time interval between laser pulses must be 5×10^4 times less than that used in the experiment in order to intercept all the atoms. Thus, to record one spectral interval 3.5×10^6 atoms flying out of the crucible were needed. To detect the total spectrum in a region of $1\,\mathrm{cm}^{-1}$ with a resolution of $0.02\,\mathrm{cm}^{-1}$ 50 steps are

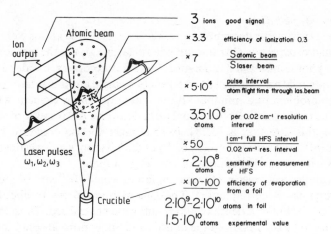

Figure 3.5 The evaluation of the absolute sensitivity of atomic photoionisation detection in a beam (from [3.28]).

required, i.e. about 2×10^8 atoms in the atomic beam. In the experiment three or four spectra were produced for each of the foil samples containing 10^{10} Yb atoms, i.e. every cycle needed $(2–3) \times 10^9$ atoms. The calculated quantity comes to 2×10^8 atoms. Hence, the coefficient of atomic yield from the crucible and foil ranges from 0.1 to 0.07. One spectral interval would be sufficient in this case to record an element content. Only $(4–6) \times 10^7$ atoms of the element would be needed here.

The experiment proved that the method of photoionisation detection in an atomic beam with dye lasers having a small pulse repetition rate makes it possible to record spectra of elements available in amounts of 10^8 to 10^{10} atoms. The sensitivity of the method can be increased radically (by 10^3 times) if the dye lasers are excited by a copper-vapour laser with a pulse repetition rate of 10 to 20 kHz. This will enable almost every hundredth atom crossing the excitation zone to be ionised. With an improved scheme geometry about 5×10^3 atoms in the sample will suffice to detect the element and 10^5 to 10^6 atoms to detect the hyperfine structure spectrum.

These results are very important for estimating the potentialities of laser stepwise photoionisation of atoms in a vacuum as a method of determining an ultra-low content of a definite element in matter. As has been shown, with a coefficient of atomic yield from the sample being about 0.1 and with a low pulse repetition rate ($f_{rep} = 12$ Hz) around 10^8 atoms in the sample will be sufficient to detect the element. The value 0.1 is characteristic for the degree of thermal atomisation for most compounds in the temperature range from 2000 to 2500 °C [3.29]. Thus, it is possible to hope for the detection of an element in a substance with the content of its atoms in the sample being about 10^8. For vacuum evaporation and atomisation a sample mass of 10 mg (the total number of atoms in the matrix is about 10^{20}) is quite reasonable. Hence, the detectable concentration will be equal to $10^{-10}\%$. For elements whose excitation transitions are suitable for dye lasers pumped by a copper-vapour laser this value may be three orders of magnitude lower. Such a low level of concentration, inaccessible for conventional analytical methods, shows that the method of laser stepwise photoionisation in a vacuum is very promising and it is necessary to develop this approach for the purpose of elaborating a new method of determining ultra-low traces of elements.

3.2.3 Selectivity of multistep excitation and ionisation

Under the conditions of real experiments on laser photoionisation the atoms to be studied are always surrounded by foreign atoms or molecules. Depending on the photoionisation approach the concentration of foreign particles N_B may vary widely—from 10^{10} cm^{-3} and less in a vacuum for the atomic beam to 10^{19} cm^{-3} in the buffer gas medium. The interaction of these particles with the laser radiation inevitably gives rise to foreign ions which can contribute to the photoionisation signal. Let the photoionisation selectivity S be defined as the ratio of the ionisation yield of atoms A of interest to the ionisation yield of foreign particles B, for example, the atoms or molecules of the matrix

$$S = \eta_A/\eta_B. \tag{3.6}$$

If no additional measures are taken to discriminate foreign particle photoions relative to the photoions of the atoms to be detected, the selectivity value places an evident limitation on the minimal detectable concentration of the atoms to be analysed: $N_A/N_B \geqslant 1/S$.

(a) Selectivity multiplication

The selectivity of the method of multistep resonant photoionisation is many orders higher than the selectivity of any spectroscopic method based on single-photon transitions as one resonant photon is absorbed. This is due to the fact that at multistep excitation there must be a resonance at each excitation step. Therefore the selectivity of the process based on a sequence of several resonant processes will be apparently multiplied. In the general case the multistep photoionisation selectivity can be expressed as

$$S = S_{exc} \times S_{ion} \tag{3.7}$$

where S_{exc} is the excitation selectivity of the atomic state to be ionised and S_{ion} is the ionising step selectivity. The excitation selectivity S_{exc}, in its turn, is the product of excitation selectivities at every step: $S_{exc} = \Pi_k S_k$, where k is the number of excitation steps. The ionising step selectivity depends on the method of ionisation. For example, when ionisation is carried out to the continuum, one can believe because of a large photoabsorption bandwidth that $S_{ion} = 1$, i.e. there is no selectivity. The situation is analogous in the case of

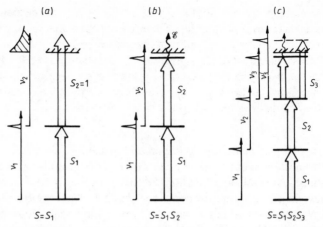

Figure 3.6 The selectivity multiplication in various schemes of laser multistep photoionisation of atoms.

electric field ionisation of Rydberg atoms because the electric field ionises any highly excited atoms. On the last ionising step a high selectivity can be realised only by photoionisation through narrow autoionisation states. Figure 3.6 gives some examples of selectivity multiplication in different schemes of multistep excitation and ionisation. It can be seen that in the simplest scheme of two-step photoionisation at the transitions to the continuum the selectivity multiplication effect is absent while in the schemes of three-step excitation of autoionisation or Rydberg states it is a maximum.

(b) *Selectivity limitation*
The use of the method of multistep photoionisation in detecting atoms allows in some cases a very high (up to 10^{19}) selectivity to be attained as the atoms of one sort are detected against the background of a large number of another sort of atoms with a high ionisation potential [3.30]. This is explained by the fact that the transition frequencies of the A and B atoms differ greatly at all steps.

There are two mechanisms which are able to limit the selectivity of multistep excitation and ionisation (figure 3.7): the excitation of unwanted B atoms at the wing of their absorption line; and the multiphoton nonresonant ionisation of unwanted B atoms by laser radiation tuned to resonance with the detected A atoms. Here we shall consider each mechanism in more detail.

When the atoms are excited in an atomic beam by monochromatic laser radiation, the absorption at the line wing is described by the formula

$$\gamma(\omega) = \frac{\Gamma}{\pi[(\omega-\omega_0)^2+\Gamma^2]} \qquad |\omega-\omega_0| \gg \Gamma \qquad (3.8)$$

where ω_0 is the central frequency, Γ is the radiative half-width of line at half

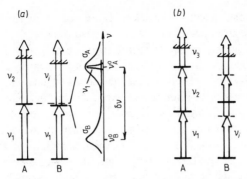

Figure 3.7 Mechanisms of selectivity limitation in laser stepwise atomic photoionisation. (*a*) The excitation of B atoms at the wing of their absorption line by laser radiation v_1 tuned to the frequency v_A^0 of A atoms. (*b*) Multiphoton nonresonant photoionisation of B atoms by laser radiation used for stepwise photoionisation of A atoms.

maximum. Since Γ is usually much smaller than the distance between the lines the A atoms are excited at the wing of absorption line of the B atoms. The distance between the centres of lines of the A and B atoms being $\delta\omega$, the maximum excitation selectivity will be equal [3.25] to

$$S=\left(\frac{\delta\omega}{\Gamma}\right)^2 \frac{\sigma_A^0}{\sigma_B^0} \qquad (3.9)$$

where σ_A^0, σ_B^0 are the absorption cross sections of the corresponding atoms at their central frequencies.

If the laser radiation width $\Delta\omega_{las}$ is larger than the absorption line radiative width, the real excitation selectivity at one step will be smaller, i.e.

$$S=\frac{2(\delta\omega)^2}{\Gamma \Delta\omega_{las}} \frac{\sigma_A^0}{\sigma_B^0}. \qquad (3.10)$$

This expression for excitation selectivity will also define the selectivity of two-step photoionisation if the quantum energy of radiation photoionising atom A is sufficient to photoionise atom B from an excited state (figure 3.7(*a*)).

Another factor that restricts the selectivity is nonlinear multiphoton ionisation of foreign atoms by laser radiation which is used for resonant stepwise photoionisation of the chosen A atom (figure 3.7(*b*)). The probability of *n*-photon atomic ionisation by laser radiation is expressed by the formula [3.31]

$$W_n=\sigma_n I^n \ (\text{s}^{-1}) \qquad (3.11)$$

where σ_n is the cross section of *n*-photon ionisation and I is the laser radiation intensity. The cross section σ_n varies within a very wide range and depends materially on the presence of intermediate resonances.

The multiphoton mechanism of limitation of multistep photoionisation selectivity is particularly important in the case of photoionisation detection of atoms with a high energy of the first resonant level which are excited using a k-multiphoton process in an intense visible wavelength laser field $E_{1\,exc} = k\hbar\omega_{las}$. In this case an intense laser pulse can quite easily ionise the atoms and molecules of the medium surrounding the atom to be detected, and the ionisation selectivity drops drastically. Such a regime of photoionisation cannot, of course, be used for detecting minute traces of elements.

Experimental realisation of high selectivity places a stringent limitation on the concentration of B atoms due to the process of nonresonant excitation transfer from the A atom at the first intermediate level to the B atom. The probability of excitation transfer during the stay of the A atom at an intermediate level τ_1 must be less than the inverse of excitation selectivity at the first step

$$\sigma_{trans} N_B v_0 \tau_1 \leqslant 1/S_1 \qquad (3.12)$$

where σ_{trans} is the cross section of excitation transfer and v_0 is the atomic velocity. For the selectivity $S_1 \simeq 10^6$ the permissible density of atoms with $\sigma_{trans} = 10^{-15}$ cm^2, $\tau_1 = 10^{-8}$ s and $v_0 = 5 \times 10^4$ cm s^{-1} will be $N_B \leqslant 2 \times 10^{12}$ cm^{-3}. It can be easily seen that the excitation transfer can decrease the selectivity of separation of different atoms but it does not reduce the total selectivity of detection of rare atoms against the background of a large number of other atoms.

Thus, the method of laser resonant multistep excitation and photo-ionisation of atoms permits in most cases detection of single atoms of a given element among 10^{14}–10^{20} atoms of other elements.

(c) *Role of selectivity in the detection of impurities*

Let us consider the effect of selectivity-limiting factors on the detection of impurities in pure substances. We shall illustrate this by an example of detection of boron impurities in germanium which is of great practical interest.

Figure 3.8 presents the simplified energy level diagrams of boron and germanium. It also shows the transitions used for the photoionisation of boron atoms through Rydberg states. It can be seen that the germanium atom has an absorption line $\lambda_{1Ge} = 249.8$ nm, very close to boron line $\lambda_{1B} = 249.68$ nm, that is conditioned by the intercombination transition $^3P_0 \rightarrow {}^1P_1^o$. The absorption of the first-step laser UV radiation at the wing of the Ge line causes the Ge atoms to be excited to the state $4s^2\,4p\,5s\,{}^1P_1^o$. The quantum energy of the second-step laser $\lambda_{2B} = 378.3$ nm used to excite the Rydberg 17p boron state is sufficient to photoionise the germanium atoms from the $^1P_1^o$ level. It is also possible to photoionise them by a quantum with λ_{1B}, but due to

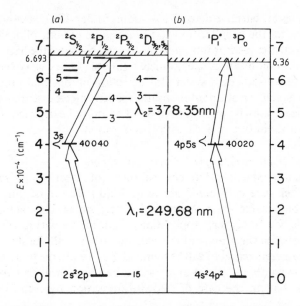

Figure 3.8 The energy level diagrams of boron (*a*) and germanium (*b*) atoms and transitions used for photoionisation (from [3.43]).

the small energy of this laser pulse the contribution of such a process will be small compared with λ_{2B}.

Let us estimate the selectivity of the boron atom photoionisation when its impurities are detected in a dense beam of germanium atoms. Substituting the value $\delta \nu = 20$ cm^{-1} (figure 3.8), $\Delta \nu_{las} = 1$ cm^{-1}, $\Gamma = 1.5 \times 10^7$ s^{-1} (5×10^{-4} cm) and $\sigma_B^0/\sigma_{Ge}^0 = 2$ [3.32] into formula (3.10) we get the value of excitation selectivity $S_{exc} \simeq 3 \times 10^6$. In the case of boron, when each of the excitation steps is saturated, $\eta_{ion\,B} \simeq 0.5$ since half of the atoms will be excited from the ground state to the 17p Rydberg state (see formula (1.23)) and every Rydberg atom can be ionised by a field pulse. With the laser parameters considered $\phi_2 \simeq 3 \times 10^{-3}$ J cm^{-2}, the ionisation efficiency of the Ge atoms for $\sigma_{ion} \leqslant 10^{-17}$ cm^2 will be equal to $\eta_{ion\,Ge} \leqslant 5 \times 10^{-2}$. So, for the ionising step selectivity we have $S_{ion} = \eta_{ion\,B}/\eta_{ion\,Ge} \geqslant 10$. The final value of total photoionisation selectivity will be equal to $S = S_{exc} \times S_{ion} = 3 \times 10^7$. This value is not sufficient for analysing high-purity germanium with boron impurities ranging from 10^{-7} to $10^{-8}\%$ by the method of conventional two-step photoionisation. Such analysis calls for additional discrimination of the ion background. The method of such discrimination is considered below in §3.3.5.

3.3 Technique of atomic photoionisation analytical spectroscopy

Having discussed the principles of laser stepwise atomic photoionisation and its potentialities for effective detection of atoms under different conditions we will now consider the peculiarities of the technique of laser photoionisation analytical spectroscopy. For this purpose we shall consider the basic elements of the laser photoionisation spectrometer: tunable lasers, a system for the atomisation of substances, a system for detection of ions and processing of signals, etc. The scheme of a photoionisation spectrometer essentially depends on the scheme of stepwise photoionisation of atoms to be detected. Therefore, the discussion should be opened with this question.

3.3.1   Choice of a universal photoionisation scheme

The pros and cons of different schemes of laser stepwise atomic photoionisation were considered above (§3.2.1(d)). The most universal and easily realisable scheme is ionisation via Rydberg states. This method is also most suitable for analytical applications. Indeed, by virtue of resonant interaction between the atom and the laser radiation at all the steps the laser energetic parameters required will be minimal. As a result, the nonselective ion background inevitable in direct analytical experiments will be two or three orders lower than in the case of stepwise ionisation by transition to the continuum.

The analysis of the spectrum of atomic energy levels for most of the elements [3.33] shows that for 19 elements of the periodic table—H, He, C, N, O, F, Ne, P, S, Cl, Ar, As, Se, Br, Kr, I, Xe, At and Rn—the wavelength of the first transition from the ground state lies in the region $\lambda < 2000$ Å. Today it is rather difficult to generate tunable laser radiation in this region. It is possible to use for these elements at the first step [3.34] two- or three-photon excitation of the lowest levels by intense UV radiation. Unfortunately, in real analytical studies this gives rise to a considerable ion background as a result of nonselective photoionisation of the matrix atoms or molecules with lower ionisation potentials. It is impossible to detect the weak selective signal under such conditions without discriminating against this background. For all the other elements of the periodic table resonant single-photon excitation can be realised at all steps using appropriate tunable lasers.

3.3.2   Laser equipment

The laser photoionisation spectrometer makes use of tunable lasers which permit resonant excitation of atomic transitions. At present, dye lasers excited by pulsed UV and visible-wavelength lasers, e.g. the nitrogen laser ($\lambda = 337$ nm), the excimer laser ($\lambda = 308$ nm, 249 nm) and second- ($\lambda = 532$ nm) and third-harmonics ($\lambda = 355$ nm) of the Nd–YAG laser, are the most suitable for the photoionisation of atoms in the region of 220 nm $< \lambda < 900$ nm where absorption lines of the most elements lie.

(a)   Dye lasers with a pulsed pumping

Dye lasers with transverse pumping by N_2-laser radiation have been rather widely used until recently [3.35]. Despite the moderate power of its pulses the N_2 laser can be used for the simultaneous excitation of several dye lasers. Such a laser spectrometer was used in analytical experiments [3.36] where high sensitivity was achieved in detecting traces of some elements in real substances (see §3.4). The characteristics of a number of commercial dye lasers with different sources of pumping have already been considered in Chapter 2 (table 2.1). As far as photoionisation analytical applications are concerned, the models made by 'Lambda Physik' and 'Molectron' are the best of the variety of available dye lasers. They make it possible to apply any of the above sources of pumping for simultaneous excitation of several dye lasers with a pulse repetition rate of up to 200 Hz. Dye lasers themselves are characterised by a high pulse power, a sufficiently narrow line and a wide range of wavelength tuning (217 to 970 nm). Thus, the laser systems considered are the optimal sources of multifrequency tunable radiation if there is no need for a high pulse repetition rate.

(b) *Dye lasers pumped by copper-vapour lasers*
As far as the fulfilment of condition (3.3) and the realisation of effective photoionisation detection of atoms in a beam are concerned, dye lasers pumped by metal-vapour laser radiation (Cu, Pb, Au) are best suited. Such lasers are characterised by a high pulse repetition rate (10 kHz and higher). The copper-vapour laser is the best developed of the pumping lasers. In the USSR, for example, the ILGI-101 type copper-vapour laser with an average power of up to 10 W ($f_{rep} = 10$ kHz, $\tau_{pul} = 18$–20 ns) is in commercial production. The characteristics of dye lasers pumped by such a copper-vapour laser were investigated in detail in [3.37]. Laser action was attained in the region of 530 to 720 nm. The laser radiation frequency was doubled over the entire tuning range with the use of a KDP crystal.

The development of lasers operating on vapours of other metals—Au ($\lambda = 628.3$ nm) and Pb ($\lambda = 723$ nm)—will probably permit widening of the generation range of dye lasers towards the long-wave band, from 850 to 950 nm.

(c) *The choice of laser equipment for an analytical spectrometer*
The most universal laser system for multistep resonant photoionisation of atoms is a combination of several dye lasers pumped by an excimer XeCl laser ($\lambda = 308$ nm) or a KrF laser ($\lambda = 249$ nm). The wavelengths of atomic transitions of more than 80% of the elements from the periodic table lie in the laser tuning range of such a system (217–970 nm, table 2.1). A relatively high pulse repetition rate (up to 200 Hz) and the long-term stability of the frequency and energetic parameters of this system make it possible to realise a high absolute sensitivity of atom photoionisation detection in prolonged analytical experiments. When the simplicity and mobility of the laser system are put in

the forefront, it is advisable to use the N_2 laser as a pumping source ($\lambda = 337$ nm).

For a considerable number of elements (mainly heavy unclosed shell elements) it is possible to choose a multistep photoionisation scheme with laser wavelengths 530 nm $\leqslant \lambda_i \leqslant 750$ nm. If in this case it is necessary to analyse a small absolute amount of a substance in a continuous atomic beam, when a high pulse repetition rate is required to intercept a significant number of the atoms to be detected, the copper-vapour laser is most advisable as a pumping source. If the mass of the substance to be analysed is unlimited, it is better, of course, to use a simpler laser system.

Thus, the choice of optimal laser equipment is defined by the specific element and, to a great extent, by the conditions of experiment, in particular by the method of substance atomisation.

3.3.3   System of atomisation

(a)   *Thermal atomisation*

Thermal atomisation of substances can be accomplished in a gas or in a vacuum. The atomisation in an inert gas medium is widely used now in atomic fluorescence spectroscopy (see §2.3.2). In most cases the atomisation process is performed in graphite crucibles, cells and tubes heated electrically up to 3000 °C. The main function of a buffer gas is to prevent thermal scattering of the substance and thus to create a considerable concentration of the atoms to be detected in the irradiation zone. This method is rather universal and is characterised by good speed of analysis. But, as for photoionisation detection, the method has material shortcomings already indicated in the introduction to this chapter.

Thermal atomisation in a vacuum which will be considered below is free of these shortcomings [3.36]. This method is characterised by rather slow evaporation and atomisation of a substance in order to form in a vacuum effusive atomic–molecular beams. In this case the interference effects between the element and the matrix taking place at a high particle density during atomisation in a buffer gas are essentially reduced due to a low vapour pressure of the substance in the crucible.

The main requirements in designing a vacuum atomiser are to achieve temperatures of up to 3000 °C, to realise prompt replacement of samples and crucibles, to ensure its compatibility with the photoionisation and ion detection systems as well as simplicity and reliability in use. For example, the atomiser used in [3.36] (figure 3.9) was based on a graphite crucible-tube heated by electric current. A special ion-suppressing system was placed above the outlet channel of the crucible to suppress a strong background of thermal ions and electrons. In the process of analysis the sample in its natural form with a mass of up to 200 mg was put into the crucible. In the analysis of solutions the fluid was first evaporated. Then the crucible was placed in the atomiser and,

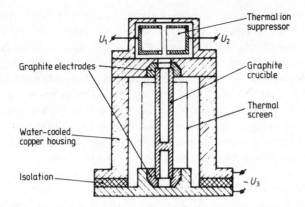

Figure 3.9 The design of the electrothermal atomiser for producing collimated atomic beams in a vacuum (from [3.49]).

after the vacuum chamber was evacuated to a residual pressure of 10^{-6} Torr, the crucible was heated by steps to the working temperature. The evaporating atoms and molecules of the substance were formed by the cylindrical channel of the crucible as an atomic–molecular beam in which photoionisation detection of the selected atoms was carried out.

(b) *Laser evaporation*
This method of evaporation and atomisation of matter began to be used in analytical studies with the appearance of powerful lasers (see §1.3). Evaporation by cw laser radiation is in fact a modification of thermal evaporation with some improvement in the locality of heating. When it is combined with photoionisation detection of atoms, the question arises as to how the laser pulse repetition rate can be optimised, how the thermal ion background can be eliminated, etc. In this respect it is more convenient to apply pulsed laser evaporation. It permits synchronisation with the pulses of detecting laser radiation that, in its turn, solves the problem of optimal consumption of the substance to be analysed. Evaporation by laser pulse can be accomplished in a buffer gas or in a vacuum. The laser radiation intensity needed for effective evaporation and atomisation of matter usually ranges from 10^7 to 10^9 W cm^{-2}. Such a wide range is conditioned mainly by a different duration of the radiation pulse (10–10^3 ns), the pulse energy being approximately the same.

(c) *Electron-beam evaporation*
This type of evaporation has been applied for a long time in analytical studies as a means of producing the highest possible temperatures (above 3000 °C), for example, to evaporate refractory metals and their oxides. The sample is heated in a vacuum by a continuous power electron beam generated by an electron-

beam gun. At present many designs of electron guns with electron energy of 5 to 30 keV, current of up to 1 A and specific beam power of 10^3 to 10^5 W cm^{-2} have been developed. With a focused electron beam, it is possible to accomplish local heating of the sample with satisfactory spatial resolution.

Electron-beam evaporation is widely used in mass spectrometric analysis of solids [3.38]. The studies [3.39] proved that the intensity of a neutral component at such evaporation is three or four orders higher than that of an ion component. With additional mass filtration of ions this makes, in principle, the use of electron-beam evaporation in combination with laser stepwise photoionisation of atoms promising for analysis of matter.

(d)   *Ion sputtering*

The sputtering of the surface of solids under the action of an accelerated ion beam is widely used in secondary-ion mass spectroscopy to study the element composition of solids and particularly to detect traces of impurities in substances. One of the problems in this method of analysis is that the secondary-ion yield depends greatly on the matrix material and, also, the ion yield is small compared with that of the neutral component. The neutral atom yield, as a rule, is two to four orders of magnitude higher than the ion yield and weakly depends on different matrix effects. All this points to ion sputtering as a promising method for atomisation of solids. The ion background arising can be discriminated using the methods of ion separation accepted in secondary-ion mass spectroscopy. The application of laser radiation enables one to obtain an increase in the degree of ionisation of the atoms of the selected element by several orders in comparison with ion bombardment.

The ion beam parameters may vary over a very wide range and are defined by a specific task. In [3.40], for example, an ion beam was applied with energy 5 to 30 keV and direct current 1 mA. The beam was modulated in order to synchronise it with the laser pulses. This modulation provided a pulse repetition rate of up to 10 kHz with pulse duration of 0.25 to 10 μs. The pulsed sputtering regime is more advantageous since the consumption of substance is minimal and it is simpler to achieve the vacuum conditions necessary for the ion separation and detection systems.

(e)   *Choice of atomisation system*

Each of the above-considered methods of atomisation of matter has its advantages and disadvantages. Thermal atomisation in a buffer gas is the simplest method to achieve in practice. But this method cannot be used to realise the ultimate sensitivity of photoionisation detection.

Pulsed laser evaporation and ion sputtering in a vacuum are best suited to local analysis of solids and thin films with a sensitivity of up to $10^{-10}\%$ and a high spatial resolution. This is conditioned by the potentialities of focusing of laser and ion beams and the application of ion energy separation and mass separation and filtration systems.

Thermal atomisation in a vacuum combines the simplicity of experimental realisation with the advantages of vacuum evaporation and selective detection of ions. This method of atomisation seems to be most universal since it can be used for direct analysis of almost any class of substance—from high-purity materials to biological objects—with a near-ultimate sensitivity. This method calls for a moderate vacuum, $\simeq 10^{-6}$ Torr, and is characterised by a good speed of operation.

This section is concerned only with those methods of atomisation which can be combined with laser stepwise atomic photoionisation and make it possible to realise the ultimate sensitivity of photoionisation detection. For this reason we have not considered here flame atomisation, one of the oldest and most widely used methods, since the ion background of flames, especially in the presence of a matrix, restricts the analysis sensitivity to a level from 10^{-5} to $10^{-7}\%$.

3.3.4   *The registration system*

There are two detectors working in essentially different conditions which are sensitive enough to detect single ions and electrons. These are the proportional counter operating at near-atmospheric gas pressure and the secondary-emission multiplier (SEM) operating in a vacuum.

In the first case (see [3.10]) the electrons formed by atomic photoionisation are detected. The electron avalanche arising can be proportionally amplified by special choice of a gas and accelerating voltage, and with a gain factor of around 10^4 it is possible to detect single photoelectrons.

In detecting the ions by SEM in a vacuum the scheme of ion detection depends essentially on the method of atomisation. The common requirement in this case is to produce a vacuum no worse than 10^{-5} Torr for normal work of the SEM and free motion of the ions from the region of the laser beam to the SEM. It is possible to use here both a unified vacuum system and differential pumping for producing a higher vacuum in the chambers of the ion source, of the mass separator, etc.

To illustrate this we shall consider a scheme for ion detection with thermal atomisation of matter in a vacuum since it is this method that has been applied in most of the analytical photoionisation experiments [3.36] described below.

The atomic molecular beam (figure 3.10) formed by atomisation of matter is irradiated at a right angle simultaneously by all the laser beams in the region between the flat metal electrodes. An ionising pulse of electric field ($\mathscr{E} = 10$–15 kV cm^{-1}, $\tau = 10$–20 ns) is fed to the electrodes 20–50 ns after the atoms are excited by lasers to a Rydberg state. The ions formed by field ionisation of Rydberg atoms gain, in the field of this pulse, a normal component to its thermal velocity in the atomic beam. The value of this component is almost two orders higher than the average velocity of atoms in the beam, and so the ions move almost normally to the atomic beam axis. After passing through the slit in the electrode the ions enter the SEM where they are detected. The

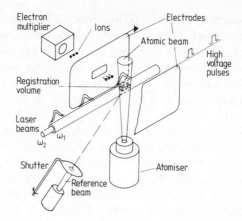

Figure 3.10   The scheme for laser photoionisation detection of atoms in an atomic–molecular beam at the thermal atomisation of a substance in a vacuum.

estimation of the ion extraction efficiency from the inter-electrode region shows that it is close to unity. Two types of SEM—SEM-2 with discrete dynodes and SEM-4 with a continuous channel dynode—were used in the experiments. The latter has better characteristics as far as the detection of single ions is concerned (the SEM-4 efficiency was 0.35–0.5 and less than 0.2 for SEM-2).

The signal from the SEM was processed in the standard manner: it was amplified, fed to the boxcar and was then registered by a recorder or by an amplitude analyser.

3.3.5   Discrimination of nonselective ion background

In direct analysis of matter the atoms of the element to be detected are photoionised among a large number of atoms or molecules of the matrix. This gives rise to a nonselective ion background related in most cases to the processes of nonlinear interaction of the laser radiation with the substance.

(a)   Sources of nonselective ions

The main sources of nonselective ions are the processes which limit the selectivity of photoionisation detection (see §3.2.3(b)). These are multiphoton nonresonant laser photoionisation of foreign atoms and molecules, and excitation and ionisation of unwanted atoms due to laser radiation absorption at the wings of their lines. Despite the small value of the cross sections of such processes the ion background may be significant since the density of the matrix particles in the beam or in the vapour exceeds the density of the atoms to be detected by many orders. As well as the ion background conditioned by laser radiation there is a residual background from the thermal ions of the atomiser which have escaped the ion suppressor. As a result, the total nonselective ion background will restrict the sensitivity of analysis at a certain step.

Consideration is now given to the methods of separating the ions of a definite element from all the ions formed after the medium to be analysed has been exposed to laser radiation.

(b) *Ion mass separation*

The ions formed by atomic photoionisation have almost the same thermal velocities as the primary atoms. This allows the formation from them, by moderate electric fields, of a collimated pulsed ion beam which can serve as a good ion source for any ion mass separator. Mass resolution and transmission coefficient are the most important characteristics of the mass separator for detecting a small number of ions and for analytical applications. Quadrupole and time-of-flight mass separators are most suitable for this purpose.

The basic advantages of quadrupole mass filters [3.41] are near 100% transmission that is of great importance for detecting weak ion flows, small dimensions which allow easy combining of the filter with the laser photoionisation sysrem, and reliability in operation. Even though quadrupole mass analysers have a moderate resolution $(M/\Delta M = 10^3 - 10^4)$, in combination with selective laser atomic photoionisation they are very promising for application in analytical spectroscopy of atomic traces.

The time-of-flight mass separator naturally matches the scheme of atomic ionisation by electric field via Rydberg states. As has been said, under the action of a field pulse the ions move at right angles to the motion of the atoms in the beam. Thus in the detection scheme given in figure 3.10 the drift in the field-free region can be realised in the same vacuum chamber with the ion detector (SEM) sufficiently removed from the electrodes. It can be easily demonstrated that such a filter is linear with respect to ion mass, and with $\mathscr{E} = 10$ kV cm^{-1}, $\tau = 10$ ns and $L = 50$ cm the interval between masses is $\Delta T = 0.5$ μs with the ion peak width being equal to 5 ns. Such an interval is convenient for temporary gating of signals from different masses [3.42].

(c) *Selection of Rydberg atoms*

Selective ionisation of Rydberg atoms under the action of two electric field pulses with definite parameters is a simpler method for suppressing the nonselective ion background and separating the ions of the element under study [3.43]. The method is based on a sharp threshold dependence of the degree of ionisation of Rydberg atoms on the amplitude of electric field pulse and its weak dependence on pulse duration. For example, for the 17p state of aluminium the critical electric field strength is equal to 6.2 kV cm^{-1} with $\tau_1 = 10$ ns and to 6.1 kV cm^{-1} with $\tau_2 = 30$ ns. It can be seen that the critical strengths differ only slightly while the pulse duration in the first case is three times shorter than in the second.

Let the atoms of the selected element in the beam or in the vapour of different atoms (figure 3.11) be excited by laser radiation pulses to a Rydberg state in several steps. After the medium is irradiated, the region between the

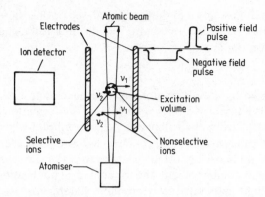

Figure 3.11 The selection of Rydberg atoms at their two-pulse ionisation by the electric field (from [3.43]).

electrodes will not only have reasonantly excited Rydberg atoms of the element being analysed but also foreign ions, namely thermal ions and ions formed by laser multiphoton or multistep photoionisation of the particles of the residual gas and the matrix. Then a negative electric pulse with strength smaller than the critical one for an excited Rydberg state is fed to the electrodes. Under its action all the ions between the electrodes acquire velocity components towards the electrode without a slit. The action of this pulse on the Rydberg atoms is negligible [3.16]. Then a positive electric pulse (the delays between the pulses and their durations are much shorter than the Rydberg state lifetime and equal 10–50 ns) with strength larger than the critical one and sufficient for full ionisation of Rydberg atoms is fed to the electrodes. The amplitudes $\mathscr{E}_i$ and the pulse durations τ_i in this case should satisfy the condition

$$\mathscr{E}_-\tau_- > \mathscr{E}_+\tau_+. \tag{3.13}$$

In this case under the action of the second pulse the nonselective ions just declerate their motion to the electrode without a slit. The ions of the element under analysis formed from Rydberg atoms will be accelerated in the field of the second pulse, pass through the slit in the electrode and be detected by SEM. Thus, if only the ions of the selected element are extracted from the excitation zone, the detection selectivity will increase compared with the selectivity in one-pulse ionisation. According to simple estimations the coefficient of suppression of nonselective background in two-pulse detection is three to four orders and is limited by the background of the scattered ions outside the interelectrode space.

3.3.6 Calibration of an analytical spectrometer

(a) *Conventional calibration methods*
The analytical calibration curve is one of the most widely used methods of

analytical signal calibration. It consists of constructing the dependence of the analytical signal on the concentration of an element in samples of the same substance with a known (standard) amount of this element. The standards are chosen so that the concentration to be found falls within the standard range in use. When the calibrating curve is linear over a wide range, it is possible, as a rule, to approximate the curve to the region of lower concentrations. This method is best suited to analysis of aqueous solutions of salts when the interference effects are minimal. It is very difficult, and sometimes impossible, to choose or prepare standard samples with the same matrix content in analysing complex objects. In these cases the addition method is advisable. The essence of this method lies in constructing a dependence of the analytical signal on the concentration of the element introduced into the substance in the form of a definite compound in a known amount and in approximating this dependence to the zero concentration level. It should be noted that this method becomes inadequate in the presence of strong interaction effects between the matrix and the added element as well as when the chemical properties of the natural and the added compounds of the element differ greatly.

These methods of calibration are also applied in laser photoionisation analytical spectroscopy. Atomisation in a vacuum in this case allows the range of application of the addition method to be widened significantly, because in the process of such atomisation all the interference effects between the element to be determined and the matrix are considerably reduced.

(b) *Reference atomic beam method*
There is another method of calibration that can be applied in vacuum photoionisation spectroscopy. An additional atomic beam of the element is introduced into an analytical vacuum chamber (figure 3.10). It can be formed, for instance, by thermal evaporation of the pure element from its metal or amorphous state in an additional oven. If this reference beam passes through the excitation zone, i.e. the region where the laser beams and the main beam intersect, the atoms of the elements are under the same conditions of laser excitation and field ionisation no matter which beam they belong to. Then, if the system of ion registration operates in a linear regime, we can write

$$I_1(t)/\Phi_1(t) = I_2(t)/\Phi_2(t) = \bar{I}_2/\bar{\Phi}_2 \qquad (3.14)$$

where $I_{1,2}$ are ion signals and $\Phi_{1,2}$ are the fluxes of the atoms to be detected in the main and the reference beams respectively. The thermodynamic properties of the element known, it is possible to calculate the flux Φ_2 in a standard way [3.44] with the geometry and the temperature of the reference oven known. The material of the reference crucible in the experiment has, of course, to be chosen so that the chemical interaction of the atoms of the element with the crucible walls is eliminated. Thus, if we know the ratio of the signals from the main and the reference beam, it is possible to determine the flux of atoms of the element in the main beam. Thus, using the reference beam we can carry out absolute calibration of the detection system.

The total analytical signal from the sample is proportional to the area $\int I_1(t)\,dt$ of the selective ion signal. Consequently, the atomic yield of the element from a sample will be expressed as

$$N = K \int \Phi_1(t)\,dt = \frac{K\bar{\Phi}_2}{\bar{I}_2} \int I_1(t)\,dt \qquad (3.15)$$

where K is the geometric factor which is equal to the ratio of the total flux of atoms from the crucible to the flux of atoms through the excitation zone. This coefficient is determined in model experiments by direct measurement of these fluxes under the same conditions of atomisation. Knowing the mass of the sample, i.e. the total number of its particles N_0, and the degree of atomisation ξ of the compounds of the element under the experimental conditions we can calculate the relative content of the element from the formula

$$C = \frac{N}{N_0\xi} = \frac{K\bar{\Phi}_2}{\bar{I}_2 N_0\xi} \int I_1(t)\,dt. \qquad (3.16)$$

The calculation of Φ_2 in this formula is somehow connected with the value of the saturated vapour pressure of the element. However, existing pressure data are rather contradictory for a number of elements. The application of laser stepwise photoionisation in a single-atom detection regime also permits absolute calibration of the reference beam, as has been done in [3.18]. Such calibration makes it possible to choose the correct values of vapour pressure from a variety of data and to use them for correct calculation of Φ_2.

3.4   The laser photoionisation method in analytical experiments

In this section consideration is given to the basic experiments on diagnostics and determination of impurities in substances in different states: from an atomic beam to biological objects. Emphasis is placed on ultra-high-sensitivity analysis of complex media.

3.4.1   *Analysis of atomic beams and vapours*

The photoionisation method is quite suitable for direct diagnostics of atomic beams with both basic and impurity atoms. In this case the laser radiation is focused into a small region of the atomic beam and the resultant photoions are detected by an SEM placed inside the vacuum chamber. Such experiments have been performed in [3.45] to diagnose the spatial structure of a Ga atomic beam.

The atoms of Ga were photoionised by one-frequency radiation in two steps via the resonant intermediate state $5\,^2S_{1/2}$ into a continuum. The Ga atoms have two different initial states, $4\,{}^2P_{1/2}$ (ground) and $4\,{}^2P_{3/2}$ ($\Delta E = 826$ cm^{-1}), so it was possible to apply radiation at two wavelengths (403.2 nm or 417.2 nm) for resonant ionisation. The sensitivity attained in this experiment was several Ga atoms in a volume under irradiation ($d \simeq 0.2$ mm). The atomic beam geometry was measured with a spatial resolution of about 10^{-2} cm as the focal point was scanned.

Photoionisation detection of Hg atoms at a low vapour pressure ($p < 10^{-4}$ Torr) was realised in [3.46]. The characteristic feature of this experiment was two-photon excitation of the transition $6\,{}^1S_0 \to 7\,{}^1S_0$ by radiation with $\lambda = 312$ nm and $\Delta\lambda = 0.05$ nm. The detection limit in this experiment was about 10^9 atoms/cm^3. It was limited by a nonselective ion background.

3.4.2 Analysis of solutions. Thermal evaporation in a buffer gas and in a vacuum

The typical state of a substance during its analysis by the methods of atomic absorption and atomic fluorescence is a specially prepared solution of the substance being analysed. As will be seen below, the photoionisation method does not need such a solution since its high sensitivity enables detection of an extremely weak flux of atoms evaporated as the sample is heated in a vacuum. Nevertheless, several experiments have been performed on photoionisation analysis of solutions. In [3.47, 3.48] Na and Yb were detected in low-concentration solutions with thermal atomisation of their compounds in a buffer gas. The atoms of the wanted element were ionised by laser radiation in three steps. The atoms of Na were ionised into the continuum using the following scheme

$$\text{Na: } 3\,{}^2S_{1/2} \xrightarrow{589\text{ nm}} 3\,{}^2P_{3/2} \xrightarrow{568.8\text{ nm}} 4\,{}^2D_{5/2} \xrightarrow{1.06\,\mu m} \text{Na}^+.$$

The atoms of Yb were ionised through the autoionisation state (AS)

$$\text{Yb: } 6\,{}^1S_0 \xrightarrow{555.6\text{ nm}} 6\,{}^3P_1 \xrightarrow{452.8\text{ nm}} 6\,{}^1D_2 \xrightarrow[(\sigma_3 \simeq 10^{-16}\text{ cm}^2)]{452.8\text{ nm}} \text{AS} \rightsquigarrow \text{Yb}^+.$$

The resulting photoelectrons were detected with a wire probe operating under a proportional amplification regime with $K = 10$–100. These studies show that atomisation in a buffer gas has certain restrictions. The lower detection limit of Na, for example, was imposed by the residual background of the Na compounds present in the gas. The presence of impurities in the buffer gas which reacted with the resultant Yb atoms was an obstacle to reaching a low detection limit of Yb.

The advantages of vacuum atomisation were demonstrated in [3.49] where the detection limit of Yb was $5 \times 10^{-2}\ \mu\text{g}\,\text{l}^{-1}$ or 1 pg in the sample. This value is smaller than the detection limits of the most sensitive analytical methods. An electrothermal atomiser was used to atomise ytterbium chlorides. The photoionisation scheme for Yb atoms was the same as described above [3.48]. The results obtained were not ultimate. The sensitivity could be improved by two orders by perfection of the atomiser design and by increasing the laser radiation power.

In analytical investigations it is often necessary to measure isotope ratios in samples with a high accuracy. Such tasks become more complicated if the interference of isobars of close elements takes place because of their incomplete chemical separation. Such a situation occurs for Lu and Yb. These two elements are similar in their properties and so the process of their chemical separation is very complicated. At the same time, laser photoionisation in a vacuum allows easy separation of these elements. The selectivity of Lu detection in the presence of Yb obtained in [3.50] was rather high. A mixture of Yb and Lu chlorides was applied on a metal filament placed in the region of the ion source of the magnetic mass spectrometer. As the filament was heated to 700 °C, the compounds of Lu and Yb were atomised and the resulting atoms of Lu were ionised in two steps by a laser pulse running parallel to the filament. The laser wavelength $\lambda = 452$ nm was tuned to the transition of Lu $^2D_{3/2}(5d\ 6s^2) \rightarrow {}^2D^o_{3/2}(5d\ 6s\ 6p)$. Thus, the atomisation system described and the ionising laser radiation were the ion source for the mass spectrometer.

Under the conditions of normal thermal ionisation the signal of Yb from an equimolar mixture of these elements is an order larger than the signal of Lu. By laser photoionisation, due to the absence of accidental one- and two-photon resonances in Yb at $\lambda = 452$ nm, the selectivity of photoionisation detection $S \geqslant 5 \times 10^4 = N_{175}/N_{174}^{1/2}$, where N_{175} is the number of photoions and N_{174} is the background signal for the 174 mass. The sensitivity of the method estimated under these conditions gave a value of the Lu content in the sample of less than 1 ng.

3.4.3   Analysis of high-purity semiconductor materials

(a)   *CdS crystals. Separation of surface and bulk impurities*
In [3.51] CdS crystals grown by the method of resublimation of CdS from the gas phase in an argon flow were analysed for Na impurities.

The Na impurities were detected in the following way. A sample of CdS crystal with a mass of several tens of milligrams was placed into a tantalum crucible with an outlet channel 1 mm in diameter and was heated in a vacuum ($p_{res} \simeq 10^{-6}$ Torr) to a temperature of 800–1050 °C. At this temperature intense sublimation of the crystal occurs ($t_{melt_{CdS}} = 1475$ °C) yielding various atoms and molecules—Cd, CdS, S_2, S_4, Na, etc. A flux of atoms and molecules was formed as a beam was irradiated by two dye lasers pumped by an N_2 laser. The wavelengths of the lasers were tuned for two-step excitation of Na atoms to the

Rydberg state 14d by the scheme:

$$3\,{}^{2}S_{1/2} \xrightarrow{\;589\,\text{nm}\;} 3\,{}^{2}P_{3/2} \xrightarrow{\;418\,\text{nm}\;} 14\,{}^{2}D_{3/2,5/2}.$$

The Rydberg atoms were then field ionised ($\mathscr{E}_{pul} = 13$ kV cm^{-1}) and the ions were detected by a SEM (see §3.3.4). The reference beam method described in §3.3.6 was used to calibrate the registered ion signals.

Figure 3.12 shows the dependence of the Na ion signal I_1 on evaporation time at $t = 940$ °C for a sample of CdS with a mass of 30 mg. The sharp maximum at the beginning of evaporation is conditioned by evaporation of Na compounds from the polluted surface of the crystal. Similar dependences were obtained for several samples of CdS crystals in the temperature range 800–1050 °C. In some cases the fluctuations of the ion signal I_1 were stronger than in figure 3.12 which was probably connected with an inhomogeneous distribution of the impurity over the volume of the sample.

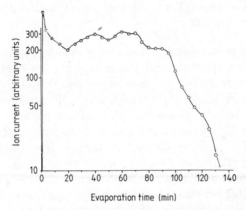

Figure 3.12 The dependence of Na ion signal I_1 on the evaporation time at a temperature $t = 940$ °C for a CdS sample with a mass of 30 mg (from [3.51]). $\eta_{Na/CdS} = 2 \times 10^{-6}\%$; $S/N \gtrsim 10^4$.

The content of Na impurity in CdS was determined using formula (3.16) for the reference beam method. The value $\int I_1(t)\,dt$ for each sample of CdS was determined from the area under the ion signal curve. At a definite temperature of the reference beam oven the ion signal $\bar{I}_2$ was measured and the flux Φ_2 for this temperature was calculated from the beam geometry [3.44]. The coefficient K was measured for the atomiser used, its value being $K \simeq 50$. The degree of atomisation of Na compounds in this experiment was taken to be equal to unity from the following considerations. Crystals with a regular structure and a low level of impurities are characterised by the impurity molecules in them being in a dissociated state [3.52]. When such crystals evaporate, the impurity molecule dissociates completely giving rise to free atoms of the impurity element. Thus, the degree of atomisation of Na impurities in CdS crystals, ξ, is approximately one.

The concentration of Na impurities obtained in the experiments for different CdS samples ranged from 2×10^{-6} to $6 \times 10^{-5}\%$. In most cases the signal-to-noise ratio was about 10^4, which would allow, in principle, reaching a registration level below $10^{-9}\%$. The ion background caused by nonselective multiphoton ionisation of the atoms and molecules of the basic substance by the second-step laser radiation is a factor limiting sensitivity. Nevertheless, the result obtained for Na—$2 \times 10^{-6}\%$ with a large value of signal-to-noise ratio (10^4)—is almost ultimate for most conventional analytical methods in the direct analysis of a substance.

The experiment has shown that with vacuum atomisation it is possible to separate surface and bulk impurities through careful evaporation of the substance below the melting point. In this case the contaminated surface layers are first to evaporate and under vacuum conditions their complete yield from the crucible takes place. Then the basic mass evaporates and the bulk impurities are determined.

(b) *High-purity germanium. Thermal evaporation in a vacuum*
Some applications of present-day technology of high-purity materials, for example the determination of aluminium, boron and other third-group acceptor elements in semiconductor materials, require control of impurities at a level of 10^{-8}–$10^{-10}\%$. Conventional analytical methods are not able to provide such sensitivity. The possibility of detection of sodium, aluminium and boron in high-purity crystals of germanium using the method of laser stepwise photoionisation in a vacuum was investigated in [3.53, 3.54, 3.43]. Sodium was chosen as a model element because the scheme of its effective ionisation via Rydberg states had been proved (see the earlier discussion in section (a) above). Under the conditions of saturation of all transitions and field ionisation every second atom of Na entering the laser beam became ionised [3.18]. The experimental technique was similar to the one described in (a). The first photoionisation experiments in a germanium atomic beam showed that, with the second-step energy flux sufficient to saturate the Rydberg transition $3p \to 14d$, a remarkable ion background occurred. It occurred because of three-photon ionisation of germanium atoms by laser radiation with $\lambda_2 = 418$ nm. To remove this background the laser beam was defocused from a diameter of 0.5 to 1.5 mm. As a result, the ionisation efficiency of sodium atoms fell to 0.1 but the ion signal was decreased only slightly due to multiple increase in the volume of the excitation zone. The signal of three-photon ionisation, however, dropped by about two orders of magnitude because of a nonlinear dependence of the rate of such ionisation on laser radiation intensity.

Figure 3.13 presents the time dependence of the Na ion signal at different temperatures. The increase of the photoionisation signal observed with every increase of the crucible temperature (figure 3.13(b)) seems to be caused by the atomisation of more and more thermostable Na compounds or by the heating of new parts of the atomiser. It must be said that this signal is conditioned mainly by the Na compounds adsorbed on the crucible surface since it was observed every time the chamber lost sealing and the crucible was withdrawn into the air. At a temperature of 1750 °C there was a gradual decrease in background signal caused by the evaporation of Na from the surfaces of the oven components under the action of thermal radiation.

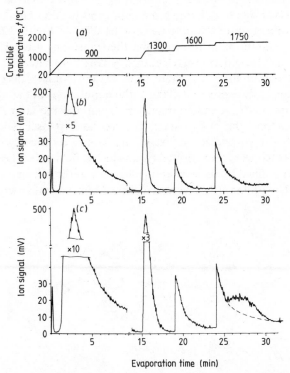

Figure 3.13 The dependence of Na ion signal on evaporation conditions: (a) temperature–time dependence; (b) empty crucible; (c) germanium sample $M = 5$ mg and $C_{Na} = 2 \times 10^{-8}\%$.

When a crystal of Ge with a mass of 5 mg and Na content of $2 \times 10^{-8}\%$ (figure 3.13(c)) was placed into the crucible, the amplitude of all the signals was increased due to an additional contribution of Na impurities from the crystal surface. The yield of Na bulk impurities took place only at 1750 °C when the diffusion rate of Na atoms from the Ge melt increased significantly. The absolute calibration of the registration system by the reference beam made it possible to evaluate the amount of Na in the sample from the additional ion signal. The result agreed quite well with the neutron activation analysis data. The detection limit of Na in Ge obtained in the experiment was about $5 \times 10^{-9}\%$.

These studies show that the method of atomisation concerned allows separate detection of surface and bulk impurities even at temperatures higher than the melting point of the substance provided that the heating regime is chosen properly. This is probably due to the fact that when evaporating in a vacuum the flux of the basic substance from the surface of the melted material prevents the surface impurities from diffusing or being dissolved in the melt.

The next stage was to determine the impurities of Al in Ge. The two-step scheme of Rydberg state excitation was also used for Al. Figure 3.14 shows the energy level diagram of an Al atom and the transitions used. The first-step laser radiation pulse with the wavelength $\lambda_1 = 3961.5$ Å excited the Al atoms to the $4\ ^2S_{1/2}$ state from the $3\ ^2P_{3/2}$ state, being $112\ \text{cm}^{-1}$ higher than the ground state. At a temperature of about $1000\ ^\circ\text{C}$ the populations of the states $3\ ^2P_{3/2}$ and $3\ ^2P_{1/2}$ (ground) are made practically equal and the laser excitation is more effective from the $3\ ^2P_{3/2}$ state. The second-step laser radiation pulse with $\lambda_2 = 4474$ Å continued to excite the atoms to the Rydberg states $15\ ^2P_{1/2,3/2}$ which were then ionised by an electric field pulse. The ionisation efficiency was defined by the excitation efficiency at the last step and came to about 10%. The Al impurities were determined by a technique similar to that

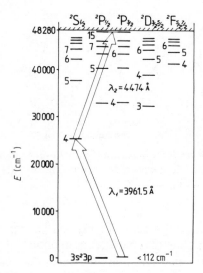

Figure 3.14   The energy level diagram of the Al atom and transitions used for excitation of Rydberg states (from [3.53]).

for Na. First the annealing of the sample was carried out at 900 °C ($t_{\text{melt}_{Ge}}$ = 937 °C). In contrast to Na, the stage of temperature increase to 1500 °C was characterised by the absence of a surface aluminium signal. Only above 1500 °C was it possible to detect an appreciable signal connected with the atomisation of surface Al impurities.

Figure 3.15 shows how the Al ion signal depends on evaporation time at 1800 °C for an empty crucible (3.15(a)) and for Ge samples differing in mass and surface area (3.15(b–d)). As it can be seen from figures 3.15(b) and 3.15(c), there are no important differences between the signals from samples of approximately the same mass but differing greatly in surface area. The signal from a sample (figure 3.15(d)), the mass of which is ten times larger than and the surface area of which is close to that of the sample 3.15(c), is much stronger than the signals from the samples 3.15(b) and 3.15(c). These results point to the

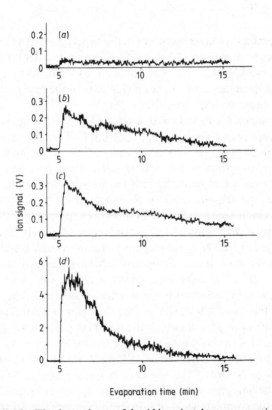

Evaporation time (min)

Figure 3.15 The dependence of the Al ion signal on evaporation time at a temperature of 1800 °C for (a) an empty crucible; (b) germanium sample $M = 2.75$ mg, surface area $S \simeq 4$ mm^2; (c) germanium sample $M = 2.5$ mg, surface area $S \simeq 18$ mm^2; (d) germanium sample $M = 28.2$ mg, surface area $S \simeq 20$ mm^2.

fact that the recorded signal is caused mainly by the Al atoms from the volume of Ge rather than from the surface impurities. In the case shown in figure 3.15(d) the selective signal-to-nonselective-noise ratio was equal to about 10^2 which gave a detection limit of Al in Ge of about $10^{-9}\%$.

The first experiments on direct laser photoionisation analysis of Ge for boron impurities revealed the presence of a strong nonselective ion background caused by Ge atoms. The analysis of this signal showed that there was two-step photoionisation of Ge atoms by the laser radiation used to ionise B atoms (figure 3.8). This process limits the selectivity of detection of B impurities in Ge to the level $S = 3 \times 10^7$ (see §3.2.3(c)).

To increase the detection selectivity the method of ion background discrimination considered in §3.3.5(c) was used. The B atoms were ionised according to the scheme

$$2\ ^2P_{1/2} \xrightarrow{\text{249.68 nm}} 3\ ^2S_{1/2} \xrightarrow{\text{378.35 nm}} 17p \xrightarrow[\text{pulse}]{\mathscr{E}\ \text{field}} B^+.$$

The energy parameters of laser radiation in the experiment [3.43] provided an ionisation efficiency of B atoms of about 5%. The electric field pulses had the parameters: $\mathscr{E}_- = 5\,\text{kV cm}^{-1}$, $\tau_- = 30$ ns, $\mathscr{E}_+ = 7\,\text{kV cm}^{-1}$, $\tau_+ = 10$ ns. The technique of ion detection and the basic elements of the experimental set-up were similar to those already described.

In the experiment a Ge sample with a mass of about 10 mg was put into a graphite crucible previously annealed at $t \lesssim 2500\,^\circ\text{C}$. Then the germanium was evaporated in a vacuum ($p \simeq 2 \times 10^{-6}$ Torr) at about 1700 °C. Since the thermodynamic properties of Ge and B differ greatly, it was mainly Ge evaporation at such a temperature. This led to a gradual increase of the B concentration in the remaining Ge melt. After 20–30 minutes the crucible temperature was increased to 2200 °C for intense evaporation of the rest of the Ge and B. Figure 3.16 shows the dependences of the B ion signal on evaporation time at $t \simeq 2200\,^\circ\text{C}$ for an empty crucible (3.16(a)) and for two Ge samples with a mass of about 10 mg containing B impurities of 2×10^{-8} (3.16(b)) and $2 \times 10^{-7}\%$ (3.16(c)) (these concentration values were obtained from electrophysical measurements of comparable samples). The comparison of the signals 3.16(b) and 3.16(c) shows that the ratio of B ion yields in these cases is close to 10, i.e. it coincides with the results of preliminary certification. Thus, the recorded signal is conditioned by the bulk impurities of B rather than by its surface impurities. The absolute calibration of the registration system was done as before by the reference beam of B atoms. This permitted estimation of the integral signal for the case 3.16(c). It was found to be consistent with the complete yield of B from the Ge sample. All this confirms the correctness of the analysis carried out.

The B impurities left in the crucible after annealing were a factor limiting the ultimate sensitivity of analysis in the experiment. Since B is widespread in nature, it is necessary to choose for crucibles graphite with a low content of B.

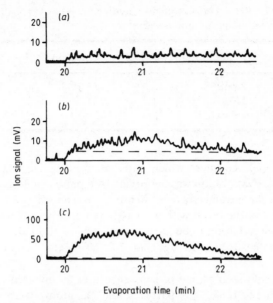

Figure 3.16 The dependence of the B ion signal on evaporation time at a temperature of 2200 °C (from [3.43]): (a) empty crucible background; (b) germanium sample $M = 9.5$ mg, $C_B = 2 \times 10^{-8}\%$; (c) germanium sample $M = 10$ mg, $C_B = 2 \times 10^{-7}\%$.

An enhancement of ionisation efficiency of B atoms by increasing the energy of laser pulses will permit realisation of the detection limit of B in Ge at $10^{-9}\%$ under the same conditions of atomisation.

The experiments carried out show that the method of laser stepwise atomic photoionisation in combination with atomisation of substances in a vacuum holds much promise for determining the content of impurities in high-purity substances. The method allows direct analysis of natural samples without any special preparation. The experiments illustrate the possibility of separate registration of surface and bulk impurities since the diffusion of surface impurities inside the melted sample decreases considerably at vacuum evaporation. The results of photoionisation determination of Na, Al and B impurities in Ge are given in table 8.1.

(c) *Silicon. Laser evaporation in buffer gas*
One of the applications of the photoionisation method for analysis was studied in [3.55] where pulsed laser evaporation (ablation) of Si was used together with synchronised laser photoionisation of Na atoms in the resulting cloud of substance (LARIS). The processes of evaporation of substance, atomic photoionisation and electron detection were carried out in a buffer gas that contained 90% argon and 10% methane. This gas was the working medium of

Table 3.1 Results of laser photoionisation analysis of germanium at the thermal atomisation in a vacuum.

Element	Concentration (wt %)	Detection limit (wt %)
Na	2×10^{-8}	5×10^{-9}
Al	10^{-7}	10^{-9}
B	2×10^{-8}	5×10^{-9}

the proportional counter which was used to detect the resultant photoelectrons. Evaporation was performed by a pulse of a flashlamp-pumped dye laser with the parameters: $\bar{\lambda} \simeq 590$ nm, $\tau_{pul} = 0.6\ \mu s$, $E_{pul} = 1$–10 mJ. Only the central part of the beam with a Gaussian intensity profile was used, and this allowed the radiation to be focused onto a spot with a diameter of up to 30 μm. The intensity in the spot of 10^7–10^8 W cm^{-2} was optimal for evaporation.

After the evaporated plume of substance was thermalised as a result of collisions with the buffer gas particles, the Na atoms were photoionised according to the scheme

$$\text{Na: } 3\ ^2S_{1/2} \xrightarrow[589\text{ nm}]{589.6\text{ nm}} 3\ ^2P \xrightarrow[568.2\text{ nm}]{568.8\text{ nm}} 4\ ^2D \xrightarrow{\lambda_1,\lambda_2} \text{Na}^+.$$

The estimated ionisation efficiency was close to 1. The sensitivity of the photoionisation system allowed detection of a concentration of Na atoms in the buffer gas of no less than 10 atoms/cm^3. Several samples of Si differing in purity were investigated. The experiments have shown that at intensities of 10^7–10^8 W cm^{-2}, 10^{-8} cm^3 of silicon crystal evaporate, which corresponds to 5×10^{14} Si atoms. The density of silicon atoms in the gas plume was estimated in this case to be 10^{14} cm^{-3}. In the process of detection one point was probed by several evaporating pulses. The first few pulses served to clean any accidental surface contamination and the others were taken into account as the bulk impurity concentrations were determined. For purest silicon the minimal signal corresponded to 500 Na atoms in the excitation volume or to 5×10^{11} cm^{-3} in the volume of Si. When the last result was obtained it was assumed that the influence of surface contamination was eliminated, the plume of substance was fully thermalised at the moment of photoionisation, and the concentration of reactive species in the buffer gas was not high enough to cause any significant decay of concentration of the resulting Na atoms.

It can be seen that certain assumptions regarding the character of all processes are needed to interpret quantitatively the results obtained. The simplest way to eliminate the need for such assumptions is to use standard samples of Si with a definite concentration of Na. But even in this case of evaporation in a buffer gas the interference effects arising between the atoms of the element to be analysed, the matrix and the impurities in the buffer gas complicate considerably the calibration of the signals and their interpretation. Under such conditions the approximation of the calibration curve to the low-concentration region may turn out to be incorrect.

(d) *Silicon. Ion sputtering in a vacuum*

Another version of the photoionisation method for analysis of solids (called 'SIRIS' by the authors) free of the shortcomings listed above was suggested in [3.40]. A pulsed beam of argon ions with parameters $E = 5$–30 keV, $\tau_{pul} = 2$ μs and $f_{rep} = 30$ Hz was used in the analytical experiment for sputtering silicon. It was formed from a continuous ion source with a special system of modulation and filtration. The ion beam intersected the target plate at $60°$ with respect to the normal forming a 2.6×3.3 mm^2 spot on the surface. The target system only provided for the escape of neutral particles into the region of detection. The resultant atomic–molecular cloud was irradiated by the laser radiation tuned for stepwise photoionisation of the atoms of the element under study. Then the resulting ions were subjected to mass filtration using a mass spectrometer with double focusing that comprised an electrostatic sector for analysing the ion energy and a magnetic sector for mass analysis. After magnetic separation the ions were detected by an channel electron multiplier equipped with a special conversion electrode. To suppress the extraneous background the electronics were time gated.

The potentialities of the method were studied by detecting gallium impurities in silicon with a concentration of $10^{-2}\%$. The atoms of Ga were photoionised according to the following scheme

$$\text{Ga: } 4p\ {}^2P^o_{1/2,3/2} \xrightarrow{2\nu_1} 4d\ {}^2D_{3/2,5/2} \xrightarrow{\nu_1} \text{Ga}^+.$$

The frequency ν_1 corresponded to three wavelengths for the first-step transitions: 5750.16 Å, 5890.07 Å and 5888.99 Å. The spectrum of the isotopes ^{69}Ga and ^{71}Ga was recorded in the experiment without extraneous mass peaks. At the same time analysis of silicon was performed by the method of secondary-ion mass spectroscopy. In this case an additional mass peak of GaH$^+$ formed in the process of ion sputtering was observed.

Although the authors [3.40] state that their experiment was preliminary and qualitative in nature it has nevertheless demonstrated the wide potential of the new method which now enables realisation of a detection sensitivity of impurities at a level of about $10^{-7}\%$. The realisation of all the technical reserves of the SIRIS apparatus will allow a sensitivity of 1 part in 10^{12}. The results of photoionisation analysis of Si for Na and Ga are given in table 3.2.

3.4.4 *Analysis of sea-water*

(a) *Aluminium in natural waters*

The development of new, highly sensitive methods of detection of

Table 3.2   Results of laser photoionisation analysis of silicon.

Element	Method of atomisation	Element concentration (atoms %)	Detection limit (atoms %)
Na	Laser ablation in buffer gas	5×10^{-9}	10^{-10}
Ga	Ion sputtering in a vacuum	10^{-2}	10^{-7}

microelements and aluminium, in particular, in natural waters is of especial importance. Aluminium, one of the most important elements in the lithosphere, is used in lithology and chemistry of the ocean as an indicator of terrigenous run-off. There are some works [3.56, 3.57] which relate aluminium dissolved in sea-water to certain biological processes. Thus, correct quantitative estimation of the participation of aluminium in the processes taking place in the ocean requires a universal method for determination of aluminium in waters with different mineralisation and different organic substance contents that can form soluble metallo-organic complexes with aluminium. River, sea and interstitial (pore) waters are waters of these types.

At present there is no direct method or universally adopted technique for determination of dissolved aluminium in natural waters [3.58]. There are many methods which differ from each other by the use of different chemical reagents which may contaminate the samples and act as an undesired background. The absence of reliable data on dissolved aluminium in sea and ocean waters and the presence of only a few data for interstitial waters point to the fact that these methods are complicated and unsatisfactory.

The merits of the method of laser stepwise atomic photoionisation as revealed by analysis of impurities in high-purity substances (see §3.4.3) has stimulated the application of this method to analysis of natural waters, which are more complex [3.59].

The procedure of aluminium determination in natural waters is as follows. A sample of water, volume 40 μl, was introduced into a tantalum or graphite crucible and evaporated in air at 90 °C to a dry residuum. Then the crucible and contents were placed in an atomiser, the vacuum chamber was evacuated to a residual pressure of 10^{-6} Torr and the crucible was heated by steps to 1750 °C. The technique of aluminium signal detection was analogous to that described in §3.4.3.

Figure 3.17(a) shows the typical dependence of the ion signal on evaporation time at 1750 °C for 40 μl of sea-water. During 0–5 minutes the crucible was heated stepwise to 1500 °C. The step heating was due to the fact that at fast ashing of the dry residuum of sea-water considerable gassing and vacuum

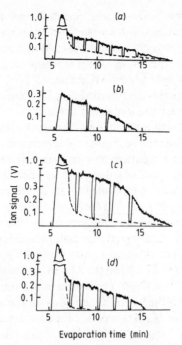

Evaporation time (min)

Figure 3.17 Dependence of the ion signal on evaporation time at a temperature of 1750 °C for a 40 μl sample (from [3.59]): (*a*) sea-water; (*b*) AlCl$_3$ calibrating solution with an Al concentration of 100 $\mu g \, l^{-1}$; (*c*) mixture of sea-water (9 parts) and AlCl$_3$ solution with an Al content of 1000 $\mu g \, l^{-1}$ (1 part); (*d*) aqueous solution with concentration of NaCl of 2% ($2 \times 10^7 \, \mu g \, l^{-1}$) and Al of 100 $\mu g \, l^{-1}$.

deterioration take place. In this temperature range the aluminium ion signal was almost absent. To discriminate the selective signal of Al from the total ion signal the first-step laser wavelength was periodically tuned off by 3–5 cm^{-1} from the Al transition $3 \, ^2P_{3/2} \xrightarrow{\lambda_1} 4 \, ^2S_{1/2}$. The dips in figure 3.17 correspond to the detuning of λ_1 which causes the ionisation efficiency of Al atoms to drop by more than two orders of magnitude (in the experiment the ionisation efficiency was equal to about 0.1). Such a procedure allowed discrimination of the nonselective ion background of the sea-water matrix from the ion signal conditioned only by aluminium. The nonselective background is probably conditioned by thermal ionisation of easily ionised sea-water components and laser photoionisation of the compounds constituting the sea-water matrix. The yield of Al in this case is defined by the area lying between the curve of the total ion signal and the broken curve.

In quantitative measurements the universally accepted technique is to construct an analytical calibration curve. The calibration solutions in this case must be prepared on the basis of the matrix being analysed. First the aluminium yield from solutions with known Al content was investigated. The solutions were prepared from aluminium chloride and deionised water. The typical time dependence of the ion signal for the calibrating solution of $AlCl_3$ with an Al content of $100 \mu g l^{-1}$ is shown in figure 3.17(b). As can be seen, for 'pure' solutions the ion signal is conditioned only by aluminium. Its total yield is defined by the area under the ion signal curve.

With the use of the addition method it was shown that the sea-water matrix composition does not distort the results of determination of aluminium content. For this purpose a solution was prepared which consisted of nine parts of sea-water and one part of calibrating solution of $AlCl_3$ with an Al content of $1000 \mu g l^{-1}$. This corresponded to an additional concentration of Al in seawater of about $100 \mu g l^{-1}$. The dependence of the ion signal on evaporation time for $40 \mu l$ of such a mixture under the same conditions is shown in figure 3.17(c). The total signal of aluminium defined by the 'selective' area under the signal curve is equal to the sum of the selective signal of aluminium from sea-water (3.17(a)) and the Al signal from the solution with Al concentration of $100 \mu g l^{-1}$ (3.17(b)). The nature of the nonselective ion signal in sea-water (3.17(a)) was studied using an aqueous solution that contained 2% NaCl (this concentration of NaCl is typical of sea-water) and aluminium at a concentration of $100 \mu g l^{-1}$ (figure 3.17(d)). For such a mixture the ion signal also consists of selective and nonselective parts. The selective signal area (figure 3.17(d)) shows that the yield of Al here is similar to that in the solution of $AlCl_3$ (3.17(b)), i.e. NaCl does not affect the yield of Al. The nonselective ion signal conditioned by the ionisation of NaCl (3.17(d)) differs from the nonselective ion signal for sea-water (3.17(a,c)) in time behaviour. This signal in the interval 5–7 minutes is caused by the ionisation of NaCl, and the prolonged nonselective signal is defined by a more complex composition of the sea-water matrix.

On the basis of $AlCl_3$ solutions in deionised water an analytical calibration curve was constructed. It turned out to be linear in the concentration range 5–$10^3 \mu g l^{-1}$. Above $10^3 \mu g l^{-1}$ the calibration curve becomes nonlinear due to saturation of the registration system. The lower limit of the calibration curve was defined by the Al content in deionised water ($5 \mu g l^{-1}$). The detection limit of Al in these experiments determined by extrapolating the analytical curve to the background level was equal to $0.01 \mu g l^{-1}$. An analytical curve was constructed on the basis of sea-water with aluminium chloride additions. The curves turned out to be parallel, which indicates that it is possible to use the calibration curve constructed on the basis of $AlCl_3$ solutions in quantitative measurements of aluminium in sea-water. The nonselective ion background affects only the value of minimum detectable concentration of Al in sea-water since it is difficult to discriminate a small selective signal against the background of a strong nonselective one. In experiments this value was equal to $1\,\mu g\,l^{-1}$ that corresponded to the absolute amount of Al in a sample of 40 pg.

The relative standard deviation was determined from twenty parallel measurements carried out for sea-water. It was equal to 5.6%. The largest contribution to this value was made by the volume fluctuations of the sample introduced in the crucible.

This method was applied to determine the content of dissolved aluminium in water samples taken from the mouth of the Kura River (the Caspian Sea). The results obtained for river water are close to the average value for the rivers throughout the world [3.58].

The concentrations of dissolved aluminium in the pore waters of the Indian Ocean were also determined. The pore water was squeezed out of sediment monoliths (10–15 cm below the sediment surface) picked with a bottom scoop and filtered through $0.45\,\mu m$ membrane filters. Figure 3.18(a) shows the location of points where the samples were taken, the ocean depths are given in brackets. The dissolved aluminium content increases in regions of subsea ridges (figure 3.18(b)) which is conditioned by the formation of metalliferous sediments at such ridges. Such an increase is particularly characteristic in the

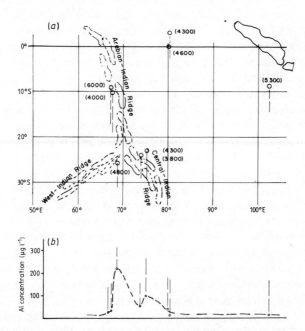

Figure 3.18 (a) Location of pore water sampling points in the Indian Ocean (the ocean depths in metres are indicated in brackets). (b) The dependence of dissolved aluminium concentration in pore waters on the longitude of sampling points.

region of a 'triple point', i.e. the point of intersection of three submarine ridges where the process of subsea orogenesis is probably still going on. The results obtained agree well with the data from [3.57].

The content of dissolved Al in the surface waters of the Mediterranean has also been measured. The value obtained $(6.5\pm0.7)\,\mu g\,l^{-1}$ is close to the average concentration of Al in the Mediterranean [3.60].

(b)   *Ruthenium in the ocean*

The distribution of the rarest elements in the ocean is of interest not only for oceanology but also for geology, geochemistry and cosmochemistry. Its study enables us to understand the processes of lithosphere formation. Among these elements is ruthenium which has not yet been studied in the ocean because of its ultralow content and the absence of sufficiently sensitive analytical methods for Ru detection.

Because of the lack of sufficient analytical data on ruthenium its abundance in the earth's crust has not been finally ascertained and, according to estimations, ranges from 10^{-8}–$10^{-9}\%$ [3.61]. The best results, with detection limits of 10^{-6}–$10^{-7}\%$ [3.62], in determining the ruthenium content in natural objects were obtained by conventional analytical methods only after preliminary chemical [3.63] or fire assay concentration [3.64] of ruthenium from large samples of mass 100–500 g. A complex multistage procedure of chemical treatment of samples and the volatility of ruthenium oxides make the method of chemical concentration labour-intensive and not absolutely reliable. The methods of fire assay concentration of platinum metals and ruthenium, particularly, into copper, lead or tin alloys are more convenient.

Laser photoionisation spectroscopy with vacuum thermal atomisation is very promising for analysis of such alloys. In this method concentration solves the problem of sample representativeness. The point is that if the sample has a typical mass of 10–100 mg used in vacuum atomisation and the content of the element in the substance is small, the amount of element in samples of such a mass fluctuates over very wide ranges. A comparatively large sample is needed to exclude these fluctuations. For example, a sample of mass 10–50 g is representative for geological objects and concentrations of 10^{-9}–$10^{-10}\%$.

Laser photoionisation spectroscopy with vacuum thermal atomisation in combination with fire assay concentration was used in [3.65] to detect ruthenium traces in a wide class of ocean species. Because of the wide-ranging variety of objects and ultralow content of ruthenium in them fire assay concentration to a lead alloy was used. This procedure [3.64, 3.66] comprised a fusion of 25 g of the substance in powder form with lead oxide and additional fluxes at 1000 °C. In this case ruthenium was collected into a metal lead alloy (lead button) of mass 30–40 g. After mechanical separation of slags the lead button was reduced to 50–100 mg by oxidation in magnesite cupels in an air flow at a temperature of 900–1000 °C. According to [3.64], such a reduction of the lead button does not lead to losses of ruthenium. The concentration factor obtained in this way ranged from 10^2 to 10^4. In analysis of waters the moisture was first evaporated and then the fire assay enrichment of the dry salt residuum was carried out.

The lead bead produced after cupellation was put into the atomiser crucible and, after the vacuum chamber was evacuated, the crucible was heated to 900 °C. At this temperature the lead was evaporated while the ruthenium was concentrated in the rest of the lead alloy because of its extremely low volatility. In this way the matrix was evaporated for 15 to 40 minutes depending on the lead mass. The rest of the sample was atomised by stepwise increase of the crucible temperature to 2400 °C.

Photoionisation detection of the ruthenium atoms in the beam thus formed was carried out by field ionisation of the Rydberg state with the effective principal quantum number $n^* \simeq 16$. The three-step excitation of Ru was performed according to the scheme (figure 3.19)

$$4d^7 5s(^5F_5) \xrightarrow{392.6\ nm} 4d^6 5s5p(^7D_4^\circ) \xrightarrow{633.05\ nm} 4d^7 6s(^5F_5) \xrightarrow{567.3\ nm} 4d^7 np\ (n^* \simeq 16).$$

The technique of ion detection is similar to that described above for aluminium.

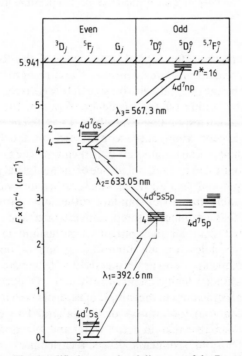

Figure 3.19 The simplified energy level diagram of the Ru atom and laser transitions for excitation of Rydberg states.

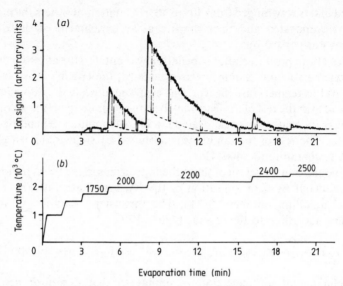

Figure 3.20 (*a*) The typical dependence of the Ru ion signal on the evaporation time and atomiser temperature for a lead bead of mass 50–100 mg. (*b*) Time dependence of the atomiser temperature (from [3.65]).

Figure 3.20(*a*) presents the typical dependence of the Ru ion signal on evaporation time and atomiser temperature. The content of ruthenium in the sample is defined by the integral selective signal. It was discriminated from the total ion signal by tuning off the frequency of one of the lasers from the resonance with a corresponding transition [3.59]. The ion signal was calibrated with a calibration curve constructed on the basis of copper standards containing a known amount of ruthenium. The possibility of such calibration was confirmed by the coincidence of the analytical photoionisation signals from copper and lead alloys in which ruthenium was collected with signals from the same amount of standard ruthenium-bearing ore.

Table 3.3 describes a number of samples investigated and results obtained. These are the first data on the content of ruthenium in these samples. Ferromanganese nodules are an interesting group of ocean formations enriched in iron, manganese and heavy metals [3.67]. One can see that they are characterised by rather a high content of ruthenium. Of particular interest is the high content of ruthenium in the sample of aerosol taken in the central part of the North Atlantic at the latitude of the Sahara. This concentration is almost three orders higher than in ocean water and four or five times higher than in pelagic bottom sediments. If such enrichment in ruthenium is characteristic of ocean aerosols as a whole, this could be taken as an argument in favour of the hypothesis of the cosmogenic origin of a considerable amount

Table 3.3 The ruthenium content in samples from the ocean environment.

Sample	Location	Ru concentration† $(\times 10^{-10}\%)$
Clayey-siliceous ooze	Southern Indian Ocean	150
Metalliferous sediment	East Pacific Rise	1700
	Red Sea, Discovery Deep	7600
Ferromanganese nodule	Northern Pacific	3500
	Southern Pacific	1600
Phosphorite (Pleistocene)	Mid Pacific Mountains	170
Fish bones	Chile shelf	56
Ocean water	Bay of Biscay	1.3
Aerosol	Northern Atlantic off Sahara coast	910

† The values are given with the accuracy $\pm 15\%$. The detection limits are equal to $10^{-10}\%$ for solid samples (25 g) and $3 \times 10^{-12}\%$ for sea-water (1 l).

of the ruthenium in the ocean since many cosmogenic materials (dust, meteorites, etc) are enriched in ruthenium [3.68, 3.69].

The measurements performed show that the combined photoionisation method has undoubted promise and, since there are no fundamental limitations on the fire assay concentration of other platinum metals, the method can be applied to detect traces of any platinum-group elements in natural materials.

3.4.5 Analysis of biological objects

One of the urgent analytical tasks important for biology and medicine is the determination of trace amounts of metals in biological objects. It is necessary to determine normal concentrations of such metals in the organism as well as the correlation between their deviations from normal levels and different functional disorders and diseases.

Biological objects are especially interesting for the application of the photoionisation method since they reveal its essential merit of insensitivity to other elements except that being analysed. This means that no preliminary treatment of the sample is needed.

This was confirmed by experiments [3.71] on photoionisation detection of Al traces in human blood. The choice of Al is connected not only with the fact that it can be easily detected, but also that it is one of the metals of interest for toxicology. The role of this element in the metabolism of living organisms remains to be determined. The point is that it is difficult to produce 'aluminiumless' diets because of its wide abundance in nature. Some diseases of the lungs and nervous system have been shown to be caused by increased content of aluminium in an organism, as in [3.72].

The analytical procedure of direct determination of the Al content in blood consisted of the following: a 40 μl sample of blood in its natural form was introduced into a tantalum crucible and was dried in air at a temperature of 90–100 °C for 3–5 minutes. The processes of ashing and atomisation of the dry residuum took place in a vacuum chamber. It is necessary to find heating regimes such that ashing does not cause the vacuum to deteriorate below 10^{-4} Torr. At the same time this process must go quickly enough in order not to lead to significant thermal evaporation of the residuum without its atomisation. The crucible temperature in the experiment during ashing was increased to 1500 °C in five steps for 10 minutes. The technique of detection of Al atoms is described in detail in the foregoing sections.

Figure 3.21(a) shows the typical time dependence of the ion signal during stepwise heating of the crucible. The procedure of wavelength detuning described in §3.4.4 was used to discriminate the selective signal of aluminium.

The total Al signal for a sample was defined by the 'selective' area under the signal curve (the difference between the total signal and the background one) (figure 3.21(a)). The value of Al concentration corresponding to such a signal was determined by the calibration curve constructed on the basis of $AlCl_3$ aqueous solutions. The possibility of such calibration was examined by the additions method. In this case 40 μl of blood and 40 μl of $AlCl_3$ solution with an Al concentration of 100 μg l^{-1} were introduced in the crucible. The

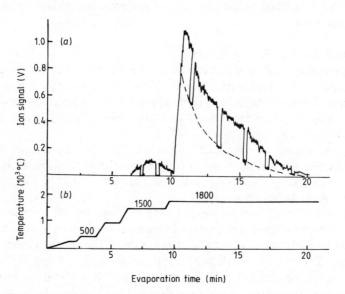

Figure 3.21 (a) The dependence of Al ion signal on evaporation time for 40 μl of human blood. (b) Stepwise increasing of crucible temperature (from [3.71]).

aluminium signal from such a mixture turned out to be equal to the sum of signals from the mixture components at their independent analysis within the measurement error (about 10%). This fact proved the absence of any effects of the blood matrix on the aluminium yield at thermal atomisation in a vacuum. The results of measurements of the aluminium content in five blood samples lie in the range $(230 \pm 50)\ \mu g\,l^{-1}$.

3.5 Conclusions and prospects

The experiments on laser photoionisation detection of trace elements in different media described above point conclusively to the merits of vacuum atomisation of matter. Table 3.4 presents the results of direct determination of some elements in different substances by the method of laser photoionisation spectroscopy in a vacuum. The results obtained under nonoptimised experimental conditions are ultimate for the most sensitive conventional analytical methods (the comparative characteristics of different analytical methods are given in table 3.5). The method of laser photoionisation spectroscopy has the potential to further reduce the obtained detection limits by one or two orders by improving the atomiser design, increasing the laser photoionisation efficiency and selectivity, removing the nonselective ion background, and so on.

Table 3.4 The results of direct laser atomic photoionisation determination of trace elements in various substances at the thermal atomisation in a vacuum.

Element	Matrix	Element concentration in matrix (atoms %)	Detection limit (atoms %)	Reference
Yb	Aqueous solution of $YbCl_3$	5×10^{-7}	2×10^{-9}	3.49
Na	CdS crystal	2×10^{-6}	2×10^{-10}	3.51
	Ge crystal	2×10^{-8}	5×10^{-9}	3.54
Al	Ge crystal	2×10^{-7}	10^{-9}	3.53
	Aqueous solution of $AlCl_3$	2×10^{-7}	2×10^{-10}	3.59
	Sea-water	5×10^{-7}	10^{-7}	3.59
	Human blood	3×10^{-5}	2×10^{-7}	3.71
B	Ge crystal	2×10^{-7}	5×10^{-9}	3.43
Ru	Sea-water	$(1-3) \times 10^{-10}$	3×10^{-12}	3.65
	Solid samples	10^{-4}–10^{-9}	10^{-10}	3.65

Table 3.5 Comparison of various methods of trace element determination.

Method	Detection limit in aqueous solutions (%)	Detection limit in real matrix (%)	Element selectivity
Atomic absorption spectrometry	$10^{-4}-10^{-9}$	$10^{-4}-10^{-7}$	Moderate
Neutron activation analysis	$10^{-5}-10^{-9}$	$10^{-5}-10^{-9}$	Moderate
Spark mass spectroscopy	$10^{-5}-10^{-8}$	$10^{-5}-10^{-7}$	High
Laser fluorescent spectroscopy	$10^{-6}-10^{-11}$	$10^{-5}-10^{-8}$	High
Laser stepwise photoionisation spectroscopy	$10^{-11}-10^{-14}$	$10^{-8}-10^{-12}$	Ultrahigh

The experiments performed confirm the universal character of the method of laser photoionisation spectroscopy in combination with vacuum thermal atomisation of substances and open up wide prospects for using it as a new analytical method. In addition, laser stepwise photoionisation of atoms in a vacuum can be combined with other methods of atomisation and allows direct combination with a mass spectrometer and different methods of discriminating selective ions.

In conclusion it should be stressed again that the merits of the method developed—the possibility of single-atom detection in the volume of interaction with laser radiation, the possibility of direct analysis of the substance in its natural form, the removal of uncontrollable contaminations by atomising the substance in a vacuum, the possibility of discriminating a selective signal against the background in one measurement and separate detection of surface and bulk impurities in solid samples—enable it to be used in determining traces of most elements in almost any matrix.

References

3.1   Winefordner J D (ed) 1976 *Trace Analysis: Spectroscopic Methods for Elements* (New York: Wiley)
3.2   Letokhov V S 1980 in *Chemical and Biochemical Applications of Lasers* vol. 5 ed C B Moore (New York: Academic) pp. 1–38
3.3   Hurst G S, Payne M G, Kramer S D and Young J P 1979 *Rev. Mod. Phys.* **51** 767
3.4   Balykin V I, Bekov G I, Letokhov V S and Mishin V I 1980 *Usp. Fiz. Nauk* **132** 293 (Engl. transl. 1980 *Sov. Phys.–Usp.* **23** 651)

3.5 Ambartsumyan R V, Kalinin V P and Letokhov V S 1971 *Pis'ma Zh. Eksp. Teor. Fiz.* **13** 305 (Engl. transl. 1980 *JETP Lett.* **13** 217)

3.6 Letokhov V S, Mishin V I and Puretsky A A 1977 *Prog. Quantum Electron.* **5** 139

3.7 Letokhov V S 1983 *Nonlinear Laser Chemistry* ed F P Schäfer (Springer Series in Chemical Physics vol. 22) (Berlin: Springer) ch 3, 7

3.8 Hurst G S, Payne M G, Nayfeh M H, Judish J P and Wagner E B 1975 *Phys. Rev. Lett.* **35** 82

3.9 Ambartsumyan R V, Apatin V M, Letokhov V S, Makarov A A, Mishin V I, Puretsky A A and Furzikov N P 1976 *Zh. Eksp. Teor. Fiz.* **70** 1660 (Engl. transl. 1976 *Sov. Phys.–JETP* **43** 866)

3.10 Hurst G S, Nayfeh M H and Young J P 1977 *Appl. Phys. Lett.* **30** 229; *Phys. Rev. A* **15** 2283

3.11 Ivanov L N and Letokhov V S 1975 *Kvant. Elektron.* **2** 585 (in Russian)

3.12 Ambartsumyan R V, Bekov G I, Letokhov V S and Mishin V I 1975 *Pis'ma Zh. Eksp. Teor. Fiz.* **21** 595 (Engl. transl. 1975 *JETP Lett.* **21** 279)

3.13 Stebbings R F, Latimer C J, West W P, Dunning F B and Cook T B 1975 *Phys. Rev. A* **12** 1453

3.14 Ducas T W, Littman M G, Freeman R R and Kleppner D 1975 *Phys. Rev. Lett.* **35** 366

3.15 Paisner J A, Solarz R W, Worden E F and Conway J G 1977 in *Laser Spectroscopy III* ed J L Hall and J L Carlsten (Springer Series in Optical Science vol. 7) (Berlin: Springer) p. 160

3.16 Bekov G I, Letokhov V S and Mishin V I 1977 *Zh. Eksp. Teor. Fiz.* **73** 157 (Engl. transl. 1977 *Sov. Phys.–JETP* **46** 81)

3.17 Bekov G I, Letokhov V S and Mishin V I 1978 *Pis'ma Zh. Eksp. Teor Fiz.* **27** 52 (Engl. transl. 1978 *JETP Lett.* **27** 47)

3.18 Bekov G I, Letokhov V S, Matveyev O I and Mishin V I 1978 *Opt. Lett.* **3** 159; *Zh. Eksp. Teor. Fiz.* **75** 2092 (Engl. transl. 1978 *Sov. Phys.–JETP* **48**)

3.19 Tursunov A T and Eshkabilov N B 1983 *Preprint* No. 12 (Institute of Spectroscopy, USSR Academy of Sciences, Troitsk)

3.20 Stebbings R F, Dunning F B and Rundel R D 1975 in *Atomic Physics IV* ed G zu Putlitz, E Weber and A Winnacker (London: Plenum) p. 713

3.21 Stebbings R F 1976 *Science* **193** 537

3.22 Bekov G I, Letokhov V S, Matveyev O I and Mishin V I 1978 *Pis'ma Zh. Eksp. Teor. Fiz.* **28** 308 (Engl. transl. 1978 *JETP Lett.* **28** 283)

3.23 Bekov G I and Vidolova-Angelova E P 1981 *Kvant. Electron.* **8** 227 (in Russian)

3.24 Bekov G I, Vidolova-Angelova E P, Ivanov L N, Letokhov V S and Mishin V I 1981 *Zh. Eksp. Teor. Fiz.* **80** 866 (Engl. transl. 1981 *Sov. Phys.–JETP* **53** 441)

3.25 Letokhov V S and Mishin V I 1979 *Opt. Commun.* **29** 168

3.26 Grossman L W, Hurst G S, Payne M G and Allman S L 1977 *Chem. Phys. Lett.* **50** 70

3.27 Grossman L W, Hurst G S, Kramer S D, Payne M G and Young J P 1977 *Chem. Phys. Lett.* **50** 207

3.28 Bekov G I, Vidolova-Angelova E P, Letokhov V S and Mishin V I 1977 in *Laser Spectroscopy IV* ed H Walther and K W Rothe (Springer Series in Optical Sciences, vol. 21) (Berlin: Springer) p. 283

3.29 Price W J 1972 *Analytical Atomic Absorption Spectrometry* (London: Heiden)

3.30   Hurst G S, Nayfeh M H, Young J P, Payne M G and Grossman L W 1977 in *Laser Spectroscopy III* ed J L Hall and J L Carlsten (Springer Series in Optical Sciences, vol. 7) p. 39

3.31   *Multiphoton Ionization of Atoms* 1980 Proceedings of the Lebedev Physical Institute, USSR Academy of Sciences, vol. 115 (Moscow: Nauka) p. 27 (in Russian)

3.32   Corliss C H and Bozman W R 1962 *Experimental Transition Probabilities for Spectral Lines of Seventy Elements* NBS Monograph 53 (Washington, DC: NBS)

3.33   Moore C E 1949, 1952, 1958 *Atomic Energy Levels* NBS Circular No. 467, vols. I, II, III (Washington, DC: NBS)

3.34   Young J P, Hurst G S, Kramer S D and Payne M G 1979 *Anal. Chem.* **51** 1050A

3.35   Hansch T W 1972 *Appl. Opt*, **11** 895

3.36   Bekov G I and Letokhov V S 1983 *Appl. Phys.* B **30** 161

3.37   Zherikhin A N, Letokhov V S, Mishin V I, Belyaev V P, Evtyunin A N and Lesnoy M A 1981 *Kvant. Elektron.* **8** 1340 (in Russian)

3.38   Kuz'min A F, Rafal'son A E, Kholin N A and Tsigel'man G E 1970 *Zh. Tekhn. Fiz.* **40** 1531 (in Russian)

3.39   Tsigel'man G E 1972 *Zh. Tekhn. Fiz.* **42** 667 (in Russian)

3.40   Parks J E, Schmitt H W, Hurst G S and Fairbank W M Jnr 1983 *Thin Solid Films* **108** 69

3.41   Slobodenyuk G I 1974 *Quadrupole Mass Filters* (Moscow: Atomizdat) (in Russian)

3.42   *Model M162 Boxcar Integrator. Operating and Service Manual* 1976 (Princeton Applied Research Corporation)

3.43   Bekov G I, Maximov G A, Nikogosyan D N and Radaev V N 1984 *Kvant. Elektron.* **11** 1262 (in Russian)

3.44   Ramsey N F 1956 *Molecular Beams* (Oxford: Clarendon)

3.45   Tursunov A T and Eshkabilov N B 1982 *Kvant. Elektron.* **9** 209 (in Russian); 1984 *Zh. Tekhn. Fiz.* **54** 166 (in Russian)

3.46   Miziolek A W 1981 *Anal. Chem.* **53** 118

3.47   Gonchakov A S, Zorov N B, Kuzyakov Yu Ya and Matveyev O I 1979 *Zh. Anal. Khim.* **34** 2312 (in Russian)

3.48   Yegorov A S 1984 *Thesis Abstract* Institute of Oceanology, USSR Academy of Sciences, Moscow (in Russian)

3.49   Bekov G I, Egorov A S and Mishin V I 1983 *Zh. Anal. Khim.* **38** 429 (in Russian)

3.50   Miller C M, Nogar N S, Gancarz A J and Shields W R 1982 *Anal. Chem.* **54** 2377

3.51   Akilov R and Bekov G I 1982 *Pis'ma Zh. Tekhn. Fiz.* **8** 517 (in Russian)

3.52   Tochitsky E I 1976 *Crystallization and Thermal Treatment of Thin Films* (Minsk: Nauka i Tekhnika) (in Russian)

3.53   Akilov R, Bekov G I, Letokhov V S, Maximov G A, Mishin V I, Radaev V N and Shishov V N 1982 *Kvant. Elektron.* **9** 1859 (in Russian)

3.54   Akilov R, Bekov G I, Devyatykh G G, Letokhov V S, Maximov G A, Radayev V N, Shishov V N and Scheplyagin E M 1984 *Zh. Anal. Khim.* **39** 31 (in Russian)

3.55   Mayo S, Lucatorto T B and Luther G G 1982 *Anal. Chem.* **54** 553

3.56   Mackenzie T F, Wollast R and Stoffyn M 1978 *Science* **199** 4329

3.57   Caschetto S and Wollast R 1979 *Geochim. Cosmochin. Acta* **43** 3

3.58   Dickinson Burrows W 1977 *CRC Critical Reviews in Environmental Control* p. 167

3.59 Bekov G I, Yegorov A S, Letokhov V S and Radayev V N 1983 *Nature* **301** 410
3.60 Caschetto S and Wollast R 1979 *Marine Chemistry* **7** 2
3.61 Reeves R D and Brooks R R 1978 *Trace Element Analysis of Geological Materials* (New York: Wiley)
3.62 Jäger H 1978 *Appl. Atomic Spectrosc.* **2** 1
3.63 Khvostova V P and Golovnya S V 1982 *Zavod. Lab.* **48** 3 (in Russian)
3.64 Kolosova L P 1982 *Zavod. Lab.* **48** 8 (in Russian)
3.65 Bekov G I, Letokhov V S, Radayev V N, Baturin G N, Egorov A S, Kursky A N and Narseyev V A 1984 *Nature* **312** 748
3.66 Haffty J, Riley L B and Goss W D 1977 *Geological Survey Bulletin* **1445**
3.67 Mero J L 1965 *The Mineral Resources of the Sea* (Amsterdam: Elsevier)
3.68 Croket H 1972 *Geochim. Cosmochim. Acta* **36** 517
3.69 Wedepohl K H 1968 in *Origin and Distribution of the Elements* ed L H Ahrens (International Series of Monographs on Earth Science, vol. 30) (Oxford: Pergamon) p. 999
3.70 Lee D S 1983 *Nature* **305** 47
3.71 Bekov G I, Letokhov V S and Radayev V N 1984 *Laser Chemistry* **5** 11
3.72 Berman E 1979 *Toxic Metals and Their Analysis* ed L C Thomas (Heiden International Topics in Science) (London: Heiden)

4 Infrared Absorption Spectroscopy with Tunable Diode Lasers

Yu A Kuritsyn

4.1 Introduction

This chapter is devoted to the applications of tunable diode lasers (TDL) to high-resolution IR spectroscopy and molecular analysis. We restrict our discussion to lasers made from $PbS_{1-x}Se_x$, $Pb_{1-x}Sn_xSe$, $Pb_{1-x}Sn_xTe$ and other lead salt compounds, for which at present laser action is obtained over the spectral range of 2.5 to 46.2 μm. Semiconductor lasers for shorter wavelengths are not now used practically in molecular spectroscopy because only high vibrational overtone bands fall in the spectral range of their emission.

The first lead salt diode laser was made from PbTe at the MIT Lincoln Laboratory in 1963 [4.1]. Its operating temperature was near liquid helium and it emitted at 6.6 μm. During the next few years lasing in various lead salt semiconductor compounds had been obtained and the range of operating temperatures was increased to 40–50 K. It was shown in 1969 that the diode laser linewidth in continuous-wave operation reaches the limit due to spontaneous radiation noise [4.2]. This means that even for lasers with a small output power of about 100 μW the emission linewidth does not exceed 1×10^{-5} cm^{-1}. The possibility of high resolution spectroscopy with TDL was first demonstrated by Hinkley in 1970 [4.3]. He used the technique of injection current tuning and measured part of an SF_6 molecule absorption spectrum in the v_3 band near 10.6 μm with the resolution restricted by Doppler broadening.

From that work, which has become a classical one, the number of publications devoted to molecular spectra recordings and other TDL applications began to increase quickly and at present number a few hundred. A few review articles on high-resolution spectroscopy with lead salt diode lasers have been published [4.4–4.8]. The latest [4.7, 4.8] contain a detailed bibliography up to 1981.

There are some unique advantages to be gained by the use of tunable diode lasers.

(1) The 3 to 50 μm wavelength range is very important for investigation of molecules and analysis of molecular gas mixtures because all molecules (excluding those of identical atoms) have intensive absorption bands in this spectral region.

(2) The linewidths of absorbing molecules at low pressures ($\lesssim 1$ Torr) are (1–5) $\times 10^{-3}$ cm^{-1}. Therefore the use of diode lasers with emission linewidths of 10^{-4}–10^{-5} cm^{-1} does not distort the spectrum recorded.

(3) The TDL can be easily tuned and modulated, for example, by changing temperature or bias current.

(4) Diode lasers have high spectral brightness and are coherent devices. These features allow the use of TDLS for long path measurements, in local analysis with a high degree of spatial resolution, and also for infrared heterodyne spectroscopy.

(5) Finally, and most importantly, commercial instruments are available from the Laser Analytics Division of Spectra-Physics [4.9].

In the Soviet Union the first lead salt diode lasers were made in the 1960s at the P N Lebedev Physical Institute of the Academy of Sciences [4.10]. At first they operated mainly under pulsed excitation. The integral (over the pulse time of 1 μs) emission linewidths of such lasers were shown to be about 0.1 cm^{-1} [4.11]. This testified to a tuning rate of 0.1 cm^{-1} μs^{-1} and it became clear that to record spectra with the 10^{-3} cm^{-1} resolution in the pulsed mode of operation a detection system with a time constant of 10 ns was necessary. The first high-resolution spectra were obtained with pulsed lasers in 1977 [4.12–4.14].

At the present time work on the applications of diode lasers to high-resolution spectroscopy and spectral analysis is carried out at the P N Lebedev Physical Institute, the Institute of General Physics, the Institute of Spectroscopy and some other organisations. Experimental commercial examples of diode laser spectrometers were manufactured in 1984 by the Leningrad Optical Mechanical Association [4.15].

Despite the fact that tunable diode lasers capable of CW operation at temperatures up to 80 K have been developed [4.16], in the majority of work being carried out in the Soviet Union preference is given to the pulsed mode of operation.

The advantages and shortcomings of different modes of operation and recording methods will be analysed in this chapter. Here we only point out that the main factor militating against the wide use of cw diode lasers for analytical applications is the necessity for them to be cooled to temperatures of 20–40 K. Through the use of pulsed mode the operational temperatures can be substantially increased and even reach the value of about 200 K obtained with Peltier thermoelectric coolers [4.17].

The purpose of this chapter is to review the experimental techniques used with lead salt tunable diode lasers for recording of spectra and measurements of gas impurity. We shall not give a comprehensive survey of the spectroscopic and analytical applications of these devices, but will use the results of some published papers to illustrate their potentialities. However, we give tables which summarise references from (mainly) the last few years. For detailed discussion of earlier work the reader is referred to the excellent reviews of McDowell [4.7] and Eng and Ku [4.8]. The proceedings of the conference on 'Tunable Diode Laser Development and Spectroscopy Applications', which was held in San Diego, California, in 1983 [4.18], can also be recommended for more detailed information about analytical and some other recent applications.

4.2   Semiconductor diode lasers

4.2.1   *Principle of operation and main characteristics*
Several authors have discussed in detail the fundamental material properties of lead salts, principles of TDL operation and characteristics of these devices [4.4, 4.8, 4.9, 4.19]. Here we shall only quote the main properties of TDLs and give a short review of recent advances in their development over the past few years.

Typical characteristics of TDLs are summarised in table 4.1. As a construction a TDL consists of a small crystal lead salt chip installed in a copper package (figure 4.1). This package is mounted on a cold finger of a cryogenic system. Lasing takes place when a forward bias current is applied.

Emission in TDLs is due to the electron–hole recombination across the energy gap E_g between conduction and valence bands of the semiconductor. Therefore the laser emission frequency is principally determined by the alloy composition. At the present time lasing is obtained in lead compounds with E_g from 500 to 26.8 meV, thus the spectral range from 2.5 to 46.2 μm is covered.

In the simplest case the diode laser cavity is a Fabry–Perot resonator formed by crystal cleaving. Ignoring the mode pulling effects the coherent

Table 4.1 Characteristics of lead salt diode lasers.

Parameter	Typical value
Overall available spectral range (by changing crystal composition)	2.5–46 μm
Width of spontaneous gain profile	5–50 cm^{-1}
Cavity mode linewidth	10^{-5}–10^{-6} cm^{-1}
Separations between longitudinal modes	1–5 cm^{-1}
Gain profile tuning rate by temperature	1–4 cm^{-1} K^{-1}
Tuning rate of a single mode	
by temperature	0.2–1 cm^{-1} K^{-1}
by current	2–20 cm^{-1} A^{-1}
with time (pulsed operation)	10^{-4}–10^{-1} cm^{-1} μs^{-1}
Range of fine continuous tuning	0.5–5 cm^{-1}
Operating temperature	10–60 K
Total temperature and current tuning range for a single crystal	$\sim$200 cm^{-1}
Effective laser linewidth	10^{-4}–10^{-3} cm^{-1}
Single-mode output power	0.01–0.5 mW

radiation frequency v coincides with the frequency of the cavity resonance v_q

$$v \simeq v_q = \frac{1}{2n^*L}\, q \qquad (4.1)$$

where q is the index of an axial mode, $n^* = n[1 + (v/n)(\partial n/\partial v)]$ index of refraction of the laser material and L the cavity length. The value of the index of refraction in lead salts varies from 4.5 to 7. Therefore the separation between the adjacent longitudinal modes for a typical cavity length of 500 μm is from 1 to 5 cm^{-1}.

Generation of coherent radiation arises when the injection current density j exceeds the threshold value j_{th}. For cw operation at 10 K j_{th} must be less than 100 A cm^{-2}. For simple homostructure TDLs the value j_{th} increases quickly with temperature. Continuous generation is usually observed below 40–50 K, but some values are even higher. In particular PbEuSeTe lasers with 147 K cw operation at 4 μm [4.20] and cw PbSnSe lasers operating at temperatures up to 118 K at 10 μm [4.31] have been made. The highest reported operating temperature for pulsed lasers (PbSSe, 5.5 μm) was 230 K [4.18].

The fundamental, quantum-phase noise-limited TDL linewidth was directly measured using an optical mixing technique for the Pb$_{0.88}$Sn$_{0.12}$Te laser at 10.6 μm [4.2] and a few PbS$_{1-x}$Se$_x$ lasers at 5.3 μm [4.22]. The minimum observed linewidth was 22 kHz (7×10^{-7} cm^{-1}).

The laser beam divergence is near the diffraction limit. Because of the small thickness of the active region the angle of divergence is considerable—from

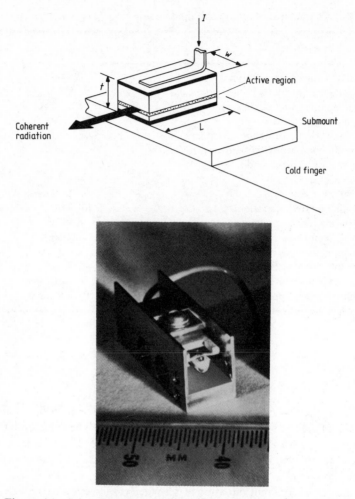

Figure 4.1 Schematic diagram of a simple homostructure lead salt tunable diode laser. Typical dimensions of the laser chip are $L \simeq 500$ μm, $t \simeq 200$ μm and $w \simeq 200$ μm. The photograph shows a laser chip mounted in a crystal package [4.9].

$\pm$ 10–20° in homolasers to 40° in some heterolasers. The TE modes are usually generated with polarisation in the p–n junction plane.

The noise spectrum of TDLS in the low-frequency region was measured by Eng *et al* [4.23]. They found that below 3 kHz there was excess noise due to instabilities of the operating conditions, mainly due to mechanical vibrations of the closed-cycle cooler in which the diode laser was housed. Above 3 kHz the excess noise from the TDL had a flat broadband spectrum with a noise power about 10 dB above the shot noise for radiation from an equivalent blackbody. In the high-frequency region excess noise resulted from the effects of mode competition; optical feedback and the spiking regime were also observed [4.24].

4.2.2 Emission frequency tuning

The most important property of a TDL is the possibility of considerable emission frequency tuning when the operating conditions are changed. Because a gain curve position v_s is determined by the bandgap energy E_g all methods of changing E_g can be used for coarse frequency tuning. E_g can be tuned by varying the temperature T of the semiconductor crystal, by applying pressure p or a magnetic field $\mathcal{H}$. The dependences of E_g on T, p and $\mathcal{H}$ for materials used for fabrication of mid-IR TDLS are given in reference [4.19].

Most conveniently, coarse emission frequency tuning is produced by varying the temperature of the cold finger where the TDL is housed. The temperature dependence of E_g for the PbSnTeSe laser is shown in figure 4.2. As may be seen, for this long-wavelength sample the wavelength (emission frequency) has been changed from 46.2 μm (216.4 cm^{-1}) at 6 K to 24.3 μm (411.5 cm^{-1}) at 78 K. The temperature tuning range of about 200 cm^{-1} is typical for a single lead salt diode. About 20 diodes fabricated from different alloys are needed to cover the total range from 3 to 46 μm by choosing crystals with different compositions and utilising the temperature tuning frequency for each separate device.

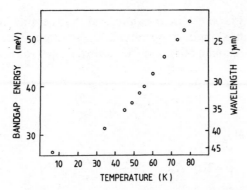

Figure 4.2 Temperature variation of bandgap energy E_g (radiation wavelength λ) for diode laser on $(PbSe)_{0.80}(SnTe)_{0.20} - PbTe_{0.68}Se_{0.32}$ [4.25].

The tuning range of a single sample can be significantly widened with the help of simultaneous temperature and pressure tuning. Ranges of up to 1000 cm^{-1} have thus been achieved [4.26].

Fine frequency tuning, which is used for high-resolution spectroscopy, is performed by shifting the frequency of a given spectral mode by control changes in the effective cavity optical length, mainly through the index of refraction. The most convenient, and therefore most often used method, is current tuning. As a matter of fact this method is the same as the temperature one, because the injection current raises the temperature of the active region due to energy dissipation.

In the general case the temperature change in a laser active layer is given by the transient equation for heat conduction

$$c\rho \frac{\partial T}{\partial t} - \nabla(k\,\nabla T) = \frac{P_T}{\delta A_j}. \tag{4.2}$$

Here c is the crystal heat capacity, ρ the sample density, k the thermal conductivity, P_T the power dissipated in the active layer, and δ and A_j are the active region thickness and area, respectively.

The solution of equation (4.2) with the heat source $P_T = IU_j(1-\eta)$ in the stationary case is

$$T - T_0 = IU_j'(1-\eta)R_T \tag{4.3}$$

where T_0 is the cold finger temperature, I the injection current, $U_j \simeq E_g/e$ the voltage drop on the p–n junction, η the external laser quantum efficiency, and $R_T = l_j/kA_j$ is the diode thermal resistance, with l_j the distance between the active layer and the cold submount. For a laser with $R_T = 200 \text{ K W}^{-1}$, $U_j = 100 \text{ mV}$ ($\lambda \simeq 10 \ \mu\text{m}$), and $I \simeq 1 \text{ A}$ the active region temperature increase will be 20 K.

This increase does not arise instantly, and thus permits frequency tuning with time in a pulsed mode of operation. The solution of equation (4.2) at a time t from the beginning of the pulse for a current step with an amplitude I is given by

$$T(t) - T_0 = \frac{IU_j(1-\eta)}{c\rho\,\delta A_j}\, t \tag{4.4a}$$

when $0 < t < \tau_\delta$, and

$$T(t) - T_0 = \frac{IU_j(1-\eta)}{kA_j}\left(\frac{D}{\pi}\right)^{1/2} t^{1/2} \tag{4.4b}$$

for $\tau_\delta < t < \tau_D$ [4.25]. Here T_0 is the active layer temperature before the current pulse, $\tau_\delta \simeq \delta^2/16D$ the characteristic time of the active layer response to the instantaneous heat pulse, $D = k/c\rho$ the thermal diffusivity, and $\tau_D \simeq l_j^2/D$ the diffusion time.

At the beginning of the pulse the heating of the active region is adiabatic and the temperature increases linearly with time.

Then, when $\tau_\delta < t < \tau_D$, the temperature rise is proportional to $t^{1/2}$ as a result of heat conductivity processes. Finally, when $t > \tau_D$, the stationary regime is established.

In the preceding analysis the thermal resistances between the crystal and its submount and the submount and cold finger have been neglected. In real conditions the time for establishment of thermal equilibrium is greater than τ_D and is usually from one to a few milliseconds. Of course this time depends on the real diode and its package construction.

The emission frequency tuning is related to the rise of the temperature of the active region by

$$v(t) = v_0 + \frac{\partial v_q}{\partial T}\, (T(t) - T_0) \tag{4.5}$$

where $\partial v_q/\partial T$ is the temperature tuning rate of a cavity mode. However, as one can see from table 4.1, the gain contour and cavity modes are tuned with different rates. Therefore the tuning curve consists of a few continuous segments separated in wavenumber by ranges where generation is absent (figure 4.3). The fine-tuning range is typically $1–2\ \text{cm}^{-1}$ wide, although continuous tunings over $6\ \text{cm}^{-1}$ have been observed for a few devices in a pulsed mode of operation [4.27]. The same ranges of continuous tuning have also been obtained in cw operation by precise variation of the temperature of the cold finger [4.21].

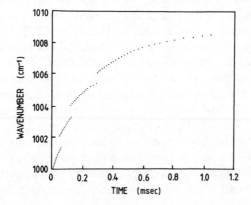

Figure 4.3 Typical frequency tuning with time for a diode laser in pulsed mode of operation. Results are shown for the PbSnSe laser. The cold finger temperature was 40 K, the injection current pulse duration was chosen to be 1.0 ms. The points indicate the transmission maxima of a Ge étalon with free spectral range of $0.0495\ \text{cm}^{-1}$.

The mode hopping effect is one of the most significant drawbacks of the TDL, but by combining different methods of tuning (current, temperature, pressure, magnetic field) it is possible to cover the whole wavelength range without gaps. A diode coupled to an external resonator also has good prospects. Another way forward is to develop distributed-feedback (DFB) lasers. For example, a continuous fine frequency tuning over 20 cm^{-1} has been obtained in a PbSnTe DFB laser [4.28].

4.2.3   Recent developments

For applications, output stability and reliability of the devices used are critical factors. Before 1980 many lasers experienced a progressive degradation of their characteristics. This degration was found to be due to the increase in contact resistance which was caused by the diffusion of indium atoms from the contact into the laser crystal. The problem was solved by depositing multiple In–Au–Pd–Au layers, which form a barrier to indium diffusion [4.29]. As a result diode lasers are fabricated which can work without any visible degradation for periods estimated to be as long as ten years.

Over the past few years some new alloy compositions with good characteristics have been developed and at present lasing is obtained in lead compounds with E_g from 500 to 26.8 meV. In particular, the PbEuSeTe lasers with 147 K cw operation at 4 μm have been made [4.20]. The heterolasers on the PbSnTeSe compounds operating in a pulsed mode gave a record value of 46.2 μm for long-wavelength generation [4.25]. Use of the mesa-stripe configuration for PbSnSe lasers has led to substantially improved performance, including cw operating temperatures of up to 118 K at 10 μm and an overall tuning range as wide as 444 cm^{-1} [4.21].

The operation of a cleaved-coupled-cavity (C^3) diode laser of PbS$_{1-x}$Se$_x$ in the 4–5 μm spectral region has been demonstrated [4.30]. The laser used consisted of two sections: the modulator section and the laser one separated by an air gap 19 μm wide. A major advantage of the C^3 configuration is that it gives a better degree of control over mode behaviour. In such a way laser frequency tuning rates can be significantly reduced (by a factor of ten) by use of modulator current tuning as opposed to laser current tuning.

PbTe lasers with a quantum well have been made [4.31]. For these lasers only single-mode operation was observed and a continuous frequency tuning with current of nearly 6 cm^{-1} was obtained. The single-ended output power was 2.5 mW.

The theory of the energy spectrum of A^{4}B^6 semiconductors has recently been developed and the analytical expressions for energy band-structure parameters and their dependences on temperature and alloy composition factor have been obtained [4.32]. It is important that this theory gives not only a qualitative but also a quantitative description of the energy spectrum and thus could be used for calculations of the electron spectra of binary lead salt semiconductors and their multicomponent solid solutions. It seems that theoretical understanding gives a good foundation for material improvement in lasers in the future.

4.3   High-resolution diode laser spectrometers

4.3.1   Main components
Despite some differences in mode of operation and optical system all diode laser spectrometers contain the following major components:

(i) a cryogenic system for diode laser cooling;

(ii) a temperature control unit for cold finger temperature tuning and stabilisation;

(iii) a diode laser power supply with the necessary modulation and sweep functions;

(iv) a grating monochromator whose functions are to isolate a single mode of the laser radiation and to give a coarse calibration of the spectrometer wavenumber scale;

(v) cells with the gases under investigation and a reference cell;

(vi) a Fabry–Perot étalon (an interferometer) for calibration of the laser tuning rate;

(vii) a cooled semiconductor infrared detector and amplifier;

(viii) a recording system for data acquisition and processing.

4.3.2   Equipment for diode laser operation
The effective laser linewidth and consequently the spectrometer resolution is directly related to the stability of the operating conditions. Since the temperature tuning rate of a single mode is of the order of $0.5 \, \text{cm}^{-1} \, \text{K}^{-1}$ a temperature stability of $\lesssim 2 \times 10^{-4} \, \text{K}$ is required to achieve an effective linewidth of $10^{-4} \, \text{cm}^{-1}$. The instabilities in the TDL driving current must not exceed $100 \, \mu\text{A}$.

In the spectrometers manufactured by Laser Analytics cryogenic cooling is provided by a closed-cycle helium microrefrigerator. This system is very convenient, for it does not require any liquid cryogens and is only electrically powered. But diode lasers housed on the cold head of the refrigerator exhibit both amplitude and frequency fluctuations resulting from mechanical vibrations and temperature oscillations. These fluctuations lead to effective TDL linewidths of about $1 \times 10^{-3} \, \text{cm}^{-1}$ or even more [4.33]. Suppression of the mechanical vibrations by an order of magnitude can be achieved by rigid coupling of the diode to an optical system and using flexible copper straps to provide thermal contact between the TDL and the cold head [4.34]. Thermal fluctuations of the cold finger can be reduced by both passive and active temperature stabilisation.

A few cryostats with continuous helium vapour flow and the necessary infrared optics have also been developed [4.35, 4.36]. These systems are free from mechanical vibration, but thermal stabilisation is needed.

Ultimate temperature stability is achieved by operating a TDL in a liquid helium or nitrogen dewar. The temperature can be varied over a small range by cryogenic vapour pumping out of the dewar tank.

A few experimental set-ups have been devised which permit tuning of the TDL frequency with applied pressure or magnetic field. An effective method of wide-band pressure variation (up to 14 kbar) has been developed utilising a fixed high-pressure chamber with a kerosene–oil liquid mixture at room temperature [4.37]. A gas module [4.38] has given the possibility of precise variation of the hydrostatic pressure up to 2 kbar at 25 K. By using the latter device the recording of high-resolution spectra with a CW PbSe laser continuously tuned by fine variation of the helium vapour pressure has been accomplished [4.39].

4.3.3   The CW diode laser spectrometer and its analytical modifications
As the laser itself is a source of tunable monochromatic radiation the optical assembly of the TDL spectrometer is a simple and compact one, especially in comparison with classical dispersive high-resolution ($\gtrsim 0.02$ cm^{-1}) instruments. Figure 4.4 shows the optical arrangement of the Laser Analytics Model SP 5000 (LS-3) basic configuration.

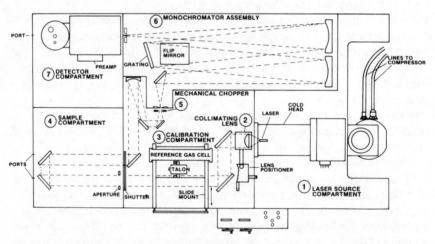

Figure 4.4   Schematic diagram of a SP 5000 (LS-3) tunable diode spectrometer [4.9].

The closed-cycle refrigerator with diode laser is mounted on the right-hand side of the scheme. TDL radiation passes through the calibration compartment with a reference gas cell and étalon, and then through the sample compartment, where a cell with the gas under study can be installed. The sample compartment is optically coupled to the grating monochromator, where the single mode of TDL radiation is selected; finally, after the monochromator, the radiation is detected by a photodetector. The detector is a cooled photovoltaic or photoconductive semiconductor element mounted in an optical dewar. Depending on the wavelength region of interest, the detector material may be InSb ($\lambda < 5.5$ μm), HgCdTe ($\lambda < 22$ μm) or Ge doped with Cu ($\lambda < 30$ μm). Liquid nitrogen is used for cooling the InSb and HgCdTe detectors, and liquid helium is needed to maintain the Ge:Cu detector at the operating temperature.

The optical configuration is quite flexible. By removing the two mirrors in the sample compartment, the laser beam can be brought out of the main frame and passed through the other sample under study. Here the Model LO-3 multipass White cell is often used, which provides adjustable pathlengths of up to 100 m, when its basic length is only 1 m.

Double-beam modifications of the SP 5000 basic system have been used in many papers. The dual channel configuration is necessary because in the cw mode of operation the frequency scan over 1 cm^{-1} takes several minutes and the reproducibility of the laser tuning curve from scan to scan is less than 1×10^{-3} cm^{-1}, therefore the simultaneous measurement of two spectra (of the gas under study and reference one or of an étalon) is needed. Several optical schemes of double-beam spectrometer have been proposed [4.40–4.42], but usually the detector was simply removed from the main frame and the laser beam after the monochromator was divided by a beam splitter, and then used with a suitable configuration of optical elements.

The recording of direct absorption spectra with the SP 5000 spectrometer can be obtained by slowly varying the TDL current while the laser beam is mechanically chopped. The Model LCM laser control module also provides modulation signals for direct spectral oscilloscope display and derivative spectroscopy. For harmonic detection lock-in amplifiers are used. The SP 5080 microprocessor-based data acquisition and control system has also been developed, which provides the diode laser current control by a 12-bit digital-to-analogue converter and is capable of automated data collection and processing.

Several modified configurations of the basic components, manufactured by Laser Analytics, have been used for spectral analysis of different gas mixtures, liquids, and solids [4.43]. The major features of these instruments are:

(i) the double-beam optical system with two detectors mounted in a single assembly, which is cooled with the help of the closed-cycle refrigerator;

(ii) the addition of a vacuum ion pump, which makes it possible to operate the closed-cycle refrigerator for maintenance-free periods of up to three months;

(iii) active frequency stabilisation by the establishment of a feedback control loop based on the first derivative of a reference absorption line;

(iv) computerised laser control and data acquisition.

Thus, fully automated TDL systems have been developed for long-term, precision measurements; specific devices have been used for monitoring nuclear reactor performance, automotive exhausts, quality of extruded plastic, chemical laser diagnostics, and other applications [4.43–4.45].

4.3.4   Fast-scan instruments

Even the early experiments [4.46, 4.47] have proved that despite the considerable increase in detection system bandwidth a good signal-to-noise ratio could be achieved in the fast-scan mode of TDL operation. However, at first the fast-scan techniques did not have widespread acceptance. We think that was due to the lack of well developed instruments to observe and to study the fast signal time evolution.

At present, two methods are used for fast laser frequency scan: (a) the sweep mode (usually called sweep integration, because sweep averaging permits signal integrations) and (b) the pulsed technique. The main difference between these methods lies in the type of TDL driving current modulation. In the sweep mode the laser current is modulated with a fast sawtooth ramp, but in the pulsed mode a TDL is supplied with rectangular pulses. Use of other shapes of TDL current pulse is also possible, thus providing laser frequency tuning rate control.

The simplest method of spectral recording is to display a detected signal on an oscilloscope screen (an ordinary oscilloscope is used if the scans are repetitive, a storage one can be utilised in a one-shot regime). To obtain the spectra in graphical or digital form sampling or digital oscilloscopes, boxcar averagers, and transient recorders can be used.

We consider that the fast-scan modes of TDL operation have some advantages over the continuous mode. Firstly the range of operating temperatures increases and, consequently, the spectral range covered by a single diode also increases. Secondly the possibility of visual control on the oscilloscope screen facilitates the choice of laser emission mode. The range of single-mode tuning and its frequency tuning rate can be controlled not only by varying the cold finger temperature and the amplitude of the laser bias current but also by changing the duration, repetition rate and shape of the current pulses. The precise wavenumber position of a particular scan can be set before recording. Thirdly in the rapid-scan regimes the wavenumber reproducibility from scan-to-scan is appreciably improved due to constant thermal conditions. Thus one can employ one beam path only for both sample and calibration measurements and so eliminate many problems associated with a double-beam optical system. Finally the use of fast-scan modes and time resolution of the detected signal eliminates modulated competitive emission from instrumental components. The time of spectral recording is considerably reduced.

The theoretical limit of TDL spectrometer resolution in the fast-scan mode is determined by the effective emission line broadening which appears due to the frequency scanning. The value Δv_{min} of the linewidth can be estimated from the uncertainty principle [4.48], which postulates that

$$\Delta v_{\min} \Delta T \simeq 1. \tag{4.6}$$

With a frequency tuning rate of dv/dt, the time

$$\Delta T = \Delta v(dv/dt)^{-1} \tag{4.7}$$

is needed to observe the spectral interval of Δv. The equations (4.6) and (4.7) yield for the laser frequency uncertainty

$$\Delta v_{\min} \simeq (dv/dt)^{1/2}. \tag{4.8}$$

The line broadening due to scanning is comparable with a fundamental, quantum-phase noise-limited linewidth of 1×10^{-5} cm^{-1} when the frequency scanning rate is approximately 3 cm^{-1} s^{-1} and reaches a value of 1×10^{-3} cm^{-1} if dv/dt is about 3×10^{-2} cm^{-1} μs^{-1}.

The real spectral resolution is often limited by the bandwidth of the detection system. In this case

$$\Delta v_{\min} \simeq (dv/dt)\tau. \tag{4.9}$$

In equation (4.9) $\tau \simeq (\tau_{det}^2 + \tau_{amp}^2)^{1/2}$ where τ_{det} is the time constant of the photodetector response and $\tau_{amp} = (2\pi \Delta f_{amp})^{-1}$ with Δf_{amp} the amplifier bandwidth. In order to obtain a spectral resolution of 1×10^{-3} cm^{-1}, when a TDL frequency scanning rate is 1×10^{-2} cm^{-1} μs^{-1}, the time constant of the detection system must not exceed 100 ns.

The line shifts and lineshape distortions also occur when the frequency tuning rate and the detector time constant are chosen incorrectly. The peak shift Δv can be estimated by an equation similar to (4.9):

$$\Delta v_{shift} \simeq (dv/dt)\tau. \tag{4.10}$$

The amount of lineshape distortion depends upon the relationship between the absorption linewidth in the time variable, $\Delta t = 2\gamma(dv/dt)^{-1}$, where γ is the half-width at half maximum (HWHM) in wavenumbers, and τ is the time constant [4.49]. In particular, in order to obtain the peak absorbance of a Gaussian profile with 1% accuracy it is necessary to fulfil the condition $\Delta t/\tau > 17$. Because detectors with a photoresponse time of 100–200 ns are usually utilised, a frequency scanning rate of not more than 1×10^{-3} cm^{-1} μs^{-1} should be used for true lineshape measurements under the Doppler-broadened conditions. When the higher rates are used the recorded spectrum can be corrected by deconvolution.

A spectrometer with sweep integration has been developed by Jennings [4.41] and at present TDL instruments with this mode of laser operation are widely used (see, for example, [4.50, 4.51]). For dual-beam sweep integration a PAR Model 4202 dual-channel signal averager is convenient, which allows 1024 data points to be recorded over one scan with 5 μs time resolution. Hence, a 200 Hz repetition rate of the injection current ramps can be used.

At the same time computer-assisted spectrometers which utilise pulsed TDLs have been developed at the Institute of Spectroscopy in Troitsk [4.35, 4.52] and at the Institute of General Physics in Moscow [4.53]. Here we give a description of the spectrometer developed at Troitsk.

An experimental set-up is shown schematically in figure 4.5. The diode lasers used have been fabricated in the Laboratory of Semiconductor Physics at the P N Lebedev Physical Institute [4.10, 4.54]. The TDLs are supplied by rectangular pulses with durations of 0.1–3 ms and repetition frequencies of 20–100 Hz. In §4.2.2 it is shown that during the pulse the fine-tuning rate decreases with time as $t^{-1/2}$, reaching values of about $1 \times 10^{-4} \, cm^{-1} \, \mu s^{-1}$ or less at the pulse end. So, the modes near the pulse end are used for obtaining the maximum resolution.

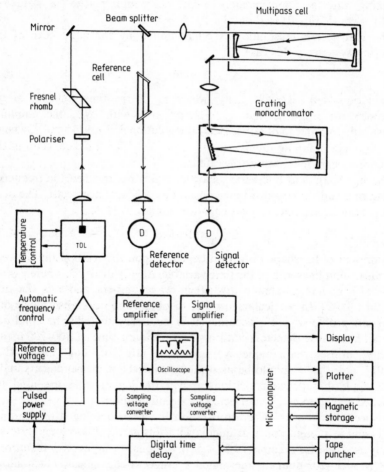

Figure 4.5 Block diagram of a pulsed diode laser spectrometer with automatic frequency stabilisation and digital data acquisition [4.52].

The laser radiation, after collimation with a KBr lens, passes through an optical isolator, which consists of a grid polariser and a Fresnel rhomb. This isolator eliminates optical feedback effects and reduces by at least one order of magnitude instabilities in laser operation caused by reflected light.

After the optical isolator the TDL radiation is divided by a beam splitter into two beams. One beam is used for spectral recording and wavenumber calibration. For this purpose the gas sample to be studied, a reference cell, or a Fabry–Perot étalon can be placed in the laser beam in front of a grating monochromator. The monochromator has a spectral resolution of about 0.5 cm^{-1} and is used as usual for TDL mode separation and coarse calibration of laser frequency. A multipass White cell is used if molecules with small absorptions are to be studied. A typical optical pathlength is 40 m. After passing through the monochromator the laser radiation falls on the photodetector. Here a boron-doped silicon photoconductive detector with time constant τ_{det} of 150 ns or a HgCdTe photovoltaic detector with $\tau_{det} < 10$ ns is used.

The signal from the photodetector is amplified by a broadband amplifier (bandwidth variable up to 20 MHz) and fed to the Model V9-5 sampling voltage converter (SVC). This instrument measures the detector signal amplitude over a fixed time-aperture of 4 ns and converts its magnitude to a near constant voltage, which is then digitised by a 10-bit analogue-to-digital converter (ADC). The readings from the ADC are then stored in the microcomputer memory.

The moment of sampling is set by the digital time-delay circuit. This circuit is a clock for the whole spectrometer and provides pulses for triggering both the laser supply unit and two sampling converters used in the main and reference channels. Time delays up to 10 ms with a 1 ns stability and the same reproducibility can be set either manually or under computer control.

As a matter of fact the data acquisition system developed acts as a dual-channel digital signal averager. The transmission spectrum of a gas cell or the étalon fringes is obtained by the detector signal pulseshape recording point-by-point. The amplitude at a single point only is usually measured over one laser pulse. Then a new delay is set and the amplitude at another moment is detected and stored. If the laser pulse repetition frequency is 50 Hz a part of spectrum consisting of 1000 data points can be recorded in this way over a time interval of 20 s.

To decrease the time of data acquisition the so-called multiplicative mode of SVC operation has been developed. In this mode the signal is sampled at many equidistant moments. Up to 100 data points over a single pulse can be stored, thus the time of obtaining the entire scan of 4000 points can be only 0.2 s.

A second laser beam is used for active laser frequency stabilisation. This can be realised as in the CW mode with a feedback loop control based on the signal from some reference line [4.56–4.58]. For this purpose, in the experimental arrangement shown in figure 4.5, the SVC in the reference channel samples the signal from the photodetector at some fixed delay after the start of the pulse when the laser frequency coincides with the slope of the reference line profile. The SVC output is then compared with a constant reference voltage and the value of the small continuous current passing through the diode in addition to the current pulses is changed proportionally to the difference between the signal and reference voltages. Results independent of the fluctuations of the TDL output power are obtained if stabilisation is based on the first derivative of the reference line.

The unique feature of the fast-scan modes is that the frequency stabilisation is conserved during the time of establishment of thermal balance. Thus pulse-to-pulse frequency reproducibility of 1×10^{-4} cm^{-1} has been attained in the whole range of TDL emission frequency tuning over a pulse (see figure 4.6). This reproducibility is conserved over a few hours, thus allowing the recording of spectra of different samples and the calibrating Fabry–Perot étalon by changing the specimen in the main channel of the spectrometer.

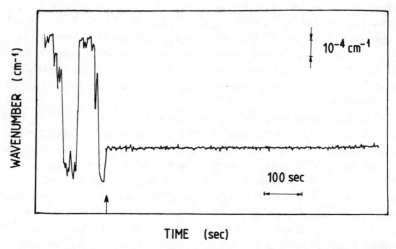

Figure 4.6 Frequency fluctuations of the pulsed diode laser at a fixed delay after the start of the pulse. The cold finger temperature stability was ± 0.002 K over 100 s. The arrow indicates the time when the active frequency control system was switched on. The pulse-to-pulse frequency reproducibility was measured at the delay which was greater than that used in a feedback loop by 100 μs. This value corresponded to the 0.3 cm^{-1} separation from the reference line [4.52].

An extensive software package has been developed. There are programs for data acquisition such as scanning the spectrum with accumulation, multiple scanning and multiplicative regimes, and measurement of the signal time evolution. The repetition of scans can be done up to a preset number of accumulations or up to a preset value of signal-to-noise ratio. A variety of programs are available for data processing: elimination of accidental overshoots, data averaging, all arithmetic functions, smoothing, differentiation, linearisation of wavenumber scale, scale expansion and shifts, determination of line positions and line intensities, and many others. There are also options for display, plotting and spectral recording on a magnetic tape.

A few examples of the employment of this spectrometer will be given later in this chapter.

4.4 Spectral and analytical techniques

4.4.1 Sensitivity limits

Theoretical considerations of the sensitivity of the laser absorption–transmission method were given by Shimoda [4.59]. Here we will cite the main expressions for the case of nonsaturation laser power and infrared photoelectric detection.

The sensitivity limit of the absorption method depends in the first place on ability of the recording system to detect small variations of laser power. In the laser shot noise limit the minimum change in the laser power which can be observed by an ideal detector is†

$$\Delta P_{min} = (2P_S h\nu \, \Delta f)^{1/2}. \tag{4.11}$$

In equation (4.11) P_S is the laser power in the absence of absorption, $h\nu$ the photon energy and Δf the bandwidth of the detection system.

For comparison of different techniques it is convenient to use the value of the limit in absorption, i.e. $\Delta P_{min}/P$. In the laser signal shot noise limited detection we have

$$\frac{\Delta P_{min}}{P} = \left(2 \frac{h\nu}{P_S} \Delta f \right)^{1/2}. \tag{4.12}$$

If a laser with photon flux $\phi_S = P_S/h\nu = 5 \times 10^{15}$ photon/s is used (for $\lambda = 10 \, \mu m$ this flux corresponds to $100 \, \mu W$ of laser power), then the minimum absorption which can be detected in 1 Hz bandwidth is 2×10^{-8}.

Table 4.2 contains the general expressions for signal and noise currents of a photoelectric IR detector in the direct photon detection scheme and in the case of optical mixing (heterodyne detection). Extra noise may be present in real detection.

† In this chapter the detection limits are defined at signal-to-noise ratio, S/N, equal to 1. In analytical spectroscopy, however, the detection limit is usually defined at an $S/N = 3$. In addition, the necessity of making the measurement on absorption resonance and off must be taken into account. Thus, the detection limits usually used are greater than those defined in this chapter by a factor of $3\sqrt{2}$.

Table 4.2 General expressions for photovoltaic and photoconductive infrared detectors.

Direct photon detection [4.60]	Optical mixing (heterodyning) [4.61]
$i = i_S + i_B = R_i(P_S + P_B)$	$|i_{IF}| = (2i_{LO}i_S)^{1/2} = R_i(2P_{LO}P_S)^{1/2}$
$\Delta i_S = R_i \, \Delta P_S$	$\Delta |i_{IF}| \simeq R_i(P_{LO}/P_S)^{1/2} \, \Delta P_S$
$\langle i_n^2 \rangle = 2mR_i \dfrac{(P_S + P_B)hv}{\eta} \Delta f$	$\langle i_n^2 \rangle \simeq 2mR_i \dfrac{P_{LO}hv}{\eta} \Delta f \qquad P_{LO} \gg P_S + P_B$
$\quad + \dfrac{4k_B T_d}{R_d} \Delta f + \dfrac{4k_B T_L}{R_L} \Delta f$	

Designations: i_S, i_B, and i_{LO} are the short-circuit photocurrents due to the detected powers of the signal laser P_S, background P_B, and laser local oscillator P_{LO} respectively; $|i_{IF}|$ the photocurrent amplitude at the IF frequency in heterodyne regime; $R_i = eG_i\eta/hv$ is the current responsivity, G_i the photogain, and η the quantum efficiency of the detector; i_n the noise current, T_d and T_L the temperatures of the detector and load resistor respectively; R_d and R_L their resistances; $m = 1$ for photovoltaic detectors, $m = 2$ for photoconductive ones.

The signal shot noise limit can be achieved theoretically in two cases: (*a*) in the direct detection regime, when the laser power P_S is greater than the detected background photon flux power P_B, and (*b*) with heterodyne detection, when the local oscillator power P_{LO} is such that $P_{LO} > P_S + P_B$. For these two cases

$$\Delta P_{min} = \left(2m \, \frac{P_S hv}{\eta} \, \Delta f \right)^{1/2} \tag{4.13}$$

where η is the detector quantum efficiency and m is introduced to take into consideration the difference between noise powers for photoconductive ($m = 2$) and photovoltaic ($m = 1$) detectors. It is essential to note that in the signal shot noise limit the minimum detectable change in power ΔP_{min} is not equal to the minimum detectable power P_{min}.

The signal shot noise limit is achieved when

$$P_S > \frac{\eta}{2hv} \, NEP_{BLIP}^2. \tag{4.14}$$

Here $NEP_{BLIP} = P_{min}/\sqrt{\Delta f}$ is the noise equivalent power for the background noise limited detector. For example, if $\eta = 0.4$, $\lambda = 10 \, \mu m$ and $NEP_{BLIP} = 1 \times 10^{-11} \, W \, Hz^{-1/2}$ then the required laser power must be not less than 1 mW. Theoretically in the signal shot noise limited case absorptions as small as 10^{-8} can be detected if $\Delta f = 1$ Hz.

In the background noise limited situation the minimum detectable change in power is given by

$$\Delta P_{min} = \left(2m \frac{P_B h\nu}{\eta} \Delta f \right)^{1/2} \tag{4.15}$$

and ΔP_{min} is equal to P_{min}.

Sensitivites of photoelectric IR detectors are high enough in the BLIP regime. For example, if P_S is only $1\ \mu$W then using a detector with $P_{min} = 1 \times 10^{-11}$ W absorptions as small as 10^{-5} can be detected.

The minimum detectable absorption coefficient κ_{min}, the minimum detectable number of molecules in the laser beam n_{min}, the minimum detectable concentration N_{min}, and minimum relative concentration c_{min} of molecular impurities in a mixture can be obtained from the value of $\Delta P_{min}/P$ using simple expressions derived from the Bouguer–Lambert–Beer law (see table 4.3).

Table 4.3 Main parameters used for characterisation of absorption method sensitivity limits.

Parameter	Expression†	Example‡
Sensitivity limit in absorption	$\Delta P_{min}/P$	10^{-5}
Minimum detectable absorption coefficient	$\kappa_{min} = \dfrac{\Delta P_{min}}{P} \dfrac{1}{l}$	10^{-7} cm^{-1}
Minimum detectable number of molecules	$n_{min} = \dfrac{\Delta P_{min}}{P} \dfrac{A}{\sigma}$	10^{12}
Minimum detectable concentration of molecules	$N_{min} = \dfrac{\Delta P_{min}}{P} \dfrac{1}{\sigma l}$	10^{10} cm^{-3}
Minimum relative concentration of impurity (analyte) in mixture	$c_{min} = \dfrac{\Delta P_{min}}{P} \dfrac{1}{\sigma l N_\Sigma}$	100 PPB

† l is the optical pathlength in the absorbing mixture; A the laser beam cross section; σ the absorption cross section at the analytical frequency; N_Σ the total concentration of molecules in mixture.
‡ The example is calculated for $\sigma = 1 \times 10^{-17}$ cm^2, $A = 1$ cm^2, $l = 100$ cm, $N_\Sigma = 10^{17}$ cm^{-3}.

TDL emission powers are 0.01–1 mW and excess noise an order of magnitude higher than shot noise is usually observed. Therefore, the situation is often intermediate between laser and background noise limits. Maximum sensitivity, $\Delta P_{min}/P$, with a TDL was measured by Reid *et al* [4.62] at 6×10^{-6}. However, this value was limited by the effects of parasitic reflections in the optical system and at least an order of magnitude improvement in sensitivity can be attained in the detector noise limited conditions.

4.4.2 Modulation techniques†

In the slow-scan mode the maximum frequency f_{max} of the signal spectrum is about 100 Hz, i.e. in the frequency region where $1/f$ noise and intensity fluctuations due to mechanical vibrations are considerable; that is why modulation techniques with phase-sensitive detection are usually used.

Amplitude modulation is the simplest method of recording a spectrum with nondistorted lineshapes and intensities. It is usually realised by chopping the laser beam with a mechanical modulator. The modulation frequency used is in the 400–500 Hz region. The signal-to-noise ratio at the output of a lock-in amplifier is no greater than 10^3–10^4.

It was shown by Eng *et al* [4.23] that the optimum frequency for TDL modulation is in the 4–10 kHz region. It is difficult to carry out amplitude modulation by mechanical choppers with such a high frequency, therefore laser source frequency modulation or molecular absorption modulation techniques are used. The signal in these cases is slightly less, but a signal-to-noise ratio improvement can be achieved, however, using the higher time constant of a lock-in detector. In addition, the sloping background signal of the laser output power and contributions from the wings of other strong lines are eliminated when harmonic detection is utilised.

Source frequency harmonic modulation is realised by modulation of the TDL injection current I so that $I = I_0 + I_1 \cos \Omega t$. In this case the laser frequency is varied as $v = v + a \cos \Omega t$, where the deviation a depends on I_1 and dv/dI_1. The absorption coefficient κ for low optical densities, when $\kappa l \ll 1$, can be written as a Fourier series [4.64]:

$$\kappa(v + a \cos \Omega t) = \sum_{k=0}^{\infty} H_k(v) \cos k\Omega t. \tag{4.16}$$

The expressions given above for detection limits are also valid in the case of signal harmonic registration, if $\kappa(v)$ and $\Delta P(v)$ are replaced by the corresponding Fourier components $H_k(v)$ and $\Delta P_k(v) = H_k(v)l P$.

When the deviation is small with respect to absorption line half-widths $(a \ll \gamma)$, the Fourier components are related to the derivatives of the absorption coefficient. The analysis of sensitivity attained with first and second derivative monitoring has been made by Hinkley *et al* [4.65]. However, the small deviation does not give the largest signal and optimum choice of the deviation value is needed. It was shown [4.64, 4.66], that a maximum peak-to-peak value of the first harmonic can be achieved if $a = 1.65\gamma$, the maximum second harmonic signal is at $a = 2.2\gamma$. These relationships for optimum modulation depths are justified for the Gaussian, Lorentzian and Voigt profiles of absorption lines.

Using frequency modulation techniques Reid *et al* [4.62] have measured

† A similar analysis of modulation techniques has been recently published by Laguna [4.63].

absorption coefficients as low as 3×10^{-10} cm^{-1}. This value was a record for TDL spectrometers. Let us enumerate the factors which were important in achieving such a high sensitivity.

(1) The technique of second harmonic detection was used. The detection frequency of 2 kHz, the time constant of the lock-in amplifier was 100 s.

(2) A multipass White cell with a distance of 5 m between mirrors was used. The optical pathlength could be adjusted up to 500 m.

(3) A TDL with high enough output power and single-mode operation was chosen. An active laser frequency stabilisation was used. The optical alignment was carefully adjusted to minimise the effects of étalon fringes from optical components in the beam and to avoid optical feedback to the laser. However, as was pointed out above, the sensitivity was limited by parasitic interference fringes formed due to reflections in the optical system (see figure 4.7 for example).

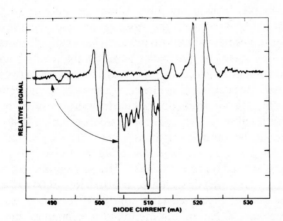

Figure 4.7 Spectrum of a standard mixture of SO$_2$ in N$_2$ equivalent to 170 PPB SO$_2$ in a multipass cell with a pathlength of 300 m. The trace was obtained by placing an additional 5 cm cell containing 1000 PPM SO$_2$ at 10 Torr in the laser beam in front of the White cell. Source frequency modulation and second harmonic detection were used. The expanded portion of the scan indicates the sensitivity to about 3 PPB SO$_2$. This corresponded to a minimum detectable absorption coefficient of 10^{-9} cm^{-1} [4.62a].

Signal fluctuations due to parasitic optical fringes can be avoided when the absorption spectrum of molecules under study is modulated. Two versions are possible: modulation of a concentration of absorbing molecules and modulation of the resonance absorption frequency.

The general method of modulation of the concentration can be realised by a mixture passing into a sample cell and then pumping out, or purging the cell by an inert gas. The same portions of the spectra should be recorded by a computer with the absorber present and with the absorber absent and the difference or ratio could be calculated. In this manner the constant étalon fringes can be removed and sensitivities $\Delta P_{min}/P$ of 10^{-5} have been obtained [4.51].

The largest specificity can be achieved for molecules which possess permanent electric or magnetic moments. In these cases modulation of the resonance frequency can be obtained by Stark or Zeeman effects [4.67] and not only is the background signal eliminated but the lines of the molecules to be detected are picked out among those of other species in the mixture.

Up to now, Stark modulation with TDLs has been used mainly for line identification (see, for example, [4.57] and some references cited in table 4.5 in §4.6.1). But we think this technique also has prospects for analytical application when small amounts of polar molecules might be detected.

Stark modulation has a few peculiarities in the infrared region [4.68]. First, due to large Doppler-broadening, observation of Stark splittings is only possible in most cases by nonlinear spectroscopy methods. On the other hand, IR transitions for molecules without permanent electric dipole moment can be modulated in some cases. In particular, considerable contributions to the Stark spectrum come from electric dipole matrix elements between different excited states with accidental coincidence. Spherical top molecules with T_d symmetry also have a vibrationally induced dipole moment in the triply degenerate vibrational states, where E components exhibit a first-order Stark effect. For example, the infrared and Stark spectra of a CF_4 molecule over the P(23) multiplet of the v_3 band are shown in figure 4.8. A typical first-order Stark effect was observed on the $E^{(4)}$ and $E^{(3)}$ lines. The value for the induced dipole moment for CF_4 was determined to be 0.036 ± 0.004 D [4.69].

Zeeman modulation is very effective for the detection of paramagnetic species (NO, radicals). A few special modulation techniques (modulation of discharge current, photolysis, Faraday rotation, drift velocity) have been developed for recording spectra of free radicals and molecular ions. Some of these techniques will be considered in §4.6.3.

4.4.3 Fast-scan techniques

When the laser emission frequency is swept at a high rate the maximum signal frequency f_{max} is usually in the range from 100 kHz to 10 MHz. Because the noise increases proportionally to the square root of a detection system bandwidth, it seems at first sight that the fast-scan techniques have no advantages for high sensitivity detection. But this is not quite so.

First of all, a shift of the signal spectrum to higher frequencies allows us to suppress low-frequency fluctuations with appropriate electrical or digital filtering. Secondly, the noise increase can be compensated in part by signal

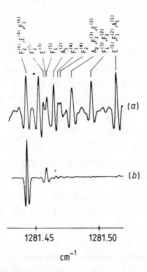

Figure 4.8   The P(23) lines in the v_3 fundamental band of CF_4 observed using a slow-scan tunable diode laser. (*a*) Second-derivative spectrum obtained by source-frequency modulation. The assignments are shown by the symmetry of the upper states. (*b*) Stark spectrum. The trace was obtained using a 0–3 kV cm^{-1} sinusoidal field at 100 kHz [4.69].

averaging. Third, the pulsed instantaneous TDL output power is often higher than in the CW mode of operation.

The simplest way to measure the concentration of particular molecular species in the fast-scan regime is to use a modification of the discrete-frequency 'on–off' differential absorption method. A two-channel boxcar averager is suitable for this measurement. When working in the ratioing log(A/B) mode the time delay in channel A is set to measure the peak absorption, and that in channel B is set to provide a signal amplitude when the absorption is absent. In this case the boxcar output is the ratio between the average value of the absorption sample and the average value of the baseline sample. Thus automatic compensation for drifts in laser beam intensity is also provided. The latter feature is very important for long path monitoring in the atmosphere [4.70].

For compensation of baseline slope fluctuations a system that measures at three different delays can be also used. In this case a value of the signal corresponding to zero absorption is obtained as an arithmetic mean of two samples on different wings of an absorption line [4.71]. Zero signal can also be measured using one of the absorption modulation techniques discussed above.

The results of references [4.70, 4.71] show that these modified techniques of differential absorption measurement combined with optimum signal filtration allow us to obtain the same detection limits in the pulsed mode of TDL operation as in the CW one.

Moreover, about 1000 data points can be recorded over one scan of 5 ms duration if the Model 4202 PAR digital signal averager is used. Thus the entire part of a spectrum, where several absorption lines of interest are located, can be recorded in one scan. Summing the results of many successive sweeps a sensitivity as low as 5×10^{-10} cm^{-1} has been obtained by Cassidy and Reid [4.50] over a 0.4 cm^{-1} region with a resolution of 4×10^{-4} cm^{-1}.

The absorption spectrum of a 200 m White cell with CO_2 at 20 Torr recorded with the use of the sweep integration technique is shown in figure 4.9. The laser frequency tuning rate was 6×10^{-5} cm^{-1} μs^{-1} and the repetition frequency was 200 Hz. Before digital quantisation the analogue presubtraction of the baseline was performed with the help of a differential amplifier. The resulting spectrum shown in figure 4.9 was obtained when a total of 32 000 sweeps were recorded with the mixture present, then a further 32 000 sweeps were taken with the White cell evacuated and subtracted from the first stored spectrum. The total time of the recording was 6 minutes. The noise level was established to be equivalent to absorption of 5×10^{-10} cm^{-1}, thus the sensitivity limit was approximately the same as the best achieved with second-harmonic detection [4.62]. However, in the last case in a time of 6 minutes only peak absorption at a single wavelength was measured.

4.4.4   *Integrative and correlation methods*

The techniques discussed above are very fruitful when light molecules are to be

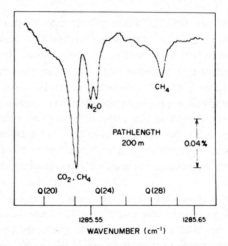

Figure 4.9   Spectrum of a sample with CO_2 at 20 Torr and a pathlength of 200 m obtained in sweep-integration mode. The absorption coefficients at the line centres of an N_2O doublet are 2×10^{-8} cm^{-1}. The detection limit is 5×10^{-10} cm^{-1} [4.50].

detected. In this case only a few individual absorption lines are observed over a range of a single TDL mode scanning. In the fast-scan regime, when the entire line profile is recorded, the sensitivity can be slightly improved by measurement of the line integral intensity. The integration technique has been realised with a pulsed TDL by Reidel [4.72], however, a sensitivity of only 10^{-2} has been obtained.

To achieve the maximum sensitivity the optimum integration interval must be chosen. The reason is, as the width of the recorded spectral region increases, the noise collected increases as its square root, while the growth rate of the integrated signal becomes slower. As was shown by Hirschfeld [4.73] an optimum integration interval for a Lorentzian peak is from $v_0 - 1.4\gamma$ to $v_0 + 1.4\gamma$, where v_0 is the centre of the line and γ its HWHM. Under optimum conditions the signal-to-noise enhancement in comparison with peak absorption measurement will be at least a factor of two.

The main drawback of the integration technique is that each of the points in the integration interval has the same weight. Further improvement is possible with the help of a cross-correlation ('matched window') technique, in which a value of

$$C_{RX}(0) = \int D_R(v)D_X(v)\,\mathrm{d}v \simeq \sum_{v_i} D_R(v_i)D_X(v_i) \qquad (4.17)$$

should be determined. In this case the various points v_i of the spectrum $D_X(v)$ of the mixture have a weight determined from a reference spectrum $D_R(v)$ of the molecule to be detected. When only a single line is recorded the gain will be at least three times [4.73]. This corresponds to a nine-fold reduction in measurement time for the same signal-to-noise ratio, when only the peak amplitude is measured.

We consider now the methods of detection of heavier molecules [4.53, 4.74–4.78]. Vibrational bands of these molecules have dense rotational structure which extends over only a small spectral range. Therefore, about one hundred or more lines can be recorded over a single TDL mode tuning range. For example this situation is observed in the spectrum of a CF_2Cl_2 molecule, shown in figure 4.10.

In this case for signal-to-noise ratio improvement it is natural to use all the information contained in the recorded spectrum. It is possible to realise this by determining the cross-correlation of two spectra, $D_X(v)$ and $D_R(v)$,

$$C_{RX}(v) = \int D_R(v)D_X(v+v)\,\mathrm{d}v \qquad (4.18)$$

where $D_X(v)$ is the recorded spectrum of a mixture under study and $D_R(v)$ is the reference spectrum of a pure individual component whose concentration in the mixture is being determined.

The cross-correlation function of two spectra for the case of CF_2Cl_2

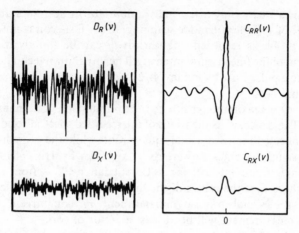

Figure 4.10 Use of a correlation technique for detection of small amounts of heavy molecules. $D_R(v)$ is the reference spectrum of the CF_2Cl_2 molecules to be determined in the studied mixture. $D_X(v)$ is the spectrum of the mixture under study (CF_2Cl_2 in air). The signal-to-noise ratio in this trace is about 0.5. $C_{RR}(v)$ designates the autocorrelation function of the reference spectrum $D_R(v)$, $C_{RX}(v)$ is the cross-correlation of $D_R(v)$ with $D_X(v)$. The signal-to-noise ratio in the latter trace at $v=0$ is equal to 15. The spectra $D_R(v)$ and $D_X(v)$ were obtained using a pulsed-diode laser spectrometer. $C_{RR}(v)$ and $C_{RX}(v)$ were calculated by a computer. (Adapted from [4.78b].)

detection in air is also shown in figure 4.10. The signal-to-noise ratio in the spectrum $D_X(v)$ of the mixtures studied was about 0.5. The autocorrelation and cross-correlation functions were calculated by a computer. It is seen that the concentration of the CF_2Cl_2 molecules could be determined from the peak value of the cross-correlation at $v=0$. A thirty-fold increase in signal-to-noise ratio has been obtained in the latter case [4.78b].

The signal-to-noise ratio advantage of cross-correlation arises due to use of all of the spectral information available from many points of the recorded spectra and is thus equivalent to data accumulation and averaging. When the line intensities are approximately equal the factor of improvement in signal-to-noise ratio is of the order of $K_{cor}\sqrt{Q}$, where K_{cor} determines the gain for one line and Q is the number of lines recorded over a given part of the spectral region. In addition, the use of cross-correlation spectral analysis reduces effects of parasitic optical étalon interference fringes by averaging their contribution at different wavenumbers [4.77, 4.78].

In the general case the mixture spectrum $D_X(v)$ can be presented as a sum of two spectra: $X(v)$, the absorption spectrum of a component being determined, and $\phi(v)$, the background one. The latter spectrum consists not only of absorptions due to other components in the mixture, but also contains additional factors which have an influence on the spectral dependence of the detected radiation (in particular, optical étalon fringes are included).

The absorption spectrum $X(v)$ can be written as

$$X(v) = c_X l_X E(v) \tag{4.19}$$

where $E(v)$ is the spectrum of absorptivity, i.e. absorbance divided by the product of the concentration c_X of the absorbing component and the sample pathlength l_X.

The reference spectrum $D_R(v)$ can be presented in the analogous form:

$$D_R(v) = c_R l_R E(v). \tag{4.20}$$

Then the value of the cross-correlation function $C_{RX}(v)$ of $D_R(v)$ with $D_X(v)$ at $v = 0$ is [4.75]

$$C_{RX}(0) = \frac{c_X l_X}{c_R l_R} \int D_R(v) D_R(v) \, dv + \int D_R(v) \phi(v) \, dv \tag{4.21a}$$

or in the vector formulation [4.74]:

$$C_{RX}(0) \equiv \boldsymbol{D}_R \cdot \boldsymbol{D}_X = \frac{c_X l_X}{c_R l_R} C_{RR}(0) + C_{R\phi}(0). \tag{4.21b}$$

In equation (4.21b) $\boldsymbol{D}_R \cdot \boldsymbol{D}_X = \Sigma_{i=1}^n D_R(v_i) D_X(v_i)$ is the dot product of the vectors $\boldsymbol{D}_R$ and $\boldsymbol{D}_X$, whose components are absorbances $D_R(v_i)$ and $D_X(v_i)$ at frequency v_i; n is the total number of points over the recorded range; $C_{RR}(0) = \boldsymbol{D}_R \cdot \boldsymbol{D}_R$ is the value of the autocorrelation function for the reference spectrum and $C_{R\phi}(0) = \boldsymbol{D}_R \cdot \boldsymbol{\phi}$ the value of the cross-correlation of the reference and background spectra, both values are taken at $v = 0$.

As equations (4.21) prove it is quite easy to determine the concentration c_X of a component in the mixture by computing the dot product of the mixture spectrum with the reference one from known concentration and pathlength. The analytical calibration curve, which presents the dependence of the function $S = C_{RX}(0)$ versus concentration c_X, should be a straight line. However, if the background spectrum partially correlates with the spectrum of the detected component, the calibration curve has no zero intercept.

If the analyte absorption is free from background interference, or if the background is constant from one measurement to the next, then the theoretical sensitivity limit would be determined as usual by detector noise. In practice, however, concentrations of interfering species can be changed from sample to sample and parasitic interference fringes have slow drift due to change with time of the temperature of the optical system elements. Nevertheless, when the properties of the background spectrum are known, a reference vector $\boldsymbol{D}_R^*$ can be chosen as orthogonal to the background vector $\boldsymbol{\phi}$, so that $C_{R\phi}(0) = \boldsymbol{D}_R^* \cdot \boldsymbol{\phi} \simeq 0$ [4.74]. The procedure of orthogonal reference vector determination is similar to analytical line selection because the elements $D_R(v_j)$

of the vector D_R at the frequencies v_j, where absorption from other species is present, must be replaced by zero. The choice of the matched vector D_R^* when a slow temperature drift is observed is equivalent to digital filtering, which excludes the contributions of frequencies where optical interference is present from the Fourier transform spectrum of the signal.

Correlation methods have been used by several authors for extracting quantitative information from spectral data obtained with TDLs (Max and Eng [4.76], Sano *et al* [4.77], Kosichkin and Nadezhdinsky [4.53] and Zasavitsky *et al* [4.78]). In most cases the functions of a digital correlator have been performed by a computer. However, it is possible to reduce significantly the time needed for data processing not by recording the spectra, but by directly recording the value of the cross-correlation function $C_{RX}(v)$ at $v=0$.

A simple method of determining the concentration of heavy molecules has been developed by Kosichkin *et al* [4.53, 4.78]. A TDL operating in a repetitive-pulsed mode was used. A boxcar averager was triggered during a single pulse by a sequence of sharp spikes corresponding to lines due to absorption in a reference cell. Thus, a value of

$$C_{RX}(0)=\sum_{i=1}^{Q} D_X[v(t_i)]=c_X N_\Sigma l_X \sum_{i=1}^{Q} \sigma(v_i) \qquad (4.22)$$

was measured over a single pulse. In equation (4.22) Q is the number of boxcar samples over the pulse, $\sigma(v_i)$ is the absorption cross section of the detected molecule at the frequency v_i (in pulsed mode v_i corresponds to time delay t_i from the start of the pulse). Then averaging over many pulses should be performed.

From the analysis given above one can see that correlation methods reduce dramatically the time of measurement and also eliminate in part the parasitic interference fringes. These parasitic optical effects are known to be the main factor limiting the sensitivity when ordinary techniques of measuring absorption at a single wavelength are used. Thus we hope that detection limits using an analysis at many wavelengths will be lower than those achieved up to now.

4.5   Calibration techniques

4.5.1   *Calibration of a wavenumber scale*
The problems of TDL spectrometer wavenumber calibration have been considered in detail in references [4.4, 4.6, 4.7]. Several methods are used to determine the frequency of laser light:

(i) method of two interfering beams with the use of a Michelson scanning interferometer and by counting numbers of interference maxima [4.81, 4.82];

(ii) method of internal standard, in which the frequency is determined from a value of absorption linewidth under Doppler-limited conditions [4.83];

(iii) calibration with a Fabry–Perot étalon and reference lines;

(iv) optical heterodyne techniques, when the frequency-stabilised CO_2 or CO laser is used as a local oscillator.

The simplest method is that in which the laser frequency dependence on a tuning parameter is determined by measurement of the Fabry–Perot étalon transmission spectrum and the absolute calibration of this dependence is performed using frequency calibration standards. The problems of obtaining frequency standards for the whole infrared region have not yet been overcome. However, over the past years many papers have been published in which new frequency standards have been obtained and those previously measured have been improved. The measurements have been made with high precision and accuracy of about $1 \times 10^{-4}\,cm^{-1}$ using Fabry–Perot interferometers (Guelachvili, Kauppinen *et al*) or heterodyne techniques (Sattler, Maki *et al*). Several recently measured spectra which are suitable for TDL frequency calibration are summarised in table 4.4 in a manner similar to that used in references [4.6–4.9].

As a result of the latest work some recommendations can be given which are useful for wavenumber calibration.

(1) First of all, we think that the repetitive, fast-scan techniques with TDL frequency stabilisation are preferable. As shown, the reproducibility of the entire frequency tuning curve from scan to scan when these techniques are used could be no worse than $1 \times 10^{-4}\,cm^{-1}$, and the procedure of spectra recording over a single mode tuning range takes only a few seconds. The sweep integration method of data acquisition is the best one, because the nonlinearities in frequency scan are small in this mode of operation, and the bandwidths of detectors and amplifiers used need not be higher than 1 MHz.

(2) Secondly, digital data recording and processing with the help of a microcomputer is very fruitful for determination of precise positions of absorption lines and étalon maxima, and for wavenumber interpolation from reference lines using the interference fringes of the calibrating étalon. With about three points over $1 \times 10^{-3}\,cm^{-1}$ and using a quadratic fit it is possible to determine the line peak positions with a relative accuracy approaching $3 \times 10^{-5}\,cm^{-1}$ [4.51].

(3) Because the refractive index of germanium (usually used as a material for étalons) has not been measured with high accuracy over the total range of interest it is desirable to calibrate the free spectral range of the étalon used directly by the reference spectrum; wavenumbers of several reference lines should be used.

(4) Baseline correction would be made using subtraction or dividing the transmission spectra of the studied gases and the étalon point by point by the

Table 4.4 Reference spectra recommended for calibration of TDL spectrometer wavenumber scale†.

Molecule	Band	Spectral region (cm^{-1})	Accuracy (10^{-3} cm^{-1})	Experimental conditions		References
				l (cm)	p (Torr)	
H_2O	Rotational spectrum	500–720	0.05	100	1.9	Kauppinen et al [4.86]
CO_2	v_2, hot bands, isotopes	584–753	0.05	100	0.05–0.5	Kauppinen et al [4.86] Jolma et al [4.87]
OCS	$10°0\leftarrow00°0$, hot bands, isotopes	823–885	0.05	100	0.3	Kauppinen et al [4.86] Jolma et al [4.87]
N_2O	v_2, hot bands, isotopes	542–633	0.05	100	0.7–7.0	Jolma et al [4.88]
$^{14}NH_3$	v_2	608–1266	0.1	60	0.5	Urban et al [4.89]
$^{15}NH_3$	v_2	652–1226	1	100	0.5	Urban et al [4.90]
OCS	$02°0\leftarrow00°0$	1025–1074	0.6–4.8	50	0.3–1	Wells et al [4.91]
H_2O, HDO	v_2, hot bands, isotopes	1066–2296	0.05	1600	0.01	Guelachvili [4.92]
SO_2	v_1	1069–1086	0.2	40	0.5–1	Sattler et al [4.93]
N_2O	13 bands	1118–1343	0.05–0.125	400	0.6	Guelachvili [4.94]
OCS	$20°0\leftarrow00°0$	1700	0.2	1700	0.2–4	Wells et al [4.95]
CO	$v=1\leftarrow0; 2\leftarrow1$	1939–2271	0.08–0.16	4200	0.05	Guelachvili [4.96]
OCS	$00°1\leftarrow00°0$	2028–2085	0.08–0.11	10	0.1–2	Guelachvili [4.97] Klebsch et al [4.98]
CO_2	$00°1\leftarrow00°0$	2272–2390	0.12			Guelachvili [4.99]

† Data from papers published in 1979–83, in addition see [4.6–4.8].

background one. Analogue or digital filtering are also effective for background elimination.

(5) To exclude systematic errors a set of spectral recordings (background, reference gas, the studied gas, and fringes of the Fabry–Perot étalon) should be made more than once at different tuning parameters.

Relative wavenumber calibration with an accuracy of $1 \times 10^{-4}\,\mathrm{cm}^{-1}$ should be attainable when these recommendations are followed [4.51, 4.53].

4.5.2 Calibration of the analytical signal

The important problem for quantitative analysis is to establish the functional dependence of a gas analyser signal on the concentration of molecules detected and to calibrate this dependence. A very sharp spectral line of a TDL ensures the validity of Beer's law. In accordance with this law, for small optical depths the signal due to absorption is linearly proportional to the number of absorbing molecules and to the power of the laser beam. It is worth mentioning that theoretically a linear dependence exists not only if the direct transmission signal is detected, but also holds when harmonic, integration or correlation methods are used. The nonlinearities may be caused by optical effects such as parasitic étalon fringes, baseline slope, interference absorption lines, or by molecular processes such as adsorption on cell walls, outgassing or chemical reactions in the mixture. In any case the linearity must be checked under real conditions.

With a sensitivity in the part per billion (PPB) range there are problems in carrying out accurate calibrations of an analyser signal because mixtures with such low concentrations are difficult to prepare and store. Fortunately, the optical absorption method enables us to calibrate in the PPB range using standard mixtures with high concentrations. For this purpose a short cell containing a calibrated mixture with a relatively high concentration should be placed in the laser beam sequentially with the multipass White cell, which is usually used to achieve high sensitivity. Because the absorption is linearly proportional to the concentration–pathlength product this method enables us to simulate an absorption signal equivalent to that from mixtures with extremely low (up to the detection limit) concentration [4.62].

Fried *et al* [4.100] have compared four independent techniques for calibration of the second-harmonic signal in the sub-PPB range of HCl concentration. Firstly, the calibration was carried out using standard mixtures of HCl in air with concentrations near the detection limit (about 0.3 PPB). These mixtures were prepared by the method based on the diffusion of the HCl molecules through a Teflon permeation membrane. Because of the chemical activity of HCl, the calibrated mixture was continuously drawn through the White cell. Pathlengths of 40 m were typically used. Secondly, the HCl concentration in the White cell was determined using a 16 cm reference cell placed in the same laser beam. The concentration of HCl in the reference cell was determined to be about 1230 PPM by calculation with the use of the analytical line parameters known from previous measurements. Thirdly, the absolute HCl concentration at the input of the White cell was also checked by neutron activation analysis of HCl collected on ultrapure nylon filters. Finally, known concentrations of CH_4 in air were passed into the White cell and the HCl concentration was calculated by taking the appropriate ratio in measurements of signal, beam intensity and absorption coefficient both for CH_4 and HCl lines. The latter three calibration techniques were in the range of statistical agreement; however, the permeation calibration resulted in HCl determinations which were outside this range and were consequently wrong. In addition, it was shown that the standard technique of the second-harmonic signal detection at a single wavelength could lead to mistakes associated with changes in the TDL output intensity. Such behaviour could be misinterpreted as a concentration change. By using repetitive scans through the analytical line that problem was avoided and the differences between a changing peak signal or changing baseline were easily identified.

Several other calibrating techniques could be used. In particular Mucha [4.101] examined the possibility of using the standard addition technique for quantitative gas analysis with derivative TDL spectroscopy. The necessity of preparing calibrated mixtures can be completely eliminated by making a measurement of the strengths and widths of the relevant absorption lines in pure gas required to be detected. The latter technique is the simplest, but in real mixtures with pressures above 1 Torr line broadening and shifting can cause systematic errors.

Comparison of calibration techniques proves that the method based upon the use of a short cell filled with a high concentration of the studied gas in a mixture at the same pressure as that in a sample cell gives an acceptable accuracy of about one per cent in the high concentration range and about 0.1–1 PPB near the detection limit.

4.6   High-resolution molecular spectroscopy

The major application of TDLs is high-resolution recording of vibrational–rotational spectra of molecules in the gas phase. The following topics, for which the TDL spectrometers have been used successfully, may be distinguished:

(i) investigation and analysis of fundamental vibrational–rotational bands of chemically stable molecules, line assignments, determination of precise spectroscopic constants and molecular structure;

(ii) study of rotational fine structure of molecular spectra;

(iii) investigation of individual line parameters—profiles, widths, intensities, and also coefficients of line broadening and shifts;

(iv) spectroscopy of transient species, in particular, free radicals and molecular ions;

(v) investigation of absorption by molecules in excited states;

(vi) study of dynamic molecular processes.

The limited space of this chapter does not allow us to survey in detail the results of all these studies. Therefore we restrict our discussion to a few examples which illustrate possibilities of TDL spectrometers and the techniques described above.

4.6.1 Fine structure of vibrational–rotational spectra

Up to the present, spectra of a large number of stable molecules from diatomic to nine-atomic (cyclopropane C_3H_6 [4.102]) and even sixteen-atomic (cubane C_8H_8 [4.103]) have been recorded. More than one hundred papers, published up to 1981, have been reviewed by Eng and Ku [4.8]. In table 4.5 we summarise the papers published during the years 1982–4 to provide the reader with a reference list which extends that given in table 2 from reference [4.8].

With the help of TDLs absorption line fine structures which correspond to interactions between different forms of molecular motion (electronic, vibrational and rotational) have been observed for many molecules for the first time. The most exciting results obtained up to 1980 were those for the spherical-top molecules such as SF_6 [4.188], UF_6 [4.189], OsO_4 [4.190] and CF_4 [4.191]. Detailed investigation of the fine structure of IR spectra of these species was necessary not only for tackling the fundamental problems of analysing the complex vibrational–rotational spectra of molecules with octahedral or tetrahedral symmetry, but also for understanding the processes of multiphoton dissociation and laser isotope separation, experiments in nonlinear optics (SF_6, UF_6, OsO_4), and for creating 16 μm lasers with optical pumping (CF_4).

Figure 4.11 shows the spectrum of the $^{32}SF_6$ molecule in the Q branch of the v_3 fundamental band [4.188]. This spectrum was obtained using a TDL operating in the CW mode. The laser beam was modulated at 700 Hz by a mechanical chopper and detected with a phase-sensitive detector. For broadening the tuning range of the TDL relatively small magnetic fields of 1.9–2.9 kG were used.

A study of the v_3 stretching band of SF_6 using a classical dispersive spectrometer (even with a resolution of 0.07 cm^{-1}) gave only the overall PQR structure. However, the spectrum obrained with a TDL shows a large number (about 10^4 per cm^{-1}) of different absorption lines. Despite the complexity of the spectrum the assignments have been made by McDowell *et al* [4.188], and the results of line identification are also presented in figure 4.11.

The fine structure of the spectrum presented is due to the splitting of rotational manifolds in the $v_3 = 1$ state of SF_6. The high degeneracy of separate rotational levels of the octahedral molecule in the excited state is removed by vibrational–rotational (Coriolis) interaction. In a dominant approximation the positions of lines can be obtained from equation (4.23), which gives the frequency of any Q-branch transition as

$$v(J, p) = m + vJ(J+1) + t_{JJ}F^{(4JJ)}_{A_1pp}. \qquad (4.23)$$

Here m is the band origin, J the total angular quantum number, v is approximately the difference between the rotational constants $B_v - B_0$ of the

Table 4.5   Molecules studied with lead salt diode lasers†.

Molecule	Spectral region‡	Special features§	Reference
Atoms			
He	919	Fine-structure transitions	4.104
	1345		
F	404	Spin–orbit splitting of ground state	4.105
Diatomic molecules			
H_2	814.4250	S(2) pure rotation quadrupole transition	4.106
CO	1–0 band at $2100 \, cm^{-1}$	Broadening by N_2 at elevated temperatures (300–600 K)	4.107
CO	2139	Dipolar broadening in solid N_2 at $T \lesssim 27$ K	4.108
CO	1–0 band	Broadening by water vapour	4.109
CO	P(20) line	Optical Stark shifting induced by Nd:YAG laser	4.110
NO	1–0 band at 5.4 μm	N_2 broadening in R branch	4.111
NO	1–0 band	Band strength, broadening by N_2, Ar and combustion gases	4.112
HCl	1–0 band P(15) line	Dicke narrowing	4.113
HD	1–0 band	Broadening by Ar, Kr, Xe	4.114
DBr	1–0 band	Heterodyne frequency measurements	4.115
BCl	1–0 and three hot bands	Spectra in MW and DC discharges	4.116
AlF	1–0, 2–4, 3–2 bands 827–55	Spectra at 1000–1190 K, Dunham constants	4.117
KF	2–0, 3–1, 4–2 bands 827–55	Spectra at 1150–1250 K	4.117
LiF	1–0 and hot bands 829–86	High temperature spectrum at 1130–1300 K	4.118
Triatomic molecules			
CO_2	4.2 μm	Line strengths and temperature dependence of half-widths	4.119
CO_2	Laser hot band	Heterodyne measurements	4.120
CO_2	2395–2680	Absolute line intensities	4.121
N_2O	7.8 μm	Line intensities and widths	4.122
N_2O	$v_1 + 2v_2$ band	Broadening by noble gases	4.123
N_2O	2450	N_2-, O_2- and air-broadened widths and intensities	4.124
H_2S	3.6–4 μm	Absorption at DF laser wavelengths	4.125
H_2S	v_2 band	Comparison of several reduced forms of rotational Hamiltonian line strengths	4.126, 4.127
OCS	v_1 band 2026–81	Band analysis for $OC^{32}S$, $OC^{34}S$, $OC^{35}S$	4.98

Table 4.5 *continued*

Molecule	Spectral region‡	Special features§	Reference
Triatomic molecules			
OCS	$20°0–10°0$ band at $2040\ cm^{-1}$	Spectrum in DC discharge	4.128
OCS	$10\ \mu m$	Heterodyne measurements of pressure broadening	4.129
OCS	$5\ \mu m$	Inelastic process induced by collisions with OCS, He, Ar	4.130
O_3	1069–104	Air-broadening	4.111
O_3	$9\ \mu m$	Air-broadening	4.131
Four-atomic molecules			
NH_3	$a2v_2–sv_2$ band	Analysis of Q branch	4.132
NH_3	v_2 band	Self-broadening and self-shifting	4.133
NH_3	$2v_2$ band	Self-broadening and self-shifting	4.134
$^{14}NH_3$	v_2 band	Calibration standard with $1 \times 10^{-4}\ cm^{-1}$ precision	4.89
$^{15}NH_3$	v_2 band	Simultaneous analysis of MW, sub-MW and IR transitions	4.90
$^{14}NH_3$, $^{15}NH_3$	v_2 band ⎱ 740–1200 ⎰	More than 500 lines with $0.0005\ cm^{-1}$ accuracy	4.135
NH_3	$s2v_2–av_2$ 620–34	Wavenumbers of Q branch transitions	4.136
$^{14}NH_3$	$2v_2, v_4$ bands	Transition dipole matrix elements from line intensities	4.137
NH_3	$10\ \mu m$	Vibrational and rotational collisional relaxation by IR–IR double resonance	4.138
$^{14}ND_3$, $^{15}ND_3$	v_2 band	Simultaneous analysis of MW and IR spectra	4.139
C_2H_2	v_4+v_5 band	R branch line strengths, band strength, N_2- and He-broadening	4.140
HCCI	v_2 and hot bands 2037–71	Hyperfine splittings due to ^{127}I nuclear quadrupole moment	4.141
HNCO	v_2 band near $5\ \mu m$	Stark modulation	4.142
HONO	v_2 and v_4 bands	Spectra of *cis*- and *trans*-HONO	4.143
$^{35}Cl_2CO$	v_1 and v_5 bands	Stark modulation, equilibrium structure and anharmonic potential function	4.144
Five-atomic molecules			
$^{12}CH_4$	v_4 band	Band strength, intensities	4.145, 4.146
CH_4	R(0) of v_3 band	N_2-, CO_2- and self-broadening at 300–1000 K	4.147
CH_4	v_2 band 1586–90	Line strengths, N_2-, air- and self-broadening	4.148

continued

Table 4.5 *continued*

Molecule	Spectral region‡	Special features§	Reference
Five-atomic molecules			
CH_4	v_3 and $v_2 + v_4$ bands 2870–83	Line strengths, N_2- and self-broadening	4.149
CF_4	v_3 band	RF–IR double resonance	4.150
$C^{35}Cl_4$	v_3 band	Stark modulation and cold jet spectrum	4.151
SiF_4	v_3 band 1023–38	Assignments of transitions, closed to CO_2 laser lines, v_3 transition dipole moment	4.152
SiF_4	v_4 band	Determination of general quadratic force field	4.153
SiF_4	v_3 band	IR–MW double resonance	4.154
GeH_4	v_4 band at $820 \, cm^{-1}$	Analysis of five isotopic species	4.155
$^{74}GeH_4$	v_2 band Q branch	Simultaneous analysis of v_2/v_4 bands	4.156
GeH_4	v_2 band at $930 \, cm^{-1}$	Analysis of v_2/v_4 interaction for isotopic species	4.157
CH_2N_2	v_2 band at $2100 \, cm^{-1}$	FTIR and TDL spectra	4.158
CDF_3	v_5 band	Heterodyne measurements	4.159
CDF_3	$v_2, v_5, v_2 + v_5 - v_5, 2v_5 - v_5$	IR double resonance	4.160
$^{12}CH_3F$	v_3 band	Assignments and frequencies of $C^{18}O_2$-pumped FIR laser lines	4.161
$^{12}CD_3H$	v_5 band	Intensities	4.162
$^{12}CD_3F$	v_2 and v_5 bands 1040–80	FTIR and TDL spectra	4.163
CH_3Br	v_6 band 929–35	Band analysis	4.164
CF_2Cl_2	v_6 band 915–30	Absorption coefficient spectra, air- and self-broadening	4.165
CF_2Cl_2	$930 \, cm^{-1}$	Air- and self-broadening	4.166
CF_3Cl	v_1 band 1099–116	More than 1000 transitions up to $J = 40$, K structure	4.167
HCCCN	CN stretching band 2240–90	l doubling and l resonance, hot bands	4.168
HCOOH	v_3 band Q branch	Simultaneous analysis with CO laser Stark spectroscopy	4.169
HNO_3	v_5 band near $880 \, cm^{-1}$	Identification of A-type transitions	4.170
H_3SiCl	v_2 and v_5 bands near $950 \, cm^{-1}$	FT, dispersive spectrometer and TDL study of $H_3Si^{35}Cl$, $H_3Si^{37}Cl$	4.172
H_3SiCl	v_1 and v_4 bands at $2200 \, cm^{-1}$	FTIR and TDL spectra	4.171

Table 4.5 *continued*

Molecule	Spectral region‡	Special features§	Reference
Five-atomic molecules			
CrO_2Cl_2	950–1020	Spectra of highly excited vibrational states	4.173
Six-atomic molecules			
CH_3CN	v_7 and hot bands near 1040 cm^{-1}	FTIR and TDL spectra	4.174
C_2H_3D	v_7 and v_8 bands 830–90	1775 lines	4.175
C_4H_2	$v_5, v_5+v_9-v_9$ 2000–37	l doubling in hot band	4.176
CD_3OH	v_4 band at 985 cm^{-1}	FTIR and TDL spectra, internal rotation in C–O stretch	4.177
PF_5	v_3 and $v_3+v_7-v_7$ bands	Cold jet spectrum	4.80
Seven-atomic molecules			
SF_6	$2v_3-v_3$	IR double resonance	4.178
$^{34}SF_6$	v_3 band 929–32	Cooling in molecular beam	4.179
SF_6	v_3 band	IR multiphoton-induced depletion of rotational sublevels	4.180
SF_6	$2v_3-v_3$	Cold jet spectrum	4.181
SF_6	$v_4+v_3-v_3$	Cold jet spectrum	4.182
SF_6	953 cm^{-1}	Air-broadening	4.166
UF_6	Nine hot bands at 16 μm	Anharmonicity constants	4.183
UF_6	v_3 band	Cold jet spectrum	4.184
Eight-atomic molecules			
C_2H_6	v_9 band Q branch 785–845	Fit of torsional splittings, determination of barrier to internal rotation	4.185
C_2H_6	v_9 band 818–50	FTIR and TDL spectra, band strength, total fit of line positions	4.186
C_2H_5Cl	v_9 band	Detailed assignments, coincidence with CO_2 R(16) line	4.187
Sixteen-atomic molecules			
C_8H_8	v_{11} at 1235 cm^{-1} $\}$ v_{12} at 850 cm^{-1}	Confirmation of octahedral symmetry, bond length	4.103

† Only papers published after 1 January 1982 are included, for earlier references see [4.7, 4.8].
‡ Vibrational band, frequency in cm^{-1}, or wavelength in μm.
§ Spectral or experimental peculiarities are noted; line positions and assignments were determined in most of the papers, spectroscopic constants were obtained if band analysis was the object.

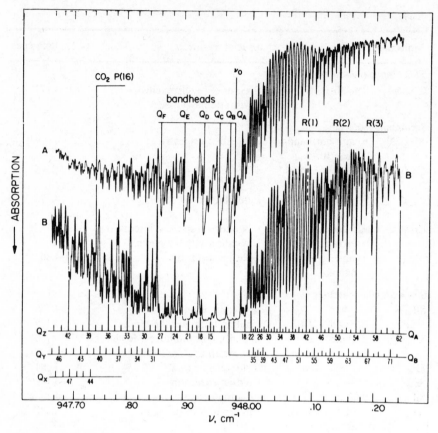

Figure 4.11   Diode laser spectrum of the ν_3 Q branch of $^{32}SF_6$, recorded at (A) 117 K and (B) 126 K; the pathlength was 10 cm [4.188].

first excited $v_3 = 1$ and ground states; t_{JJ} is the J-dependent coefficient and $F^{(4JJ)}_{A_1pp}$ is the eigenvalue of the octahedrally invariant fourth-rank tensor operator. A value of v determines the frequency spread of the Q branch. The term $t_{JJ}F^{(4JJ)}_{A_1pp}$ describes splitting of a single rotational J manifold, thus produces overlapping of many absorption lines. The Q branch can be represented as a series of subbranches that are designated $Q_A, Q_B, \ldots, Q_Z$, where Q_A and Q_Z correspond to the largest positive and negative values, respectively, of the term $t_{JJ}F^{(4JJ)}_{A_1pp}$ for each J manifold.

Although the spectrum of the SF_6 molecule is very dense, it could be analysed from readings obtained in direct absorption. A significantly more complicated example is the CCl_4 molecule.

Carbon tetrachloride has five isotopic species, three of them ($C^{35}Cl_4$, $C^{35}Cl_3{}^{37}Cl$, $C^{35}Cl_2{}^{35}Cl_2$) have approximately the same abundances. Since the vibration–rotation transitions of the isotopic species and several hot bands are heavily overlapped with one another, the lines cannot be resolved and assigned even by Doppler-limited spectroscopy. Despite these difficulties a diode laser spectrum of the v_3 band of $C^{35}Cl_4$ has been obtained and analysed recently by Yamamoto *et al* [4.151]. The assignments were possible because two techniques for extracting the lines belonging to this molecule and for suppression of the hot bands were used. The first one was a Stark modulation technique. It was mentioned above (§4.4.2) that tetrahedral molecules have a first-order Stark effect based on a vibrationally induced dipole moment in a triply degenerate vibrational state and molecules with lower symmetry have only a second-order Stark effect. Hence the absorption lines originating from the tetrahedral species could be extracted by Stark modulation. The second technique was the cooling of molecules using supersonic free jet expansion. Under this condition, the high-J transitions which were observed in the absorption spectra at room temperature vanished and the $\dot{P}$, Q and R branch transitions with low J were observed clearly (figure 4.12).

TDLS operating in a pulsed mode were used for recording the spectra of NH_3 [4.13, 4.39], CF_4 [4.192], OsO_4 [4.193], GeH_4 [4.156, 4.157] and CCl_2F_2 [4.165]. The technique of pulsed supersonic jet cooling combined with pulsed

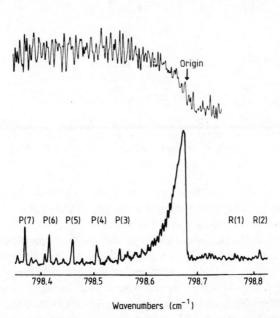

Figure 4.12 Infrared absorption spectrum of CCl_4 molecules cooled in a pulsed jet (lower trace) and at room temperature (upper trace). The cold-jet spectrum was recorded by phase-sensitive detection synchronised with the pulsed jet. The estimated rotational temperature in the jet was 20 K. (Adapted from [4.151].)

lasers was applied to analysis of complicated spectra of heavy molecules such as SF_6, WF_6 and UF_6 [4.194].

As an example of spectra recorded using the computer-assisted pulsed TDL spectrometer described in §4.3.4 a spectrum of a $^{74}GeH_4$ molecule in the 932.4–935 cm^{-1} region is presented in figure 4.13. This spectrum contains more than 10 000 data points. Despite the large number of points the time of one-scan recording was only 15 s. This was attained using the multiplicative mode of data acquisition. A TDL operating at 20 Hz with pulses 3 ms long has been used. During a single pulse 42 points were digitised and stored by a microcomputer.

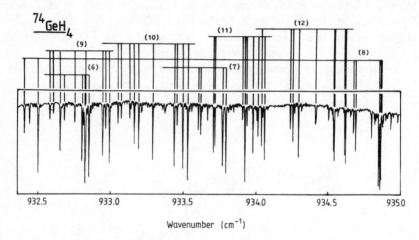

Figure 4.13   Spectrum of a portion of the v_2 Q branch of $^{74}GeH_4$ recorded using the computer-assisted pulsed-diode laser spectrometer. All of the lines have been identified and the J values are presented above the trace [4.156a].

The recording of the spectrum shown in figure 4.13 required four scans: the gas under study, the empty cell, the reference gas, and the Fabry–Perot étalon. The whole time of recording was a few minutes.

During this time the pulse-to-pulse reproducibility at any point of the laser tuning curve was never poorer than 1×10^{-4} cm^{-1}. This was achieved using automatic frequency control as described in §4.3.4. The cell with the natural GeH_4 mixture was placed for this purpose in the channel of stabilisation.

After recording, the spectra of the gas samples and that of the étalon were corrected for the background. The wavenumber sacle was linearised using the laser tuning function obtained from the Ge-étalon transmission spectrum. Absolute calibration was performed with the use of the $^{14}NH_3$ frequencies in the v_2 band [4.89] as standards.

The $^{74}GeH_4$ spectrum presented corresponds to part of the v_2 Q branch. It is known that the v_2 absorption band is forbidden in tetrahedral molecules of the XY_4 type in a harmonic oscillator–rigid rotor approximation; but it appears in the IR spectrum due to Coriolis interaction between the v_2 and active v_4 vibrations. A detailed comparison between positions of the $v_2 = 1$ energy levels for the CH_4, SiH_4 and GeH_4 molecules has been given in reference [4.156]. It was proved that Coriolis interaction between the v_2 and v_4 states leads not only to the activation of the v_2 band but is also the main factor determining the shifts and splittings of the $v_2 = 1$ energy levels.

By the use of computer data processing the effective spectral resolution can be achieved below the Doppler limit after deconvolution. This resolution enhancement is demonstrated in figure 4.14, where the portions of the observed and deconvoluted spectra for the $^RQ_3$ branch of the v_9 band of ethane are shown [4.185]. The spectrum E was recorded using a single-beam TDL spectrometer with sweep integration. The digitised spectrum was deconvoluted using a Gaussian lineshape function with $2\gamma_D$ (FWHM) = 2.2×10^{-3} cm^{-1}. The effective resolution achieved after the deconvolution was $(0.5–1.0) \times 10^{-3}$ cm^{-1}. It can be seen from spectrum F that the deconvolution reveals the small splittings of the low-J lines, which appear only as shoulders in the spectrum E. These torsional splittings in the v_9 level caused by a Coriolis interaction with the close lying $3v_4$ state have been analysed by Susskind *et al* [4.185] and Henry *et al* [4.186]. A global least-squares fit of 2206 Fourier transform lines and 58 values of splitting obtained with a TDL has been performed in reference [4.185].

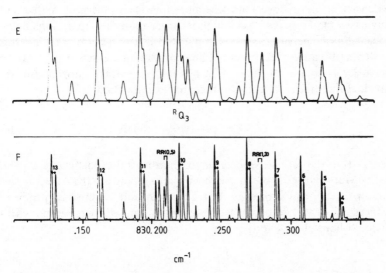

Figure 4.14 Observed (E) and deconvoluted (F) spectra of the v_9 $^RQ_3$ branch of ethane [4.185].

4.6.2 Absorption line parameters

It is important for analytical applications not only to find the most suitable spectral region, but also to establish the relations between the analytical signal and the concentration of molecules and to determine the influence of conditions such as temperature, pressure and composition of the mixture. The information about line parameters is necessary for monitoring the atmosphere, for optimising the conditions of IR gas lasers with optical pumping, and for investigations of flames and combustion. Data on spectral line broadening and shift contain information about interactions between molecules and relaxation processes.

The general procedure for measurement of intensity, width, and determination of profile of an isolated vibrational–rotational line has been stated in [4.4, 4.7]. The intensity of a separate line can be obtained by measurement in direct absorption of the area under the absorption contour or, if the functional form of the contour is known, by calculation from the peak absorbance and width.

Absorption line strengths can also be measured from harmonics or the integral signal by calibrating against lines of known strength.

Although derivative spectroscopy is very sensitive, the lineshape parameters are often difficult to obtain using this technique. Nevertheless, the true width for a separate line can be estimated from the second derivative spectrum using sets of precomputed model line profiles [4.195].

Up to now, in most papers the parameters of light molecules such as CO, NO, CO_2, CH_4, O_3 and similar ones have been obtained. For heavier molecules the problem of obtaining information about linewidth is more serious, because the density of energy levels of these molecules is considerably higher and in many cases the Doppler broadening is larger than the separations between adjacent lines. The situation is complicated if it is necessary to investigate broadening because overlapping of line contours increases. It was shown by Zasavitsky *et al* [4.166] that it is possible to extract information about individual linewidths from the autocorrelation function

$$C'_{XX}(v) = \int D'_X(v) D'_X(v+v) \, dv \qquad (4.24)$$

where $D'_X(v)$ is the derivative spectrum of the absorption coefficient. To obtain true linewidths it is possible to use the distance Δ between the autocorrelation function $C'_{XX}(v)$ zero crossings. For Gaussian and Lorentzian profiles the distances Δ_D or Δ_L, respectively, have simple analytical relations with the individual linewidth [4.166]:

$$\Delta_D = 2\gamma_D / \sqrt{\ln 2} \simeq 2.40\gamma_D \qquad (4.25a)$$

$$\Delta_L = 2\gamma_L \sqrt{4/3} \simeq 2.31\gamma_L \qquad (4.25b)$$

where γ_D and γ_L are the HWHM of individual lines with Gaussian and Lorentzian profiles, respectively.

The value of Δ can also be used to measure linewidths in intermediate situations, thus the broadening of lines for a heavy molecule with a dense spectrum can be investigated. For example, the spectrum of a CF_2Cl_2 molecule broadened by air and the corresponding autocorrelation functions are shown in figure 4.15.

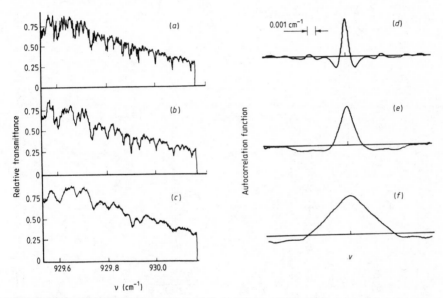

Figure 4.15 Absorption spectra of CF_2Cl_2 and their autocorrelation functions. Trace (a) is a spectrum of 0.5 Torr CF_2Cl_2. Traces (b) and (c) have 0.5 Torr CF_2Cl_2 in addition to air with pressures of 12 and 55 Torr, respectively. Curves (d), (e) and (f) are the computed autocorrelation functions $C_{XX}(v)$ of the spectra (a), (b) and (c), respectively. (Adapted from [4.166].)

In recent years TDLs have been used as a powerful tool for measurement of the gains in IR gas lasers [4.196, 4.197]. The advantages of using TDLs for these studies are the possibilities of probing both lasing and absorbing transitions, obtaining information about populations of different levels, and study of the dynamics of generation. The high sensitivity of the TDL technique allows measurement of gains as low as 0.001% per centimetre.

Reid *et al* [4.197] have recently used a TDL operating at about 12 μm to investigate the transitions in NH_3 when they were pumped with the R(30) CO_2 laser. Figure 4.16(a) illustrates the energy levels in NH_3 and transitions associated with the CO_2 laser pump for this case. Large changes in

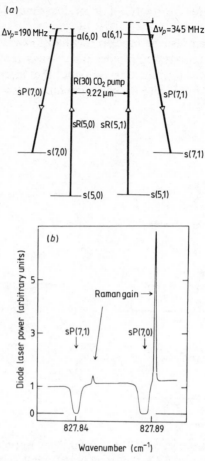

Figure 4.16   (a) Simplified energy level diagram of NH_3. (b) Diode laser scan of the sP(7, 0) and sP(7, 1) transitions of NH_3 in the presence of CO_2 pumping at 9.22 μm. The scan demonstrates Raman gain offset from absorption line centres [4.197a].

transmission that are due to the action of the CO_2 pump beam are clearly seen in figure 4.16(b). The gain spikes are offset from the sP(7, 0) and sP(7, 1) lines of the v_2 band of NH_3 by 200 and 341 MHz respectively, thus confirming the Raman nature of the gain.

These measurements were very important for the design of cw NH_3 lasers. By use of two acousto-optical modulators Rolland *et al* [4.198] succeeded in shifting the pump frequency up to 180 MHz and thus the resonance pumping scheme has been realised. In this way the cw NH_3 laser has been developed operating on 20 lines in the 10.7–13.3 μm region with output powers as high as 10 W.

4.6.3 Infrared spectroscopy of free radicals and molecular ions

Recently great interest has been evoked by the high-resolution recording of vibrational–rotational spectra of radicals in the gas phase. No direct absorption measurements have been made previously in the mid IR due to the lack of sensitivity and a suitable narrow linewidth tunable spectral source. But the analysis of IR spectra of transient molecules is important for understanding the physical properties and molecular structure of these species and also for monitoring chemical reactive interactions.

Many of the transient intermediate chemical species have an open-shell electronic structure. This structure causes additional complexities in vibrational–rotational spectra due to the interaction of electronic spin and electronic orbital angular momentum with rotation and vibration. Analysis of fine and hyperfine (due to the nuclear spin and molecular angular momentum coupling) structures of lines provide information which would not be obtained for molecules without unpaired electrons. This information and measurements of electric dipole moments can be used, in particular, to deduce the charge-density distribution in a molecule and therefore to test bonding models.

The only method used until recently for obtaining such information was laser magnetic resonance (LMR) with CO_2, N_2O and CO lasers. This method has a high sensitivity (as few as 3×10^8 paramagnetic molecules/cm^3 can be detected in mid IR), but its possibilities are limited due to the requirement of near-coincidence between laser and molecular transition frequency. Furthermore, many unstable molecules often have Zeeman effects that are not large enough for LMR studies.

Recent investigations performed mainly in Japan [4.199] and Great Britain [4.200] have demonstrated that vibrational–rotational spectra of free radicals could be recorded with high resolution and high sensitivity by diode laser spectroscopy. The merits of the TDL techniques are that they permit direct measurement of radical frequencies at zero magnetic field and therefore assignments can be made more easily. The TDLs can also give spectra of molecules in essentially nonmagnetic states such as the $^2\pi_{1/2}-^2\pi_{1/2}$ transitions. While resolution and sensitivity are not as high as in the LMR, diode laser spectroscopy yields structural information previously unattainable by that method and also by fluorescence and microwave spectroscopy.

Table 4.6 provides a list of free radicals studied with the help of diode lasers. In some cases the concentration of radicals created reaches 10^{12} to 10^{14} cm^{-3}. Under these conditions the spectrum of radicals can even be displayed upon an oscilloscope screen using the fast-scan technique. But the absorption lines of other molecules (mainly the starting components), which always present in the mixture, are often observed in the same spectral region.

Various modulation methods are used for detection of radicals. Because many states of radicals are paramegnetic the first method useful for their study is Zeeman modulation. Since Zeeman coefficients of vibrational–rotational

Table 4.6 Vibration–rotation spectra of free radicals studied with tunable diode lasers.

Radical†	Method of preparation	Spectral region (cm^{-1})	Recording technique‡	Number of lines	Results§	Reference
CN ($^2\Sigma^+$)	Discharge in N_2/C_2N_2	1994–2131	SI	10	Spin-rotational fine structure	4.201
CF($^2\pi_{1/2}$, $^2\pi_{3/2}$)	60 Hz discharge in CF_4 or C_2F_4	1250–312	ZM	15 ($^2\pi_{3/2}$), 8 ($^2\pi_{1/2}$)	Band analysis, Λ doubling in $^2\pi_{1/2}$	4.202
CCl ($^2\pi_{1/2}$, $^2\pi_{3/2}$)	60 Hz discharge in CCl/Ar	865–80; $v=1\leftarrow0, 2\leftarrow1$	ZM	24 (C^{35}Cl), 12 (C^{37}Cl)	Band analysis, Λ doubling in $^2\pi_{1/2}$	4.203
HO ($^2\pi_{1/2}$, $^2\pi_{3/2}$)	Photolysis in O_3/H_2O	3407.8; 3422	MP	2	Line strengths	4.204
FO ($^2\pi_{3/2}$)	Mw discharge in CF_4/O_2	1007–63	ZM	14	Constants for $v=0, 1$ states	4.205
PO ($^2\pi_{1/2}$, $^2\pi_{3/2}$)	Mw discharge in $H_2/He/O_2$ with red P	1226–54	ZM, SM	7 ($^2\pi_{3/2}$), 10 ($^2\pi_{1/2}$)	Constants for $v=1$ state, Λ doubling in $^2\pi_{1/2}$	4.206
ClO	Mw discharge in $Cl_2/He/O_3$	840–70	SM	10 (^{35}ClO), 7 (^{37}ClO)	Band centre, rotational constants, line strengths, Λ doubling in $^2\pi_{1/2}$	4.207
ClO	Mw discharge in $Cl_2/He+O_3$	845–75	ZM	24 (^{35}ClO), 11 (^{37}ClO)	Rotational constants, Λ doubling in ^{35}ClO ($^2\pi_{1/2}$), line strengths	4.208
BrO ($^2\pi_{3/2}$)	60 Hz discharge in Br/O_2	700–60	ZM	20 (^{79}BrO), 22 (^{81}BrO)	Band analysis	4.209

Species	Production	Range; bands	Technique	Number	Description	Ref.
NF ($^3\Sigma^-$)	H (discharge in H_2)$+NF_2$ (thermodissociation of N_2F_4)	1092–148	SI	51	Band centre, equilibrium rotational parameters	4.210
NF ($a^1\Delta$)	$H+NF_2$	1114–207	SI	31	Constants for $v=0$, 1 states, fluorine nuclear hyperfine splittings	4.211
NS ($^2\pi_{1/2}$, $^2\pi_{3/2}$)	Mw discharge in N_2/S_2Cl_2	1185–238	SM–SH	12 ($^2\pi_{1/2}$), 11 ($^2\pi_{3/2}$)	Band analysis, Λ doubling in $^2\pi_{1/2}$	4.212
PH ($^3\Sigma^-$)	DC discharge in H_2/P (solid)	2107–308	SI	20	Band origin, rotation, spin–rotation, spin–spin parameters	4.213
SF ($^2\pi_{1/2}$, $^2\pi_{3/2}$)	Glow discharge in $OCS/CF_4/He$	821–38; $v=1\leftarrow0$, $2\leftarrow1$	ZM, DCM	33	Equilibrium constants	4.214
BO_2 ($^2\pi_{1/2}$, $^2\pi_{3/2}$)	60 Hz discharge in $BCl_3/O_2/N_2$	1238–323; v_3, $2v_3-v_3$, $v_2+v_3-v_3$	SM–FH, ZM		Renner effect in $v_2=1$	4.215
CH_2, CD_2 ($^3B_1$)	$F+CH_2CO$ (CD_2CO)	709–18; 892; v_2	ZM	1 (CH_2), 5 (CD_2)	Molecular parameters for CD_2	4.216
CF_2 ($^1A_1$)	mw discharge in CF_2CFCl/Ar	1242–74; v_1	SI	~70	Constants for $v_1=1$ state, asymmetry doubling	4.217
CF_2 ($^1A_1$)	mw discharge in CF_4/H_2O	1080–140; v_3	SI	~200	Constants for $v_3=1$ state, v_1/v_3 Coriolis coupling, asymmetry doubling	4.218
FCO	60 Hz discharge in C_2F_4/O_2, SF_6/CO, C_2F_4/CO	1013–46; 1852–75; v_1, v_2	SM, ZM	171 (v_1), 260 (v_2)	Band analysis, spin–rotation splitting	4.219
HO_2 ($^2A_1$)	60 Hz discharge in allyl alcohol/O_2/He	1370–416; v_2	ZM	153	Band analysis, spin–rotation interaction and K doubling	4.220

continued

Table 4.6 *continued*

Radical†	Method of preparation	Spectral region (cm^{-1})	Number of lines	Recording technique†‡	Results§	Reference
HO_2	Flash photolysis in $Cl_2/CH_3OH/O_2$ or $Cl_2/HCHO/O_2$	1117.5; ν_3	1	VD	Band strength $S = 30$ cm^{-2} (STP atm)$^{-1}$	4.221a
HO_2	F (discharge in F_2) $+H_2O_2$	1081; ν_3	4	SM	$S = 35 \pm 9$ cm^{-2} (STP atm)$^{-1}$	4.221b
FO_2	F (MW discharge in F_2/He)$+O_2$	571–84; ν_2	>100	SM, ZM	Band analysis, spin–rotation and K splittings	4.222
NF_2 ($^2B_1$)	Thermal dissociation in N_2F_4	918–1145; ν_1, ν_3		SM-FH, SI	Band analysis, spin–rotation, ν_1/ν_3 Coriolis interaction	4.223
CH_3 ($^2A_2$)	Pyrolysis or 60 Hz discharge in $[(CH_3)_3CO]_2$	600–750; ν_2, $2\nu_2-\nu_2$, $3\nu_2-\nu_2$		AM, SM, ZM	Constants for $v=0,\dots,3$ states, potential function	4.224
CF_3	60 Hz discharge in CF_3I	1246–76; ν_3		ZM	Band analysis, Δl, $\Delta K = 2$, 2 and 2, -1 interactions, ζ_3 Coriolis constant, molecular structure	4.225

† Radicals in ground electronic state were studied if another is not pointed.

† AM = amplitude modulation, DCM = discharge-current modulation, MP = modulated photolysis, SI = sweep integration, SM = source modulation (FH = first harmonic detection, SH = second harmonic detection), VD = video detection, ZM = Zeeman modulation.

§ The band origin, rotational constants, centrifugal distortion constants and parameters of fine structure were determined in most papers if band analysis was the object.

transitions of paramegnetic molecules are of the order of $0.1–1\,kHz\,G^{-1}$ magnetic fields as high as $1\,kG$ are necessary to modulate Doppler-broadened infrared lines.

An experimental set-up for recording spectra of radicals produced by a 60 Hz discharge in a multipass cell is shown in figure 4.17 and has been described in detail by Hirota [4.199]. An example of the spectra recorded is presented in figure 4.18. This figure shows the transmission spectra of the discharge cell with di-tert-butilperoxide obtained by Yamada *et al* [4.224] with the help of ordinary source modulation and with Zeeman modulation. It can be seen that Zeeman modulation allows us to pick up transitions due to paramagnetic particles among many interfering lines of stable molecules with diamagnetism. The fringe pattern that arises from reflections in the optical scheme is also suppressed.

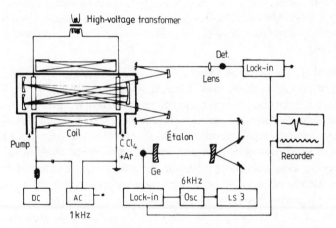

Figure 4.17 Schematic diagram of a TDL spectrometer with Zeeman modulation of paramagnetic molecule absorption [4.203].

The lines observed in the spectrum obtained with Zeeman modulation were due to methyl radicals and were assigned as transitions in the Q branch of the v_2 fundamental band of CH_3. The concentration of the radicals was estimated to be $2 \times 10^{13}\,cm^{-3}$ (about 5×10^{-4} of the total cell pressure) under steady state conditions, when peroxide was pumped continuously through the cell. The lifetime of CH_3 was about 1 ms and pathlengths of 10 m were typically used.

The effective vibrational temperature was high enough (approximately 600 K) to allow investigation of both the v_2 fundamental band and the hot bands arising from the $v_2 = 1$ and $v_2 = 2$ levels. More than 100 lines have been assigned by Yamada *et al* [4.224] and the rotational and centrifugal distortion constants for the ground and three excited vibrational states have been

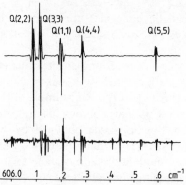

Figure 4.18 Transmission spectra taken through a discharge cell with di-tert-butylperoxide. The lower trace was obtained by source-frequency modulation and second-harmonic detection at 10 kHz. The upper trace is the first harmonic spectrum recorded using Zeeman modulation at 1 kHz. An AC magnetic field of 700 G p–p superimposed on a DC field of 350 G was used. The lines due to the CH_3 radicals are clearly observed and identified [4.224].

determined. The molecular structure parameters and the force field of CH_3 have been obtained from this detailed information [4.226]. It has been confirmed that the CH_3 radical is completely planar in the ground state and its potential function has no double minimum. Since CH_3 is planar it has no large permanent electric dipole moment, and therefore it will only possess a very weak pure rotational spectrum due to magnetic dipole transitions. Thus IR spectroscopy is a promising tool for detection of the interstellar methyl radicals. However, a value of the transition dipole moment is needed in calculating the concentration of CH_3 from the observed lines.

By using a TDL Yamada and Hirota [4.227] estimated the transition dipole moment of the CH_3 v_2 band by observing the spectral intensity decay with time. This decay was due to recombination of the radicals after the discharge generating the radicals was switched off. The transition moment of the CH_3 $v_2 = 1 \leftarrow 0$ band was calculated to be 0.280 ± 0.049 D. This value is very large; for comparison the transition moment of the NH_3 v_2 band, one of the most intense infrared bands, is 0.24 D.

The ground electronic state of most of the diatomic radicals is $^2\pi$. Analysis of the molecular Hamiltonian in an external magnetic field shows that the magnetic momentum of a molecule in the $^2\pi_{1/2}$ state is very small, because the g factor due to the electron spin momentum is compensated by the contribution from the electron orbital momentum. In the $^2\pi_{3/2}$ state the Zeeman effect also decreases rapidly with the increase of the rotational quantum number J. Therefore, the Zeeman technique has limited application.†

† One of the exceptions is the CF radical, in which the mixing of $^2\pi_{1/2}$ and $^2\pi_{3/2}$ states is so large that the $^2\pi_{1/2}$ spectra have been observed by Zeeman modulation [4.202].

Endo *et al* [4.124] have developed a technique of discharge current modulation which is, in essence, a variation of the concentration modulation method. The experimental arrangement is shown schematically in figure 4.19. The SF radicals were generated directly in a cell by a glow discharge in a mixture of OCS and CF_4. The SF fundamental absorption band is overlapped by the C–S stretching band of OCS; the latter molecule was one of the main components in the mixture. Therefore, it was practically impossible to identify the SF lines obtained by ordinary source modulation (figure 4.20(a)). In contrast, discharge current modulation allows us to pick up not only the high-J lines in the $^2\pi_{3/2}$ state such as the $v = 2 \leftarrow 1$, R(16.5) line in figure 4.20(b), but also to detect even the $^2\pi_{1/2}$ lines.

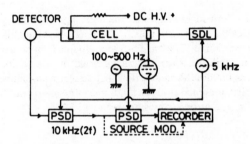

Figure 4.19 Schematic diagram of a TDL spectrometer for recording free radical spectra by discharge-current modulation [4.214].

Modulation of the radical concentration can also be effected by other methods. In particular, Podolske and Johnston [4.204] have used the modulated photolysis of an O_3/H_2O mixture for creation of the OH radicals and measurement of the OH vibrational–rotational line strengths and absolute OH concentrations.

To record vibrational–rotational spectra of molecular ions produced in an AC discharge cell Gudeman *et al* [4.228] have developed a technique of modulating the drift velocities of these charged particles, which leads to Doppler shifts of their spectral lines and thus discriminates the ion lines from those of neutral molecules. This technique has been successfully used with TDL spectroscopy for high-resolution study of the H_3O^+ [4.229], HCO^+ [4.230, 4.231], NH_2^+, DN_2^+, DCO^+ [4.232] and some other ions [4.233]. Diode lasers have also been fruitfully used for measurements *in situ* of the ArH^+ ion drift velocity and mobility in a DC glow discharge [4.234].

4.6.4 *Observation of transient molecular effects*
Systematic investigations of vibrational–rotational spectra of unstable molecules, which are being carried out now, should be considered as a first step

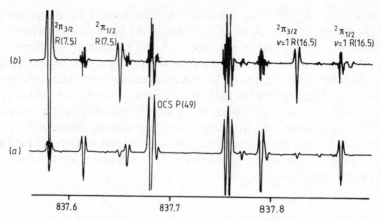

Figure 4.20 Transmission spectra taken through a discharge cell with a mixture of OCS (50 mTorr) and CF_4 (150 mTorr) diluted in He (a total pressure was 1 Torr). The trace (a) was obtained by source-frequency modulation, while the trace (b) was recorded by source-frequency and discharge-current double modulation. The assignments of lines due to the SF radical are presented in trace (b). Lines of $v = 1 \leftarrow 0$ and $2 \leftarrow 1$ in both $^2\pi_{1/2}$ and $^2\pi_{3/2}$ electronic states are seen [4.124].

in the detailed study of the different gas-phase chemical and photochemical reactions by IR high-resolution spectroscopic methods. The first experiments on this problem, in which TDLs have been used for monitoring, are summarised in table 4.7.

High spectral power density and the possibility of fast frequency tuning are those features of TDLS which make them unique for studying nonstable molecular phenomena connected with the change of populations in vibrational–rotational levels, such as molecular excitation and relaxation, interaction between molecules, energy exchange and transfer, and also to record the spectra of rapid processes. When a TDL is operating in the pulsed mode the emission frequency tuning rate can be in the range 10^{-4}–10^{-1} cm^{-1} μs^{-1}. This mode of operation allows both the recording of the spectra of the single-shot processes with duration from 10 μs to 10 ms in a spectral range of about 1 cm^{-1} and investigation of the kinetics of repeated transient processes with time resolution which depends only on the time constant of the data acquisition system [4.52].

The possibility of displaying spectra over a single laser emission pulse on a storage oscilloscope has been used to observe the Zeeman effect in a NH_3 molecule [4.241]. This molecule, as almost all stable ones, does not possess an electron magnetic moment and therefore a magnetic field as high as 500 kG is necessary to produce the Zeeman splittings of about 1×10^{-2} cm^{-1}. Such a high magnetic field can be obtained only in a pulsed mode of a solenoid.

Table 4.7 Diagnostics of chemical reactions with tunable diode lasers.

Topic	Molecules under study	Spectral region (μm)	Reference
Determination of dissociation constant $H_2SO_4 \rightarrow SO_3 + H_2O$	H_2O, SO_3	8.2, 11.3	4.235
Determination of rate constants $HNO_3 + NO \rightarrow HNO_2 + NO_2$ $HNO_2 + O_3 \rightarrow HNO_3 + O_2$	HNO_2, HNO_3	5.9	4.236
UV laser photolysis $UF_6 \rightarrow UF_5 + F$	UF_5, UF_6	16–17	4.237
Recombination of HO_2 radicals produced by lamp flash photolysis $HO_2 + HO_2 \rightleftharpoons H_2O_4 \rightarrow H_2O_2 + O_2$	HO_2	8.95	4.238
Reactions of CH_3 radicals, produced by uv laser photolysis $CH_3 + CH_3 \rightarrow C_2H_6$ $CH_3 + O_2 + M \rightarrow CH_3O_2 + M$	CH_3	16.5	4.239
Recombination of CH_3 radicals. Measurement of transition dipole moment of the v_2 band $CH_3 + CH_3 \rightarrow C_2H_6$	CH_3	16.5	4.227
Reaction of vibrationally excited CH_3F molecules $Br + CH_3F^* \rightarrow HBr + CH_2F$	HBr		4.240

Similar recording techniques have been used to investigate the depletion of rotational levels of the SF_6 ground vibrational state which has been induced by CO_2 laser excitation [4.180]. To understand the physics of excitation it was important to perform experiments under collisionless conditions, therefore cooling in a pulsed free jet was explored. The molecules were probed by a diode laser 120 μs after the excitation pulse. The time of flight of excited molecules through the probing area was about 35 μs. Using a laser frequency tuning rate of $(2-3) \times 10^{-3}$ cm^{-1} μs^{-1} the 0.3 cm^{-1} region could be scanned during a single pulse of the jet (see figure 4.21). The major result of this investigation was that the depletion of all rotational levels is effective even at considerable (up to 11 cm^{-1}) detunings of pumping frequency from the linear absorption spectrum of the molecule. The depletion produced by the P(16) CO_2 laser line excitation at $v_{ex} = 947.74$ cm^{-1} (figure 4.21) can be explained by resonance transitions in the v_3 Q branch, since when the exciting pulse energy density is more than 0.4 J cm^{-2} the power broadening of the laser line exceeds the

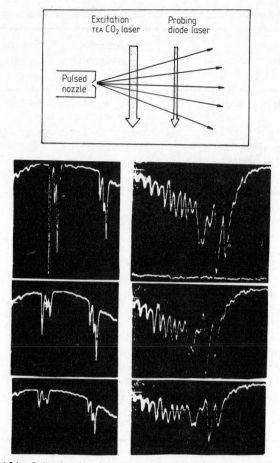

Figure 4.21 Scheme of excitation and probing of the SF_6 molecules in a pulsed jet. Below are oscillograms showing the decrease in absorption of the v_3 P(10) and P(9) multiplets (left-hand traces) and the Q branch (right-hand traces) of SF_6 under the excitation with the P(16) line (947.74 cm^{-1}) of a CO_2 laser. The decrease is due to depletion of rotational sub-levels of the ground vibrational state of the SF_6 molecules under the action of CO_2 laser. The top traces show absorption of diode laser radiation in a jet without excitation. The middle and lower oscillograms were obtained with excitation laser energy fluxes of 0.5 and 2.3 J cm^{-2} [4.180].

distance between v_{ex} and the transition frequencies in the Q branch of the v_3 fundamental [4.242].†

A novel method for investigation of highly excited vibrational states in molecules has been developed by Yevseyev *et al* [4.173]. The CrO_2Cl_2

† For discussion of the results of paper [4.180] see also [4.243, 4.244].

molecules were prepared in high vibrational states by electronic excitation with a Nd–YAG laser followed by nonradiative electronic–vibrational energy conversion. A diode laser has been used to probe the induced IR absorption and its time evolution. The FWHM of the spectrum of the molecules in the quasi-continuum ($v = 18$–19) has been measured to be approximately 15 cm^{-1} at a delay of 100 ns after the exciting pulse [4.173].

The method of probing the time-dependent changes in population of different vibrational–rotational levels by TDL radiation has recently been used by Chu *et al* [4.245] to study vibrational excitation in CO_2 caused by hot H atoms produced by excimer laser photolysis.

4.7 Analytical applications of diode lasers

The possibility of fine frequency tuning over the required spectral range, high monochromaticity of radiation, and high spectral brightness allows us to use TDLS for the solution of most problems in molecular analysis in the gas phase when it is necessary to determine mixing ratios at the level from 1 PPM to 1 PPB and, according to calculations, even to 1 PPT (10^{-12}). The use of TDL spectrometers is justified in those cases when rapid, highly selective and high-sensitivity analysis is necessary. When particular molecules are to be detected measurements should be made only in a selected spectral range or at a single wavelength. The development of gas analysers based on TDLS operating at a fixed liquid nitrogen temperature seems promising for measurements of individual species. The design of such sensors may be very simple and reliability of TDLS already permits their continuous operation over periods of one to three years.

Two main directions may be distinguished where the application of TDLS gives an advantage over other methods of molecular analysis:

(i) a fast (with sensor response time of the order of 10 ms), highly selective analysis of gaseous mixtures for content of molecular components with milli- and microconcentrations, i.e. within the range from 1000 PPM to 1 PPM;

(ii) detection of trace amounts of molecular impurities whose concentrations are at the level of 1 PPM or less.

Generally speaking the solution of these problems requires different experimental equipments and data acquisition techniques. When milli- and microconcentrations are to be detected fine adjustment of the optical system and elimination of all parasitic fluctuations are not required since owing to the high spectral brightness of lasers, even in the fast-scan mode of operation, the signal-to-noise ratio is about 10^3. At a pathlength of 1 m and a pressure of 10–30 Torr this permits detection of molecules with concentrations as small as 10 PPM. By the use of a rather simple multipass cell and second harmonic signal detection technique the sensitivity limit may be improved down to 10 PPB.

Detection of trace amounts of impurities requires, first of all, a laser diode with high spectral brightness, high stability of frequency and output power, and also with a single-mode character of generation over the selected spectral range. In addition, it is necessary to eliminate all sources of parasitic noise, to choose the optimal recording mode and use a multipath cell to increase the absorbing layer thickness. The measuring time for the detection of impurities with concentrations of the order of 10 PPB and less may be several minutes.

4.7.1 Detection of pollutants

The range of problems solved in atmosphere monitoring includes the control of the composition, detection and quantitative determination of impurities, the study of photochemical processes in the atmosphere and the influence of anthropogenic and natural factors. Evaluations show that even very small concentrations of some impurities may appreciably change physical and chemical properties of the atmosphere. The list of such impurities is rather broad and their background concentrations vary from 1 PPM (CH_4) to 1–10 PPB (HCl, H_2S, SO_2) [4.246].

Several methods of determining concentrations of atmospheric impurities may be distinguished in which diode lasers are used:

(i) control by sampling when the gas being investigated flows through a cell (usually a multipath one) and detection is carried out at reduced pressure (of the order of several torr);

(ii) open-path remote measurements with a double-ended instrument (laser source and photodetector at different ends of an air path) or with a single-ended sensor using a retroreflector;

(iii) heterodyne measurements.

As is noted above, the record value of minimum detectable absorption coefficient $\kappa_{min} = 3 \times 10^{-10}$ cm^{-1} has been obtained by Reid *et al* [4.62] using source-frequency harmonic modulation for detection of some atmospheric impurities in a multipass White cell with a basic length of 5 m. The possibility of detecting SO_2, N_2O, NH_3, O_3 and NO_2 in air has been demonstrated using that set-up [4.62, 4.247–4.249]. The detection limits were from 1 PPB to 10 PPT depending on the intensity of analytical lines. If a line strength is 1–5 cm^{-2} atm^{-1} sensitivity at the same level may be attained with a multipass cell of a shorter basic length of about 1 m.

One of the problems in connection with the use of air sampling is the possible change of sample composition at the input into the cell or during the measurements. For direct nonperturbing atmosphere control open-path measuring devices are used. The successes achieved in remote monitoring of the atmosphere with the use of TDLS are illustrated in table 4.8.

Long-path atmospheric gas analysers are used mainly to determine carbon monoxide concentrations. Thus, Astakhov *et al* [4.70] have developed a long-path CO sensor, which uses $PbS_{1-x}Se_x$ pulsed diode lasers, operating at fixed

Table 4.8 Remote detection of atmospheric pollutants with tunable diode lasers.

Molecular pollutant	Spectral region (μm)	Method	Path-length l (m)	Detection limits c_{min} (PPB)	$c_{min}l$ (PPB m)	Reference
CO	4.7	First derivative	610	5	3	4.250
CO	4.7	First harmonic	400	100	40	4.251
CO	4.7	First harmonic	400	60	24	4.252
CO	4.7	Integrative, pulsed laser	200	10–100	2–20	4.72
CO	4.7	Second harmonic	250	0.25	0.06	4.253
CO	4.7	Differential absorption, pulsed laser	600	10	6	4.70
N_2O	8.73	Second harmonic	250	20	5	4.253
SO_2	9	Correlation, pulsed laser	120	100	12	4.76

liquid nitrogen temperature. After passing through the atmosphere laser radiation is recorded by the differential absorption technique. The use of the pulsed mode of TDL operation allowed the elimination of the influence of turbulence in the atmosphere and of meteorological factors. The instrument sensitivity $c_{min}l$ is 6 PPM m, which permits confident detection of the CO background concentration at a pathlength of one hundred metres. The analyser response time is 0.1 s.

In the papers cited in table 4.8 mirror retroreflectors have been used to provide a return signal from a pathlength up to 1 km. The use of topographic targets is undoubtedly of interest; however, TDLS in the mid IR region have output powers which are too low to obtain a stable return signal. Therefore, the possibility of measurements when a portion of the TDL radiation scattered from a target is detected, which has been demonstrated by Webster and Grant [4.254], should be considered at present as somewhat academic.

The gas analysers described above are intended for the study of the atmosphere in the layer near to the Earth. To understand the processes which determine the climate of our planet monitoring of the upper atmosphere is desirable. However, an investigation of photochemical reactions in the stratosphere requires not only high-sensitivity gas analysis, but also simultaneous control of several 'key' species families, NO_x, HO_x and ClO_x. A multispecies analysis is necessary in order to determine the reaction paths, to check the agreement of the relative content of species and the general composition of the stratosphere with the existing theoretical models, and to determine the contribution of each of the processes in ozone balance.

Menzies *et al* [4.255] have described the Balloon-borne Laser In-Situ Sensor (BLISS) designed for such a multispecies analysis of the stratosphere. The BLISS instrument seems to be the most complicated and technically perfect set-up with diode lasers developed up to now. It is sufficient to note that the optical system of the instrument consists of more than one hundred elements which include four TDLs for different spectral ranges, four liquid helium cryostats each with 9.5 l of cooling agent, five HgCdTe detectors, reference cells and other components. The electronic part provides spectral recording with a resolution better than 1×10^{-3} cm^{-1} in the automatic regime or by commands from the ground. Short-time temperature instabilities were measured to be ± 0.04 mK over a three minute period. The estimated RMS value of the TDL injection current noises was 0.02 mA.

BLISS includes, in fact, four TDL spectrometers placed on one platform (1.5 m × 2 m). The spectrometers are designed according to a double-beam scheme with common measuring and four separate reference channels (figure 4.22). The measuring laser beam is obtained from a retroreflector lowered 300–500 m below the instrument gondola. The contribution of fluctuations caused by instabilities in the optical arrangement are diminished and sensitivity is increased by recording the ratio of the 4 kHz second-harmonic signal to the total return signal, which is simultaneously measured by laser beam amplitude modulation at 128 Hz.

The first flight test of BLISS was carried out in October 1983 [4.255b]. Figure 4.23 shows the results of the detection of nitrogen oxide and water vapour at a

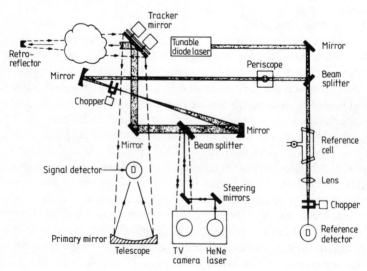

Figure 4.22 Optical scheme of the BLISS instrument. The optical path for a single TDL only is shown [4.255a].

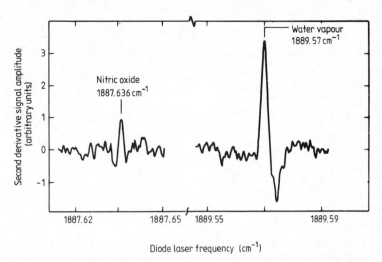

Figure 4.23 Second-derivative scans of an open-air path of 0.6 km obtained at an altitude of 36 km during the first BLISS flight [4.255b].

height of 36 km. The NO concentration calculated from these data was 17 ± 5 PPB.

4.7.2 Isotope gas analysis

Promising results have been obtained in experiments on isotope analysis of molecular gases. The possibility of such analysis is based on isotopic effects which arise in infrared spectra due to variation of molecular moments of inertia and vibrational frequencies upon isotope substitution, and also of orientations of the main molecular axis in the case of nonsymmetrical isotope substitution. The potential to determine abundances of widespread stable isotopes such as ^{13}C in CO_2 and ^{15}N in NH_3 has been shown in references [4.39, 4.256]. The relative content of ^{13}C in natural samples is 1.1×10^{-3}, the natural content of ^{15}N is 3.65×10^{-2}. The detection of isotope-substituted molecules with such concentrations is not difficult at a pressure of about 10 Torr in a cell 10 cm long. In reference [4.257] the development of an automated system based on diode lasers is reported which permits high-precision measurements of relative intensities of absorption lines of light gases. By the use of this system the amount of ^{13}C in CO_2 may be found with the accuracy of 0.1% in a sample which contains only 5×10^{-10} g of $^{13}CO_2$. Though the precision of the absorption method is below that attained with the mass spectrometer, the rate of analysis with diode lasers, and in some cases the selectivity, may be appreciably higher.

In the analysis of rare isotopes the main characteristic required is the detection limit. Labre and Reid [4.258] discussed the possibility of the determination of the ^{14}C isotope in CO_2. It is known that in natural samples

Table 4.9 Analysis of different gas mixtures using tunable diode lasers.

Topic	Component and mixture	λ (μm)	Experimental conditions			Detection Limit (PPM)	Reference
			l (m)	p (Torr)	T (K)		
Shock tube diagnostics	CO in $CO + H_2 + Ar$	4.76		148	3340		4.259
Combustion research: gas species concentrations, temperature distribution, spectral line parameters	CO in flames, sooting flame	4.7–5	0.1		2000		4.259 4.260
Mineral prospecting	SO_2 in air	7.3	2				4.76
Dynamic characteristics of engines and catalytic converters	CO, benzene in exhausts	4.7 10	100	200 10	400	1	4.261 4.43
Tobacco blend and cigarette filter developments	HCN; H_2O, CH_4, C_2H_6 in smoke	13.02 6.44	1.25–5 5	3–12 12			4.262a 4.262
Integrated circuit reliability	H_2O in IC packages	6.05	0.16	10		2	4.263
Chemical laser diagnostics	F in HF/DF	24.7	0.1				4.45, 4.105 4.264
Analysis of technological gases used in semiconductor production	PH_3 in GeH_4	9.10	1	10		0.25	4.71
Utilisation of industrial waste	HCl in air	3.7	64	750		0.1	4.265
Heat exchanger design for nuclear reactors; permeability to HD and HT	H_2O, HDO D_2O in He	6.96	100			0.01 0.05 0.2	4.43

Diagnostics of reactor rod degradation by isotope doping	$^{14}CO_2$ in CO_2	4.52	100	10	0.1	4.43
Optimising of semiconductor surface dry etching	F, CF_4 and Hg in etching gas	24.75, 15.87 and 25 respectively	0.1	10	10	4.45
Corrosion effects	HF in UF_6	25	0.1		10	4.45
Pollutants in aluminium smelting sheds	HF in air	25	1		0.5	4.45

the content of this radioactive isotope is at present 1.3×10^{-12}. At such concentrations the absorption coefficient at the centre of the strongest P(20) line of the v_3 band of $^{14}CO_2$ at $v = 2209.124$ cm^{-1} at a pressure of 30 Torr and temperature of 296 K is 2.2×10^{-10} cm^{-1}. It is not difficult to calculate that for $^{14}CO_2$ detection with an S/N ratio equal to 3 a gas analyser is required with the following characteristics: $(\Delta P/P)_{min} = 10^{-6}$ at a pathlength of 150 m. The possibility of such an instrument was shown in reference [4.258].

4.7.3 Examples of spectral analysis of natural gaseous mixtures

As noted above, the use of diode lasers in industry is partially hampered by the necessity for cryogenic cooling. Nevertheless, these lasers have already been successfully applied in some industrial laboratories and directly in technological processes. Table 4.9 gives a summary of papers in which lead salt TDLS were used to analyse various natural gas mixtures.

As an example of detection of impurities in the milliconcentration range we consider the problem of determination of moisture in hermetically sealed integrated circuit packages [4.263]. Water is known to be one of the most active chemical compounds and even in small amounts leads to corrosion of metals and changes the rates of physical and chemical processes. Thus, the residual concentration of water vapour which may considerably decrease the lifetime of an integrated circuit (IC), is of the order of 5×10^{-3}.

In figure 4.24 time dependences of the concentration of water vapour in a Pyrex cell are shown. The dependences were recorded at the injection into the cell of a standard mixture of 0.1% H_2O in N_2 at the moment designated in the figure by an arrow. The signal was obtained using a $PbS_{1-x}Se_x$ diode laser frequency modulation and the first harmonic detection. The time constants of the lock-in amplifier were chosen to be 1, 0.3 and 0.03 s for cases designated by A, B and C, respectively. Also shown is a signal for a continuous flow of the standard mixture through the cell. It can be seen, in particular, that the use of

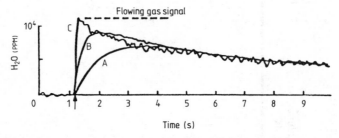

Figure 4.24 Time dependences of the first derivative TDL signal obtained at 1662 cm^{-1} from a simulator volume release (A–C) and continuous flow of a standard mixture with 10^4 PPM of water vapour. The arrow indicates the time when mixture was released into test cell. The time constants of the lock-in amplifier were (A) 1 s, (B) 0.3 s and (C) 0.03 s. (From [4.44].)

any analytical procedure with time response of the order of 1 s gives incorrect results because of strong adsorption of water vapour by the instrument. In order to obtain correct results the sensor response time must not exceed 30 ms.

Mucha [4.265] has described a gas analyser based on a diode laser which is designed for the output control of integrated circuits for water vapour content. The analyser consists of a commercial TDL spectrometer (Model SP 5000), a 16 cm Pyrex sample cell with a special arm where the IC package is punctured, and a vacuum system for the standard gas inlet and exhaust. The optical arrangement of the gas analyser is a double-beam one. The second channel serves for TDL frequency stabilisation by the reference line. Source modulation and 2 kHz second-harmonic detection were employed to detect the second derivative of water absorption. The analytical signal was calibrated by the method of standard additions [4.101].

The IC package volume V_0 is of the order of 0.01–0.1 cm^3; consequently, the task of the determination of water vapour content in the IC package amounts to the detection of 10^{14}–10^{16} molecules of H_2O. Since in the case being considered the task is to detect a definite number of molecules, the mixture ratio detection limit c_{min} depends on the dimensions of the cell in which the analysis is carried out:

$$c_{min} = \left(\frac{\Delta P}{P}\right)_{min} \frac{1}{\sigma n_\Sigma} \frac{V_{cell}}{l}. \qquad (4.26)$$

Here $n_\Sigma = N_\Sigma p_0 V_0$ is the total number of all molecules in the analysed sample and V_{cell} is the volume of the cell where measurements are made. One can easily see that at a cell cross section area $A_{cell} = V_{cell}/l = 1$ cm^2, $(\Delta P/P)_{min} = 1 \times 10^{-4}$ and $\sigma = 1 \times 10^{-17}$ cm^2 the threshold sensitivity is about 0.4 PPM at $p_0 V_0 = 1$ atm cm^3. The detection limit obtained in reference [4.265] without optimisation of the cell dimensions was 2 PPM per atm cm^3. The precision was found to be of the order of 1% during an eight-hour period. The analysis throughput of the laboratory instrument is 50 packages per day and may be increased up to several hundred.

Among other examples of high-speed analysis by TDLs one should note the determination of the content of toxic impurities in automotive exhausts [4.43, 4.261] and diagnostics of flames and combustion gases [4.259, 4.260]. In all these cases the time of measurement was of the order of 10 ms.

A problem whose solution requires high sensitivity may be exemplified by the analysis of high-purity hydrides and chalcogenides of germanium and silicon used for manufacturing semiconductor devices. Thus, the content of phosphine PH_3, arsine AsH_3 and diborane B_2H_6 in gaseous germanium and silicium hydrides (GeH$_4$ and SiH$_4$) must be controlled at the 1–100 PPB level. Such an analysis requires an instrument with a minimum detectable absorption coefficient of the order of 10^{-8} cm^{-1}.

In reference [4.71] determination of PH_3 concentration in GeH$_4$ was used to demonstrate that sufficiently low detection limits may be obtained not only with CW diode lasers, but also in the pulsed mode of operation. The sensitivity was increased using photodetector signal filtration and data accumulation by a microcomputer.

Figure 4.25(a) shows part of the transmission spectrum of a standard mixture of 10 PPM PH_3 in GeH_4 near the PP(2, 2, E) absorption line of the v_4 band of PH_3. This line was chosen as the analytical one. The spectrum was recorded with a pulsed PbSnSe diode laser, whose radiation frequency was tuned at a rate of 5×10^{-3} cm^{-1} μs^{-1}. The change in the TDL radiation power due to absorption produced by the PH_3 molecules after passing through a 1 m cell with the mixture at 10 Torr was only 4×10^{-4}. With data accumulation the signal-to-noise ratio was improved proportionately to the square root of the number of observations. In figure 4.25(b) the spectrum over the same spectral range is shown, but already after 1000 accumulations at each of the

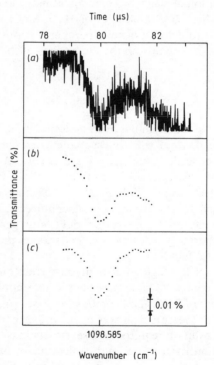

Figure 4.25 Portion of the transmission spectrum taken through a mixture of 10 PPM PH_3 in GeH_4 at a total pressure of 10 Torr and pathlength of 1 m. The traces were obtained with the pulsed-diode laser spectrometer. Trace (a) was recorded without accumulation, while each point in trace (b) was obtained by averaging 1000 data points. Trace (c) is the result of baseline subtraction with the help of a microcomputer. The detection limit (S/N = 1) is 5×10^{-8} cm^{-1} [4.71].

presented points. It is seen that in this case the sensitivity is limited by the background line slope. The final form of the spectrum obtained after the subtraction of the background is presented in figure 4.25(c).

Further measurements were made at three points, two of which were chosen to be on the different wings of the line and one at the absorption maximum. At a pathlength of 1 m and a time of 1 min for a single measurement the minimum detectable absorption coefficient was 3×10^{-8} cm^{-1}. This corresponds to 30 PPB PH$_3$.

For many industrial applications continuous monitoring of one or several molecular components is required. The efficiency of analyser operation in such cases depends to a large extent on its reliability. Fully automated TDL systems have been developed by Laser Analytics (see §4.3.3). These systems allow continuous measurements over three-month periods [4.43].

One of these systems is employed, in particular, for the determination of H_2O, HDO and D_2O content in helium used as a heat exchanger in gas-cooled nuclear reactors [4.43, 4.44]. Detection of all the three components is performed by tuning the laser emission frequency within one mode near 1436 cm^{-1}. The value of the TDL injection current is set by a computer so that the laser output was tuned sequentially to an absorption due to H_2O, HDO and D_2O. Fine adjustment is provided by an automatic frequency stabilisation circuit. The sensitivities of a pathlength of 100 m are 10–60 PPB for H_2O, 50–100 PPB for HDO and 200–700 PPB for D_2O. The system can also be used to determine the permeability of heat exchangers to HT and for diagnostics of reactor rod degradation [4.43].

4.7.4 Analysis of liquid and solid samples

The absorption lines in liquids and solids are usually of the order of 10 cm^{-1} wide; therefore the determination of concentrations of components in such samples does not require high spectral resolution. However, the use of classical thermal sources of IR radiation (a Nernst glower or a globar) for transmission measurements in some cases gives rise to serious difficulties caused primarily by the low spectral density of their radiation power.

High spectral brightness (about 10 W cm^{-1}) of semiconductor diode lasers allows their use in analysis of optically dense samples up to very high absorbances ($A_{max} = (\kappa l)_{max} = 12$) [4.6]. The most striking example is the analysis of room-temperature vulcanised silicone rubber (RTV) for the catalyst (methyltriacetoxysilane) content [4.266]. During the manufacturing process this kind of rubber is a gel-like fluid consisting of a variety of components, monomers, crosslinkers, fillers, etc. The minimum thickness required to maintain a continuous flow of rubber through the sample cell is 500 μm. Even at such a small optical pathlength the absorbance by the catalyst in the centre of the acetoxy carbonyl absorption band at 1750 cm^{-1} reaches 10, therefore the analysis is carried out at 1781 cm^{-1}, the frequency which corresponds to the shoulder of that absorption band. Owing to a narrow laser line the linear dependence of absorbance versus concentration (Beer's law) is preserved over a wide range of sample thicknesses and catalyst concentrations.

A TDL system designed for determination of the catalyst content in silicone rubber was developed in 1978 and has been used for more than five years for continuous real-time control directly in the RTV manufacturing process. An analogous system has been used also to measure the concentration of polychlorinated biphenyls in transformer oil [4.44].

One of the recent interesting applications of TDLs is the investigation of the properties of molecules adsorbed on surfaces. The spectrum can be obtained by Stark modulation of the resonant vibrational frequencies of such molecules and measuring the surface reflectivity using a tunable diode laser. This method was used by Lambert [4.268] to observe the first-order Stark effect for the CO molecules adsorbed on a Ni surface.

Tunable lead salt diode lasers have been shown to be a convenient source for spatial and temporal investigations. For example, Ohsawa *et al* [4.269] have used a TDL emitting at 9 μm to control oxygen impurities in silicon wafers with a spatial resolution of 200 μm. The pulsed TDL spectrometer described in §4.3.4 allows measurements under computer control with a time resolution up to 4 ns [4.52].

4.8   Conclusions

At present, two decades after the development of the first lead salt laser, the interest in TDL application for high-resolution spectroscopy and molecular analysis is not decreasing. On the contrary, TDLs have become widely used devices and the number of works using TDLs increases year after year.

In this review we have considered only one method, namely, the absorption–transmission method, in which the use of TDLs has allowed an appreciable improvement of sensitivity and selectivity as compared with classical IR gas analysers. The possibility of TDL frequency tuning in a broad spectral region draws one's attention to the use of these lasers in other methods of linear absorption spectroscopy, opto-acoustic [4.270], optothermal [4.271] and optogalvanic [4.272]. Remote heterodyne measurements with the use of a TDL as a local oscillator are of great interest [4.61, 4.273]. However, output powers of mid IR semiconductor lasers are at present about an order of magnitude lower than the optimum for these techniques.

Recent improvements in the manufacturing technology have resulted in considerable broadening of the spectral range and increase of stability and reliability of tunable diode lasers. One should not think that all possibilities for improving the characteristics of this type of laser have been exhausted. In particular one of the main drawbacks is the existence of 'dead zones', i.e. spectral ranges where generation is lacking. This drawback can, apparently, be eliminated by the development of compact constructions with an external resonator. Generation of TDLs coupled to an external cavity also paves the way for intracavity spectroscopy that will permit high-sensitivity molecular analysis in such cases when it is not possible to provide the required thickness of an absorbing layer by means of multipath cells.

TDLS are among the best IR sources for monitoring processes which involve vibrational–rotational excitation. The investigations of spectra and kinetics of transient molecular phenomena are, in our opinion, very important. The sensitivity and selectivity of the TDL absorption method should make this a powerful technique for future studies of gas-phase chemical and photochemical reactions.

At the same time the investigations of nonstable processes are impossible without knowledge of spectroscopic characteristics of molecules (including intermediate ones) in the ground and excited states. These data are also of interest for the study of the fundamental properties of molecules and the solution of numerous applied problems. Therefore, in future diode lasers will be used as before to study fine structure of spectra, investigation of individual line parameters, etc.

One should also expect both further application of TDL spectrometers to high-sensitivity and highly selective multispecies analysis of various molecular mixtures and development of TDL sensors operating at sufficiently high temperatures intended for determination or continuous control of individual molecular impurities.

Acknowledgments

I would like to thank Professor V S Letokhov and Dr V G Koloshnikov for their encouragement and interest during the work. I am indebted to my colleagues at the Institute of Spectroscopy and wish to acknowledge especially useful discussions with V R Mironenko and the contributions of V M Krivtsun, E P Snegirev and I Pak. I thank Professor A P Shotov, Dr I I Zasavitsky and Dr A D Britov for supplying diode lasers. Special thanks are due to G V Vedeneeva, Z I Yarikina and A I Lazareva who helped greatly in the technical production of the manuscript.

References

4.1	Butler J F, Calawa A R, Phelan R J Jnr, Harman T C, Strauss A J and Rediker R H 1964 *Appl. Phys. Lett.* **5** 75
4.2	Hinkley E D and Freed C 1969 *Phys. Rev. Lett.* **23** 277–80
4.3	Hinkley E D 1970 *Appl. Phys. Lett.* **16** 35
4.4	Hinkley E D, Nill K W and Blum F A 1976 in *Laser Spectroscopy of Atoms and Molecules* ed H Walther (Berlin: Springer) pp. 125–96
4.5	Nill K W 1977 *Laser Focus* **13**(2) 32–7

4.6   Eng R S, Butler J F and Linden K J 1980 *Opt. Eng.* **19** 945–60

4.7   McDowell R S 1981 in *Vibrational Spectra and Structure* vol. 10 ed J R Durig (Amsterdam: Elsevier) pp. 1–151

4.8   Eng R S and Ku R T 1982 *Spectrosc. Lett.* **15** 803–929

4.9   Spectra-Physics, Laser Analytics Division, Bedford, Massachusetts 01730

4.10   Shotov A P *Proc. 4th Conf. on Solid State Devices, Tokyo, 1972* (Suppl. *J. Jpn. Soc. Appl. Phys.* **42** 282–8 (1973))

4.11   Ageikin V A, Zasavitsky I I, Koloshnikov V G, Lihter A I, Pel E G and Shotov A P 1974 *Opt. Spectrosc.* **36** 808–11 (in Russian)

4.12   Britov A D, Karavaev S M, Kalyuzhnaya G A, Kurbatov A L, Gorina Yu A and Sivachenko S D 1978 *Sov. J. Opt. Technol.* **45** 415

4.13   Vedeneeva G V, Zasavitsky I I, Koloshnikov V G, Kuritsyn Yu A and Shotov A P 1978 *Sov. Techn. Phys. Lett.* **4** 927–31 (in Russian)

4.14   Anzin V B, Glushkov M V, Gorina Yu A, Kalyuzhnaya G A, Kosichkin Yu V and Nadezhdinskii A I 1978 *Sov. Phys.–Lebedev Institute Reports* No. 7 18–21 (in Russian)

4.15   Somsikov A I, Britov A D, Zasavitsky I I, Koloshnikov V G, Kurbatov A L, Kuritsyn Yu A, Lebedev E I and Shotov A P 1983 *Abstracts of the XIX All-Union Congress on Spectroscopy, Tomsk, 1983* part VI pp. 115–17

4.16   Shotov A P and Vyatkin K V 1980 *Sov. Techn. Phys. Lett.* **6** 1199–202 (in Russian)

4.17   Preier H, Bleicher M, Riedel W, Pfeiffer H and Maier H 1977 *Appl. Phys.* **12** 277–81

4.18   Lo W ed 1983 *Tunable Diode Laser Development and Spectroscopy Applications* (*Proc. Soc. Photo-Opt. Instrum. Eng.* (SPIE) **438**)

4.19   Preier H 1979 *Appl. Phys.* **20** 189–206

4.20   Partin D L 1983 *Appl. Phys. Lett.* **43** 996–7

4.21   Butler J F, Reeder R E and Linden K J 1983 *IEEE J. Quantum Electron.* **QE-19** 1520–5

4.22   Freed C, Bielinski J W and Lo W 1983 *Appl. Phys. Lett.* **43** 629–31

4.23   Eng R S, Mantz A W and Todd T R 1979 *Appl. Opt.* **18** 1088–91

4.24   Harward C N and Sidney B D 1980 *NASA Conf. Pub.* CP-2138 part 1 129

4.25   Kurbatov L N, Britov A D, Karavaev S M, Sivachenko S D, Maksimovsky S N, Ovchinnikov I I, Rzaev M M and Starik P M 1983 *JETP Lett.* **37** 422–4 (in Russian)

4.26a   Engeler W and Garfinkel M 1965 *Solid State Electron.* **8** 585–604

4.26b   Zasavitsky I I, Kosichkin Yu V, Perov A N, Polyakov Yu A, Shirokov A M and Shotov A P 1981 *P N Lebedev Physical Institute Preprint* No. 150

4.27   Kuritsyn Yu A, Britov A D, Koloshnikov V G, Shotov A P, Snegirev E P, Vedeneeva G V and Zasavitsky I I 1982 *Seventh Int. Conf. on High Resolution Infrared Spectrosc., Liblice, 1982* (Prague: Czechoslovak Spectroscopy Society) p. 66

4.28   Hsieh H H and Fonstad C G 1980 *IEEE J. Quantum Electron.* **QE-16** 1039–44

4.29   Lo W 1981 *J. Appl. Phys.* **52** 900

4.30   Linden K J and Reeder R E 1984 *Appl. Phys. Lett.* **44** 377–9

4.31   Freed C, Bielinski J W, Lo W and Partin D L 1984 *J. Opt. Soc. Am.* B **1** 544–5

4.32   Volkov B A, Pankratov O A and Sazonov A V 1983 *Sov. Phys.–JETP* **85** 1395–408 (in Russian)

4.33 Reid J, Cassidy D T and Menzies R T 1982 *Appl. Opt.* **21** 3961–5

4.34 Jennings D E and Hillman J J 1977 *Rev. Sci. Instrum.* **48** 1568–9

4.35 Vedeneeva G V, Zasavitsky I I, Koloshnikov V G, Kuritsyn Yu A, Pak I, Snegirev E P and Shotov A P 1981 *Proceedings of the Second French–Soviet Symposium on Optical Instruments, Moscow, 1981* pp. 47–55

4.36 Kosichkin Yu V, Kotlov Yu N, Kryukov P V, Kuznetsov A I, Nadezhdninskii A I, Pelipenko V I, Perov A N and Stepanov E V 1983 *Pribory i Technika Eksperimenta* No. 4 228–31 (in Russian)

4.37 Anzin V B, Glushkov M V, Kosichkin Yu V, Nadezhdinskii A I, Perov A N and Shirokov A M 1980 *Appl. Phys.* **22** 241–4

4.38 Zasavitsky I I 1978 *Pribory i Technika Eksperimanta* No. 5 191–4 (in Russian)

4.39 Gorshunov B M, Shotov A P, Zasavitsky I I, Koloshnikov V G, Kuritsyn Yu A and Vedeneeva G V 1979 *Opt. Commun.* **28** 64–8

4.40 Dubs M and Günthard H H 1978 *Appl. Opt.* **17** 3593–7

4.41 Jennings D E 1980 *Appl. Opt.* **19** 2695–700

4.42 Traenkle G and Huettner W 1982 *Appl. Opt.* **21** 4151

4.43 Forrest G T and Wall D L 1983 *Proc. SPIE* **438** 84–91

4.44 Forrest G T, Wall D L and Mantz A W 1982 *Photonics Spectra* **16** 68–71

4.45 Forrest G T 1983 *Lasers and Applications* No. 6 63–7

4.46 Nill K W, Strauss A J and Blum F A 1973 *Appl. Phys. Lett.* **22** 677–9

4.47 Preier H and Riedel W 1974 *J. Appl. Phys.* **45** 3955–8

4.48 Britov A D, Karavaev S M, Kalyuzhnaya G A, Kurbatov A L, Maksimovskii S N and Sivachenko S D 1977 *Sov. J. Quantum Electron.* **7** 1138

4.49 Dmitrievskii O D, Neporent B S and Nikitin V A 1958 *Sov. Phys.–Usp.* **64** 447–92 (in Russian)

4.50 Cassidy D T and Reid J 1982 *Appl. Opt.* **21** 2527–30

4.51 Davies P B, Hamilton P A, Lewis-Bevan W and Okumura M 1983 *J. Phys. E: Sci. Instrum.* **16** 289–94

4.52 Kuritsyn Yu A, Vedeneeva G V, Koloshnikov V G, Krivtsyn V M, Pak I, Snegirev E P, Britov A D, Zasavitsky I I and Shotov A P 1983 *Institute of Spectroscopy Report* No. 14

4.53 Kosichkin Yu V and Nadezhdinskii A I 1983 *Izv. Akad. Nauk* **47** 2037–45 (in Russian)

4.54 Zasavitsky I I, Matsonashvili B N, Pogodin V I and Shotov A P 1974 *Sov. Phys.–Semicond.* **8** 467

4.55 Flicker H, Aldridge J P, Filip H, Nereson N G, Reisfeld M J and Weber W H 1978 *Appl. Opt.* **17** 851–2

4.56 Ohi M 1980 *Jpn. J. Appl. Phys.* **19** L541–3

4.57 Weber W H, Leslie D H and Peters C W 1981 *J. Mol. Spectrosc.* **89** 214–22

4.58a Kosichkin Yu V, Kuznetsov A I, Nadezhdinskii A I, Perov A N and Stepanov E V 1982 *Sov. J. Quantum Electron.* **9** 822–5 (in Russian)

4.58b Zasavitsky I I, Kuznetsov A I, Kosichkin Yu V, Kryukov P V, Nadezhdinskii A I, Perov A N, Stepanov E V and Shotov A P 1982 *Sov. Tech. Phys. Lett.* **8** 1168–71 (in Russian)

4.59 Shimoda K 1973 *Appl. Phys.* **1** 77–86

4.60 Long D 1977 in *Optical and Infrared Detectors* ed R J Keyes (Berlin: Springer) pp. 101–43

4.61 Menzies R T 1976 in *Laser Monitoring of the Atmosphere* ed E D Hinkley

(Berlin: Springer) pp. 298–353

4.62a Reid J, Shewchun J, Garside B K and Ballik E A 1978 *Appl. Opt.* **17** 300–7

4.62b Reid J, Garside B K, Shewchun J, El-Sherbiny M and Ballik E A 1978 *Appl. Opt.* **17** 1806–10

4.63 Laguna G A 1984 *Appl. Opt.* **23** 2155–8

4.64 Reid J and Labrie D 1981 *Appl. Phys.* B **26** 203–10

4.65 Hinkley E D, Ku R T and Kelley P L 1976 in *Laser Monitoring of the Atmosphere* ed E D Hinkley (Berlin: Springer) pp. 238–97

4.66 Wilson G V H 1963 *J. Appl. Phys.* **34** 3276–85

4.67 Townes C H and Schawlow A L 1955 *Microwave Spectroscopy* (New York: McGraw Hill) chap. 10

4.68 Malov L R and Mukhtarov R T 1984 *Sov. J. Appl. Spectrosc.* **40** 211–17 (in Russian)

4.69 Takami M 1979 *J. Chem. Phys.* **71** 4164–5

4.70 Astakhov V I, Galaktionov V V, Zasavitsky I I, Kosichkin Yu V, Nadezhdinskii A I, Perov A N, Tishchenko A Yu, Trofimov V T, Khattatov V U and Shotov A P 1982 *Sov. J. Quantum Electron.* **9** 531–6 (in Russian)

4.71 Krivtsyn V M, Kuritsyn Yu A, Snegirev E P, Zasavitsky I I and Shotov A P 1985 *Sov. J. Appl. Spectrosc.* **43** 571–6

4.72 Riedel W 1977 *Proc. SPIE* **99** 17–21

4.73 Hirschfeld T 1976 *Appl. Spectrosc.* **30** 67–8

4.74 Morgan D R 1977 *Appl. Spectrosc.* **31** 404–15

4.75 Lam R B 1983 *Appl. Spectrosc.* **37** 567–9

4.76 Max E and Eng S T 1977 *Opt. Quantum Electron.* **9** 411–18; 1979 *Opt. Quantum Electron.* **11** 97–101

4.77 Sano H, Koga R, Kosaka M and Shinohara K 1981 *Jpn. J. Appl. Phys.* **20** 2145–53

4.78a Zasavitsky I I, Kosichkin Yu V, Nadezhdinskii A I, Stepanov E V, Tishchenko A Yu and Shotov A P 1983 *Sov. Phys.–Lebedev Institute Reports* No. 9 13–17 (in Russian)

4.78b Zasavitsky I I, Kosichkin Yu V, Kryukov P V, Nadezhdinskii A I, Prokhorov A M, Stepanov E V, Tishchenko A Yu and Shotov A P 1984 *Sov. J. Tech. Phys.* **54** 1542–51 (in Russian)

4.79 Kataev D I and Mal'tsev A A 1973 *JETP* **64** 1527–37

4.80 Takami M and Kuze H 1984 *J. Chem. Phys.* **80** 2314–18

4.81 Nagai K, Kawaguchi K, Yamada C, Hayakawa K, Takagi Y and Hirota E 1980 *J. Mol. Spectrosc.* **84** 197–203

4.82 Arkhipov V V, Kurbatov A L and Shubin M V 1983 *Sov. J. Opt. Tech.* No. 12 46–7 (in Russian)

4.83 Chang T-Y, Morris R N and Yeung E S 1981 *Appl. Spectrosc.* **35** 587–91

4.84 Sattler J P, Worchesky T L, Ritter K J and Lafferty W J 1980 *Opt. Lett.* **5** 21–3

4.85 Wells J S, Petersen F R and Maki A G 1983 *Proc. SPIE* **438** 110–18 and references therein

4.86 Kauppinen J, Jolma K and Horneman V-M 1982 *Appl. Opt.* **21** 3332–6

4.87 Jolma K, Kauppinen J and Horneman V-M 1983 *J. Mol. Spectrosc.* **101** 300–5

4.88 Jolma K, Kauppinen J and Horneman V-M 1983 *J. Mol. Spectrosc.* **101** 278–84

4.89 Urban Š, Papoušek D, Kauppinen J, Yamada K and Winnewisser G 1983 *J. Mol. Spectrosc.* **101** 1–15

4.90 Urban Š, Papoušek D, Belov S P, Krupnov A F, Tret'yakov M Yu, Yamada K and Winnewisser G 1983 *J. Mol. Spectrosc.* **101** 16–29

4.91 Wells J S, Petersen F R and Maki A G 1979 *Appl. Opt.* **18** 3567–73

4.92 Guelachvili G 1983 *J. Opt. Soc. Am.* **73** 137–50

4.93 Sattler J P, Worchesky T L and Lafferty W J 1981 *J. Mol. Spectrosc.* **88** 364–71

4.94 Guelachvili G 1982 *Can. J. Phys.* **60** 1334–47

4.95 Wells J S, Petersen F R and Maki A G 1983 *J. Mol. Spectrosc.* **98** 404–12

4.96 Guelachvili G 1979 *J. Mol. Spectrosc.* **75** 251–69

4.97 Guelachvili G 1979 *Opt. Commun.* **30** 361–3

4.98 Klebsch W, Yamada K and Winnewisser G 1983 *Z. Naturf.* **A38** 157–62

4.99 Guelachvili G 1980 *J. Mol. Spectrosc.* **79** 72–83

4.100 Fried A, Sams R and Berg W W 1984 *Appl. Opt.* **23** 1867–80

4.101 Mucha J A 1982 *Appl. Spectrosc.* **36** 393–400; 1984 *Appl. Spectrosc.* **38** 68–73

4.102 Weber W H, Leslie D H, Peters C W and Terhune R W 1980 *J. Mol. Spectrosc.* **81** 316–26

4.103 Pine A S, Maki A G, Robiette A G, Krohn B J, Watson J K G and Urbanek Th 1984 *J. Am. Chem. Soc.* **106** 891–7

4.104 Nagai K, Tanaka K and Hirota E 1982 *J. Phys. B: At. Mol. Phys.* **15** 341–5

4.105 Laguna G A and Beattie W H 1982 *Chem. Phys. Lett.* **88** 439–40

4.106 Jennings D E and Brault J M 1982 *Astrophys. J.* **256** 29–31

4.107 Lowry S H and Fisher C J 1982 *J. Quant. Spectrosc. Radiat. Transfer* **27** 585–91

4.108 Dubost H, Charneau R and Harig M 1982 *Chem. Phys.* **69** 389–405

4.109 Wills R E, Walker H C Jnr and Lowry H S III 1984 *J. Quant. Spectrosc. Radiat. Transfer* **31** 373–8

4.110 Knapp K and Hanson R K 1983 *Appl. Opt.* **22** 1980–5

4.111 Lundqvist S, Margolis J and Reid J 1982 *Appl. Opt.* **21** 3109–13

4.112 Falcone P K, Hanson R K and Kruger C H 1983 *J. Quant. Spectrosc. Radiat. Transfer* **29** 205–21

4.113 Wegdam G H and Sondag A H M 1984 *Chem. Phys. Lett.* **111** 360–5

4.114 Nazemi S, Javan A and Pine A S 1983 *J. Chem. Phys.* **78** 4797–805

4.115 Wells J S, Jennings D A and Maki A G 1984 *J. Mol. Spectrosc.* **107** 48–61

4.116 Maki A G, Lovas F J and Suenram R D 1982 *J. Mol. Spectrosc.* **91** 424–9

4.117 Maki A G and Lovas F J 1982 *J. Mol. Spectrosc.* **95** 80–91

4.118 Maki A G 1983 *J. Mol. Spectrosc.* **102** 361–7

4.119 Malathi Devi V, Fridovich B, Jones G D and Snyder D G S 1984 *J. Mol. Spectrosc.* **105** 61–9

4.120 Petersen F R, Wells J S, Siemsen K J, Robinson A M and Maki A G 1984 *J. Mol. Spectrosc.* **105** 324–30

4.121 Malathi Devi V, Rinsland C P and Benner D C 1984 *Appl. Opt.* **23** 4067–75

4.122 Da-Wun Chen, Niple E R and Poultney S K 1982 *Appl. Opt.* **21** 2906–11

4.123 Sondag A H M, van der Liet H A and Wegdam G H 1983 *J. Mol. Spectrosc.* **97** 353–61

4.124 Hawkins R L and Shaw J H 1983 *J. Quant. Spectrosc. Radiat. Transfer* **29** 543–8

4.125 Pokrowsky P 1983 *Appl. Opt.* **22** 2221–3

4.126 Strow L L 1983 *J. Mol. Spectrosc.* **97** 9–28

4.127 Strow L L 1983 *J. Quant. Spectrosc. Radiat. Transfer* **29** 395–406

4.128 Klebsch W, Yamada K and Winnewisser 1983 *J. Mol. Spectrosc.* **99** 479–81

4.129 Leavitt R P and Sattler J P 1983 *J. Quant. Spectrosc. Radiat. Transfer* **29** 179–81

4.130   Picard-Bersellini A and Whitaker B J 1984 *J. Mol. Struct.* **115** 347–50
4.131   Hoell J M Jnr, Harward C N, Bair C H and Williams B S 1982 *Opt. Eng.* **21** 548–52
4.132   Baldacchini G, Marchetti S, Montelatici V, DiLonardo M, Leavitt R R and Sattler J P 1982 *J. Mol. Spectrosc.* **95** 30–4
4.133   Baldacchini G, Marchetti S, Montelatici V, Buffa G and Tarrini O 1982 *J. Chem. Phys.* **76** 5271–7
4.134   Baldacchini G, Marchetti S, Montelatici V, Sorge V, Buffa G and Tarrini O 1983 *J. Chem. Phys.* **78** 665–8
4.135   Job V A, Patel N D, D'Cuhna R and Kartha V B 1983 *J. Mol. Spectrosc.* **101** 48–60
4.136   Baldacchini G, Marchetti S and Montelatici V 1984 *J. Mol. Spectrosc.* **103** 257–61
4.137   Urban Š, Papoušek D, Malathi Devi V, Fridovich B, D'Cuhna R and Rao K N 1984 *J. Mol. Spectrosc.* **106** 38–55
4.138   Kuze H, Jones H, Tsukakoshi M, Minoh A and Takami M 1984 *J. Chem. Phys.* **80** 4222–9
4.139   Urban Š, Papoušek D, Bester M, Yamada K, Winnewisser G and Guarnieri A 1984 *J. Mol. Spectrosc.* **106** 29–37
4.140   Loewenstein M, Podolske J R and Varanasi P 1983 *Proc. SPIE* **438** 189–96
4.141   Tanimoto M, Yamada K, Winnewisser G and Christiansen J J 1983 *J. Mol. Spectrosc.* **100** 151–63
4.142   Lemoine B, Yamada K and Winnewisser G 1982 *Ber. Bunsenges. Phys. Chem.* **86** 795–7
4.143   Maki A G and Sams R L 1983 *J. Mol. Struct.* **100** 215–21
4.144   Yamamoto S, Nakanaga T, Takeo H, Matsumura C, Nakata M and Kuchitsu K 1984 *J. Mol. Spectrosc.* **106** 376–87
4.145   Jennings D E and Robiette A G 1982 *J. Mol. Spectrosc.* **94** 369–79
4.146   Restelli G and Cappellani F 1982 *Chem. Phys. Lett.* **92** 439–42
4.147   Walker H C Jnr and Phillips W J 1983 *J. Appl. Phys.* **54** 4729–33
4.148   Devi V M, Fridovich B, Snyder D G S, Jones G D and Das P P 1983 *J. Quant. Spectrosc. Radiat. Transfer* **29** 45–7
4.149   Malathy Devi V, Fridovich B, Jones G D and Snyder D G S 1983 *J. Mol. Spectrosc.* **97** 333–42
4.150   Takami M 1982 *J. Chem. Phys.* **76** 1670–5
4.151   Yamamoto S, Takami M and Kuchitsu K 1984 *J. Chem. Phys.* **81** 3800–4
4.152   Patterson C W, McDowell R S, Nereson N G, Krohn B J, Wells J S and Petersen F R 1982 *J. Mol. Spectrosc.* **91** 416–23
4.153   McDowell R S, Reisfeld M J, Patterson C W, Krohn B J, Vasquez M C and Laguna G A 1982 *J. Chem. Phys.* **77** 4337–43
4.154   Takami M and Kuze H 1983 *J. Chem. Phys.* **78** 2204–9
4.155   Das P P, Malathy Devi V, Narahari Rao K and Robiette A G 1982 *J. Mol. Spectrosc.* **91** 494–8
4.156a Vedeneeva G V, Krivtsyn V M, Kuritsyn Yu A and Snegirev E P 1983 *Opt. Spectrosc.* **54** 941–4 (in Russian)
4.156b Cheglokov A E, Kuritsyn Yu A, Snegirev E P, Ulenikov O N and Vedeneeva G V 1984 *J. Mol. Spectrosc.* **105** 385–96
4.157   Cheglokov A E, Kuritsyn Yu A, Snegirev E P, Ulenikov O N and Vedeneeva

G V 1984 *Mol. Phys.* **53** 287–94

4.158 Vogt J, Winnewisser M, Yamada K and Winnewisser G 1984 *Chem. Phys.* **83** 309–18

4.159 Leavitt R P, Sattler J P and Worchesky T L 1984 *J. Mol. Spectrosc.* **106** 260–79

4.160 Harradine D, Foy B, Laux L, Dubs M and Steinfeld J I 1984 *J. Chem. Phys.* **81** 4267–80

4.161 Davies P B, Ferguson A H, Hamilton P A, Amyes T L and Van Laere I M R 1983 *Int. J. Infrared Millimeter Waves* **4** 1029–36

4.162 Cappellani F, Restelli G and Tarrago G 1984 *J. Mol. Spectrosc.* **103** 262–7

4.163 Caldon G L and Halogen L 1982 *Mol. Phys.* **46** 223–37

4.164 Baldachini G, Marchetti S and Montelatici V 1982 *Lett. Nuovo Cim.* **33** 267–71

4.165 Zasavitsky I I, Kosichkin Yu V, Nadezhdinskii A I, Stepanov E V, Tishchenko A Yu and Shotov A P 1983 *Sov. J. Appl. Spectrosc.* **41** 396–401 (in Russian)

4.166 Zasavitsky I I, Kosichkin Yu V, Nadezhdinskii A I, Stepanov E V, Tishchenko A Yu and Shotov A P 1984 *Sov. J. Quant. Electron.* **11** 2443–51 (in Russian)

4.167 Giorgianni S, Visinoki R, Gambi A, Ghersetti S, Restelli G and Cappellani F 1983 *J. Mol. Spectrosc.* **101** 245–57

4.168 Yamada K, Best R and Winnewisser G 1983 *Z. Naturf.* **A38** 1296–308

4.169 Kuze H, Amano T and Shimizu T 1982 *J. Chem. Phys.* **77** 714–22

4.170 Maki A G and Wells J S 1984 *J. Mol. Spectrosc.* **108** 17–30

4.171 Bürger H and Schulz P 1983 *J. Mol. Spectrosc.* **102** 160–73

4.172 Bürger H, Schippel G, Ruoff A, Essig H and Cradock S 1984 *J. Mol. Spectrosc.* **106** 349–61

4.173 Yevseyev A V, Krivtsyn V M, Kuritsyn Yu A, Makarov A A, Puretzky A A, Ryabov E A, Snegirev E P and Tyakht V V 1984 *Sov. Phys.—JETP* **87** 111–24 (in Russian)

4.174 Mori Y, Nakagawa T and Kuchitsu K 1984 *J. Mol. Spectrosc.* **104** 388–401

4.175 Herbin P, Blomquet C, Walraud J, Courtoy C P and Fayt A 1984 *J. Mol. Spectrosc.* **104** 262–70

4.176 Pasternack L and McDonald J R 1984 *J. Mol. Spectrosc.* **108** 143–52

4.177 Weber W H and Maker P D 1982 *J. Mol. Spectrosc.* **93** 131–53

4.178 Dubs M, Harradine D, Schweitzer E, Steinfeld J I and Patterson C 1982 *J. Chem. Phys.* **77** 3824–39

4.179 Baldacchini G, Marchetti S and Montelatici V 1982 *J. Mol. Spectrosc.* **91** 80–6

4.180 Apatin V M, Krivtsun V M, Kuritsyn Yu A, Makarov G N and Pak I 1983 *Opt. Commun.* **47** 251–6

4.181 Baronov G S, Britov A D, Bronnikov D K, Karavaev S M and Kurbatov L N 1984 *Sov. J. Quantum Electron.* **11** 371–4 (in Russian)

4.182 Baronov G S, Bronnikov D K, Zasavitsky I I, Karavaev S M, Razumov A S and Shotov A P 1984 *Opt. Spectrosc.* **56** 5–7 (in Russian)

4.183 Krohn B J and Kim K C 1982 *J. Chem. Phys.* **77** 1645–8

4.184 Takami M, Oyama T, Watanabe T, Namba S and Nakane R 1984 *Jpn. J. Appl. Phys.* **23** 288–90

4.185 Susskind J, Reuter D, Jennings D E, Daunt S J, Blass W E and Halsey G W 1982 *J. Chem. Phys.* **77** 2728–44

4.186 Henry L, Valentin A, Lafferty W J, Hougen J T, Malathi Devi V, Das P P and Narahari Rao K 1983 *J. Mol. Spectrosc.* **100** 260–89

4.187 Zhu Qingshi, Francisco J S and Steinfeld J I 1982 *J. Mol. Spectrosc.* **92** 257–65

4.188   McDowell R S, Galbraith H W, Cantrell C D, Nereson N G and Hinkley E D 1977 *J. Mol. Spectrosc.* **68** 288–98

4.189   Jensen R J, Marinuzzi J G, Robinson C P and Rockwood S D 1976 *Laser Focus* **12**(5) 51–63

4.190   McDowell R S, Radziemski L J, Flicker H, Galbraith H W, Kennedy R C, Nereson N G, Krohn B J, Aldridge J P, King J D and Fox K 1978 *J. Chem. Phys.* **69** 1513–21

4.191a Patterson C W, McDowell R S, Nereson N G, Begley R F, Galbraith H W and Krohn B J 1980 *J. Mol. Spectrosc.* **80** 71–85

4.191b McDowell R S, Reisfeld M J, Galbraith H W, Krohn B J, Flicker H, Kennedy R C, Aldridge J P and Nereson N H 1980 *J. Mol. Spectrosc.* **83** 440–50

4.192   Alimpiev S S, Zasavitsky I I, Karlov N V, Kosichkin Yu V, Kryukov P V, Nabiev Sh Sh, Nadezhdinskii A I, Sartakov B G and Shotov A P 1980 *Sov. J. Quantum Electron* **7** 1885–94 (in Russian)

4.193   Glushkov M V, Kosichkin Yu V, Nadezhdinskii A I, Zasavitsky I I, Shotov A P, Gerasimov G A and Fomin V V 1980 *Sov. J. Quantum Electron.* **7** 908–11 (in Russian)

4.194   Baronov G S, Britov A D, Karavaev S M, Karchevskii A I, Kulikov S Yu, Merzlyakov A V, Sivachenko S D and Scherbina Yu I 1981 *Sov. J. Quantum-Electron* **8** 1573–6 (in Russian)

4.195a Olson M L, Grieble D L and Griffiths P R 1980 *Appl. Spectrosc.* **34** 50–5

4.195b Grieble D L, Olson M L, Sun J N-P and Griffiths P R 1980 *Appl. Spectrosc.* **34** 56–60

4.196a Dang C, Reid J and Garside B K 1983 *Appl. Phys.* B **31** 163–72

4.196b Dang C, Reid J and Garside B K 1983 *IEEE J. Quantum Electron.* **QE-19** 755–64

4.197a Rolland C, Reid J, Garside B K, Morrison H D and Jessop P E 1984 *Appl. Opt.* **30** 87–93

4.197b Sinclair R L, Reid J, Garside B K, Rolland C and Morrison H D 1984 *J. Opt. Soc. Am.* B **1** 439–40

4.198   Rolland C, Reid J and Garside B K 1984 *Appl. Phys. Lett.* **44** 725–7

4.199   Hirota E 1980 in *Chemical and Biochemical Applications of Lasers* vol. 5 ed C B Moore (New York: Academic) chap. 2

4.200   Davies P B and Russell D K 1980 *J. Mol. Struct.* **60** 201–4

4.201   Davies P B and Hamilton P A 1982 *J. Chem. Phys.* **76** 2127–8

4.202   Kawaguchi K, Yamada C, Hamada Y and Hirota E 1981 *J. Mol. Spectrosc.* **86** 136–42

4.203   Yamada C, Nagai K and Hirota E 1981 *J. Mol. Spectrosc.* **85** 416–26

4.204   Podolske J R and Johnston H S 1983 *J. Chem. Phys.* **79** 3633–8

4.205   McKellar A R W, Yamada C and Hirota E 1983 *J. Mol. Spectrosc.* **97** 425–9

4.206   Butler J F, Kawaguchi K and Hirota E 1983 *J. Mol. Spectrosc.* **101** 161–6

4.207a Menzies R T, Margolis J S, Hinkley E D and Toth R A 1977 *Appl. Opt.* **16** 523–5

4.207b Margolis J S, Menzies R T and Hinkley E D 1978 *Appl. Opt.* **17** 1680–2

4.208   Rogowski R S, Bair C H, Wade W R, Hoell J M and Copeland G E 1978 *Appl. Opt.* **17** 1301–2

4.209   Butler J E, Kawaguchi K and Hirota E 1984 *J. Mol. Spectrosc.* **104** 372–9

4.210   Davies P B and Rothwell W J 1983 *Proc. R. Soc.* A **389** 205–12

4.211   Davies P B, Hamilton P A and Okumura M 1981 *J. Chem. Phys.* **75** 4294–7

4.212   Matsumura K, Kawaguchi K, Nagai K, Yamada C and Hirota E 1980 *J. Mol. Spectrosc.* **84** 68–73
4.213   Anacona J R, Davies P B and Hamilton P A 1984 *Chem. Phys. Lett.* **104** 269–71
4.214   Engo Y, Nagai K, Yamada C and Hirota E 1983 *J. Mol. Spectrosc.* **97** 213–19
4.215   Kawaguchi K, Hirota E and Yamada C 1981 *Mol. Phys.* **44** 509–28
4.216   McKellar A R W, Yamada C and Hirota E 1983 *J. Chem. Phys.* **79** 1220–3
4.217   Davies P B, Lewis-Bevan W and Russell D K 1981 *J. Chem. Phys.* **75** 5602–8
4.218   Davies P B, Hamilton P A, Elliott J M and Rice M J 1983 *J. Mol. Spectrosc.* **102** 193–203
4.219   Nagai K, Yamada C, Endo Y and Hirota E 1981 *J. Mol. Spectrosc.* **90** 249–72
4.220   Nagai K, Endo Y and Hirota E 1981 *J. Mol. Spectrosc.* **89** 520–7
4.221a Buchanan J W, Thrush B A and Tyndall G S 1983 *Chem. Phys. Lett.* **103** 167–8
4.221b Zahniser M S and Stanton A C 1984 *J. Chem. Phys.* **80** 4951–60
4.222   Yamada C and Hirota E 1984 *J. Chem. Phys.* **80** 4694–700
4.223a Davies P B, Handy B J and Russell D K 1979 *Chem. Phys. Lett.* **68** 395–8
4.223b Davies P B, Hamilton P A, Lewis-Bevan W and Russell D K 1984 *Proc. R. Soc.* A **392** 445–55
4.223c Davies P B and Hamilton P A 1984 *Proc. R. Soc.* A **393** 397–408
4.224   Yamada C, Hirota E and Kawaguchi K 1981 *J. Chem. Phys.* **75** 5256–64
4.225   Yamada C and Hirota E 1983 *J. Chem. Phys.* **78** 1703–11
4.226   Hirota E and Yamada C 1982 *J. Mol. Spectrosc.* **96** 175–82
4.227   Yamada C and Hirota E 1983 *J. Chem. Phys.* **78.**669–71
4.228   Gudeman C S, Begemann M H, Pfaff J and Saykally R J 1983 *Phys. Rev. Lett.* **50** 727–31
4.229   Haese N N and Oka T 1984 *J. Chem. Phys.* **80** 572–3
4.230   Foster S C, McKellar A R W and Sears T J 1984 *J. Chem. Phys.* **81** 578–9
4.231   Davies P B, Hamilton P A and Rothwell W J 1984 *J. Chem. Phys.* **81** 1598–9
4.232   Foster S C and McKellar R W 1984 *J. Chem. Phys.* **81** 3424–8
4.233   Lemoine B and Destombes J L 1984 *Chem. Phys. Lett.* **111** 284–7
4.234   Haese N N, Fu-Shin Pan and Oka T 1983 *Phys. Rev. Lett.* **50** 1575–8
4.235   Eng R E, Petagna G and Nill K W 1978 *Appl. Opt.* **17** 1723
4.236   Streit G E, Wells J S, Fehsenfeld F C and Howard C J 1979 *J. Chem. Phys.* **70** 3439–45
4.237   Kim K C, Reisfeld M J and Person W B 1980 *J. Mol. Struct.* **60** 205–13
4.238a Thrush B A and Tyndall G S 1982 *J. Chem. Soc. Faraday Trans.* **78** 1469–75
4.238b Thrush B A and Tyndall G S 1982 *Chem. Phys. Lett.* **92** 232–5
4.239   Laguna G A and Baughcum S L 1982 *Chem. Phys. Lett.* **88** 568–71
4.240   Kleinermanns K and Wolfrum J 1983 *Laser Chemistry* **2** 339–59
4.241   Koloshnikov V G, Kuritsyn Yu A, Pak I, Ulitskiy N I, Kharlamov B M, Britov A D, Zasavitsky I I and Shotov A P 1980 *Opt. Commun.* **35** 213–17
4.242   Makarov A A and Tyakht V V 1985 *Opt. Commun.* **54** 270–2
4.243   Hodgkinson D P and Taylor A J 1984 *Opt. Commun.* **50** 214–18
4.244   Tosa V, Deac I, Mercea D, Gulácsi Zs and Mercea V 1985 *Appl. Phys.* B **36** 55–7
4.245   Chu J O, Wood C F, Flynn G W and Weston R E Jnr 1984 *J. Chem. Phys.* **80** 1703–4
4.246   Hinkley E D (ed) 1976 *Laser Monitoring of the Atmosphere* (Berlin: Springer)
4.247   Reid J, Garside B K and Shewchun J 1979 *Opt. Quantum Electron* **11** 385–91

4.248   El-Sherbiny M, Ballik E A, Shewchun J, Garside B K and Reid J 1979 *Appl. Opt.* **18** 1198–203

4.249   Reid J, El-Sherbiny M, Garside B K and Ballik E A 1980 *Appl. Opt.* **19** 3349–54

4.250   Ku R T, Hinkley E D and Sample J O 1975 *Apr¨. Opt.* **14** 854–61

4.251   Cappellani F, Melandrone G and Restelli G 1977 in *Lasers in Chemistry* (Amsterdam: Elsevier) pp. 61–9

4.252   Chaney L W, Rickel D G, Russwurm G M and McClenny W A 1979 *Appl. Opt.* **18** 3004–9

4.253   Cassidy D T and Reid J 1982 *Appl. Opt.* **21** 1185–90

4.254   Webster C R and Grant W B 1983 *Appl. Opt.* **22** 1952–4

4.255a Menzies R T, Webster C R and Hinkley E D 1983 *Appl. Opt.* **22** 2655–64

4.255b Webster C R and Menzies R T 1984 *Appl. Opt.* **23** 1140–2

4.256   Lehmann B, Wahlen M, Zumbrunn R, Oeschger H and Scnell W 1977 *Appl. Phys.* **13** 153–8

4.257   Sams R L and DeVoe J R 1983 *38th Symposium on Molecular Spectroscopy, Abstracts* (Columbus: The Ohio State University)

4.258   Labrie D and Reid J 1981 *Appl. Phys.* **24** 381–6

4.259   Hanson R K 1983 *Proc. SPIE* **438** 75–83 and references therein

4.260   Hanson R K 1980 *Appl. Opt.* **19** 482–4

4.261   Sell J A 1983 *Proc. SPIE* **438** 67–74

4.262a Forrest G and Vilcins G 1980 *Appl. Spectrosc.* **34** 418–19

4.262b Forrest G T 1980 *Appl. Opt.* **19** 2094–6

4.262c Vilcins G, Harward C N, Parrish M E and Forrest G T 1983 *Proc. SPIE* **438** 48–54

4.263   Mucha J A 1983 *Proc. SPIE* **438** 55–60

4.264   Stanton A C and Kolb C E 1980 *J. Chem. Phys.* **72** 6637–41

4.265   Pokrowsky P and Herrmann W 1981 *Proc. SPIE* **286** 33–8

4.266   Schweid A N and Hardman B B 1983 *Proc. SPIE* **438** 61–6

4.267   Wall D L 1984 *Laser Anal. Lett.* No. 1

4.268a Lambert D K 1983 *Phys. Rev. Lett.* **50** 2106–9

4.268b Lambert D K 1983 *Proc. SPIE* **438** 158–64

4.269   Ohsawa A, Honda K, Ohkawa S and Ueda R 1980 *Appl. Phys. Lett.* **36** 147–8

4.270   Vansteenkiste T H, Faxvog F R and Roessler D M 1981 *Appl. Spectrosc.* **35** 194–6

4.271   Gough T E and Scoles G 1984 in *Laser Spectroscopy V* ed A R W McKellar, T Oka and B P Stoicheff (Berlin: Springer) pp. 337–40

4.272   Webster C R and Menzies R T 1983 *J. Chem. Phys.* **78** 2121–8

4.273   Glenar D A 1983 *Proc. SPIE* **438** 125–36

4.274   Weber W H and Terhune R W 1983 *J. Chem. Phys.* **78** 6437–46

5 Laser Opto-acoustic Spectroscopy in Chromatography

V P Zharov

5.1 Introduction

The opto-acoustic (OA) method is based on a sequence of several physical processes. These are optical excitation of the medium, nonradiative relaxation of excited particles, heating of the medium, formation of acoustic vibrations and their detection.

Before the advent of lasers the OA method was most widely used for quantitative and qualitative analysis of gas media using nondispersive infrared systems and for measuring weak radiation using nonselective OA detectors.

With the advent of lasers with their unique properties the scope of the OA method has been widened considerably primarily due to a large increase of sensitivity. Up to the present the efficiency of the OA method has been demonstrated in the study of both high- and weak-absorption media of different aggregate states including solid, liquid and gaseous phases in a wide temperature range with the use of radiation of different spectral ranges, from ultraviolet to radiofrequency. The applications of the OA method with laser sources include high-resolution spectroscopy of weak-absorption media, detection of ultra-low concentrations of impurities, investigation of nonlinear effects, analysis of surfaces, studies of photoactive media, microscopy, measurement of thermodynamic parameters of matter, etc. These applications are covered in a number of review papers and monographs [1.15, 1.43–1.44, 5.1–5.4].

One of the most important and valuable analytical applications of the OA method is the detection of impurities. In practice the highest sensitivity of the OA method can be reached, as a rule, when there is just one type of compound to be analysed in a nonabsorbing or weakly absorbing medium. If several types of compounds are analysed in a mixture, their detection threshold may be impaired by possible overlapping of absorption bands of different compounds. Therefore in analysis of multicomponent mixtures it is very important to find methods to increase the selectivity of the OA method.

The natural solution to this problem is a combination of the powerful and highly developed technique of chromatography for separation of a multicomponent mixture and the laser spectroscopy technique, particularly the laser opto-acoustic method for sensitive detection of separated molecular components. A review of this type of laser analytical technique is the subject of this chapter.

5.2   Analytical opto-acoustic spectroscopy of simple mixtures

This section considers briefly the technique of laser OA spectroscopy in the analysis of gas and condensed media and its possibilities as the impurities in simple mixtures are detected.

5.2.1   General measuring scheme

Figure 5.1 shows a generalised block diagram of a laser OA spectrometer. The basic elements of this scheme are a tunable laser, a modulator, an optical system and the OA cell with an acoustic transducer. The OA signal from the transducer enters the phase-sensitive amplifier and then the recording unit. In linear absorption the OA signal amplitude is proportional to the laser radiation power. Thus, to eliminate the effect of radiation fluctuations on the accuracy of measurement, it is necessary either to stabilise the laser parameters or, more simply, to apply a second reference channel for independent recording of radiation power with subsequent normalisation of the OA signal to it. As the laser frequency is scanned, the OA spectrum, which under definite conditions is identical to the standard absorption spectrum, is measured. The type of admixture absorbing the radiation is identified from the OA spectrum and then the concentration of this admixture is determined using the calibration data.

The formation of OA signals in the mixture occurs as a result of modulation of the parameters of the radiation or the medium itself. Amplitude and frequency modulations of the laser radiation are widely used. Amplitude modulation is usually realised with a mechanical or electro-optical modulator. There is no need for amplitude modulation under pulsed laser operation. Frequency modulation is realised by periodic tuning of the laser wavelength or by Stark and Zeeman modulation of absorption lines.

The optical system is designed to provide the transfer of radiation energy

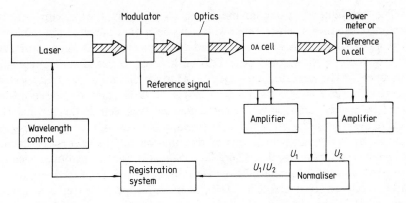

Figure 5.1 Generalised structural scheme of a laser OA spectrometer.

from the laser to the medium. The sensitivity of the OA method increases with the radiation intensity in the OA cell so that, with the laser parameters fixed, improvement can be obtained by focusing the radiation, increasing the number of passages of the beam through the OA cell or placing the OA cell inside the laser cavity. The OA cells can be classified as resonant or nonresonant. The difference is that in the first case acoustic resonances are formed in the measuring chamber.

The electronic treatment of OA signals is achieved using the technique of phase-sensitive detection since this provides the highest sensitivity.

5.2.2 Opto-acoustic gas spectroscopy

(a) *Technique of opto-acoustic gas spectroscopy*
According to the spectral range of the radiation sources used, different energy levels can be excited in gases: electronic levels in the UV and visible spectral regions, vibrational–rotational levels in the IR region and purely rotational ones in the submillimetre range. The IR spectral range, approximately 2 to 15 μm, which contains intense absorption bands of most molecules to be detected, holds much promise for analytical purposes. The laser radiation power for the OA method must be rather high, at least 10^{-2}–1 W, so that a high sensitivity of OA spectroscopy may be realised. The radiation sources most widely used in OA gas spectroscopy are low-pressure gas lasers: the HF laser (tuning range from 2.7 to 2.9 μm), He–Ne laser (3.39 μm), DF laser (3.8–4 μm), CO laser (5.2–6 μm) and CO_2 laser (9.2–10.8 μm) as well as isotopic modifications of these molecules. Such lasers usually generate a number of lines spaced at several tenths to several units of cm^{-1}. The generation frequency is usually tuned by turning the intractivity diffraction grating.

To detect some molecules, like NO_2, SO_2 and others, one can use lasers of UV and visible spectral ranges, for example dye lasers.

The technique of OA measurement consists of placing the gas sample to be analysed into the OA cell chamber which has been pre-evacuated or foreblown with a neutral gas. It is sometimes called the spectrophone or OA detector. It is also possible to allow continuous flow of the analysing medium through the chamber. At the same time the modulated laser radiation of required spectral range is passed through the chamber. When the laser radiation line coincides with the absorption band of the molecules under detection, the gas absorbs part of the laser radiation power, is heated and induces an increase in pressure, that is, an OA signal, which carries information on the absorption factor (concentration) of molecules. Then the OA signal is transformed to an electric signal with the use of a microphone placed in the OA cell. The best operation in laser OA cells, as far as simplicity of design and high operational efficiencies are concerned, has been demonstrated for the most part by capacitor and electret microphones of plane geometry. The sensitive element in such microphones is a thin (from 1 to 10 μm) elastic membrane made of Mylar, Teflon or a metal. As the membrane deforms under pressure, the capacity between the plates comprised of the surface of a fixed electrode and the metallised surface of the membrane, internal or sometimes external with respect to the former, undergoes changes. Most practical tasks can be solved if commercial types of microphone one or half an inch in diameter and with sensitivity ranging from 5 to 50 mV Pa^{-1} are used. However, the maximum possible sensitivity is provided by special microphones with optimised characteristics, particularly when the gas elasticity matches with the elasticity of the membrane.

The change of microphone capacity can be recorded by using two basic types of electronic circuit: the microphone connected directly to the preamplifier in series with a high-resistance resistor and polarising voltage; or a circuit with frequency conversion. The first circuit has received wide use in laser OA spectroscopy together with high-resistance (up to 10^{10} Ω), low-noise, field-effect transistor preamplifiers.

The OA cell is made in the form of a cylindrical metal chamber, to the ends of which are fitted transparent windows. Typical dimensions of the chambers of nonresonant OA cells are: internal diameter 5 to 10 mm and length 5 to 20 cm. The internal diameter of some resonant cells is as large as 10 cm. The modulation frequencies in nonresonant OA cells are from one to hundreds of hertz. The acoustic resonance frequencies in resonant OA cells are generally of the order of several kilohertz.

The theoretical sensitivity limit of the OA method can be reached by increasing the radiation intensity until the absorption in the medium is saturated and by restricting it below with the Brownian noise threshold. With respect to the absorption factor this limit is about 10^{-13} cm^{-1} [5.4] which corresponds to a detection limit of concentration for many molecules of 10^{-1} to 10^{-2} PPT. However, it is difficult to realise such a high sensitivity in practice due to different backgrounds and noise.

(b)   *Sources of background and methods of its discrimination*

The background signal level in most cases can be estimated in the form of equivalent background absorption in a medium, κ_B, the OA signal of which is equal to that of the background in amplitude. Independently of their physical nature and time character all background signals can be classified in this way [5.4, 5.5].

Random nonoptical background connected with the effect of electromagnetic, thermal and acoustic noise uncorrelated with the process of radiation modulation. The source of such a background may be the usual acoustic noise in the laboratory, the gas flow turbulence through the OA cell, etc.

This type of background is excluded by using standard methods of thermo-, acoustic-, electro- and vibro-isolation of the OA cell, by imparting a laminar character to the gas flow and by choosing the modulation frequency on the basis of maximisation of the signal-to-noise ratio.

Correlated nonoptical background caused by the process of modulation but not related to the radiation's passing through the OA cell. The source of such a background may be the laser itself, where electrical and acoustic noise coincident in phase with the laser pulses is present, and also the modulator.

The effect of the electrical noise caused by the laser can be removed by time or phase selection by introducing the optical delay line, for example, providing additional transmission of the radiation before entering the OA cell through a long optical fibre or an acoustic delay line. In some cases it is possible to introduce additional low-frequency 'noiseless' modulation with suppression of the noise of high-frequency modulation at the second stage of synchronous detection (see [5.5]).

Correlated optical instrumental background arising as the radiation passes through the OA cell but not related to the absorption in the medium. This background may be caused by radiation absorption by the walls and windows of the OA cell chamber as well as by the scattered radiation falling onto the acoustic transducer. The windows and walls give rise to a background due to their thermal expansion as well as to the transfer of heat to the medium. Radiation falling onto the elements of the OA cell may be caused by misadjustment, touching of the Gaussian beam tail, diffraction and scattering by the windows, modulator, etc, reflections in the windows, Rayleigh and Mie scattering and by fluorescence. The background signals in the transducer arise due to thermal deformations of the sensitive element or due to the pyroeffect.

The background from the windows can be suppressed almost completely with pulsed excitation of the medium on the basis of spatial and time selection because the background OA signals reach the transducer with a time delay relative to the absorption OA signals in the medium. It is rather efficient to introduce buffer zones with diaphragms in the path of the background OA signals from the windows [5.6–5.9]. Such diaphragms in resonant OA cells reduce the effect of window background by about two orders [5.6, 5.10].

Absorption by the walls is decreased when they are polished or made of an optically transparent material. The fraction of radiation reflected from the windows can be reduced if they are placed at the Brewster angle and fixed to flanges made of optically transparent material. It is also efficient to use differential schemes with two OA cells [5.11].

The value $\kappa_B \simeq 10^{-9}$ cm^{-1} obtained under time selection and monopulsed irradiation of the medium [5.12] and $\kappa_B = 3.3 \times 10^{-9}$ cm^{-1} obtained with the use of a differential scheme with two identical OA cells and with the medium being irradiated by CW laser modulated radiation [5.11] have so far been the best results on instrumental background suppression.

Correlated optical background from the medium due to the radiation's passing through the medium. The resonant or quasiresonant background in this case may be caused by extraneous impurities, the effect of intense line wings, hot bands, nonlinear effects, etc. The nonresonant background may be conditioned by electrostriction, optical breakdown and other effects showing themselves at high intensities. The suppression of such a background is the most difficult problem in OA measurements. Some possible solutions are given below.

Table 5.1 shows approximate levels of background OA signals in the form of the parameter κ_B.

Table 5.1 Background level in gas.

Sources of background signals	Background level, κ_B (cm^{-1})	Notes
Gas flow turbulence	10^{-7}	For flow rate 50 cm^3 min^{-1}
Absorption in windows	10^{-6}–10^{-7}	For NaCl and ZnSe windows on CO_2 laser lines
Reflection from the output window	10^{-7}	For Brewster windows
Scattering in windows and in gas	10^{-8}	For CO_2 lasers
Scattered radiation falling on the microphone	10^{-7}	For cylindrical capacitor microphone

(c) *Detection limits*

To detect just one type of molecule in a 'pure' medium it is necessary to choose a laser having a wavelength coinciding with the intense absorption band of the molecule and to suppress the instrumental background. Some information on the limits of detection of some molecules using the OA method can be obtained from table 5.2 where the results of some experiments in this field are listed.

Table 5.2 Limits of OA detection of molecules.

Molecules	Threshold of detection (PPB)	Laser and wavelength, λ (μm)	Background level, κ_B (cm^{-1})	Type of OA cell	Reference
SO$_2$	0.12	Dye, 0.3	7.5×10^{-9}	Cylindrical, multi-pass, resonant with azimuthal modes	[5.13]
NO$_2$	10	Dye, 0.598	—	Cylindrical, resonant with longitudinal modes	[5.14]
HF	10^3	HF, 2.7	10^{-9}	Nonresonant with time selection	[5.12]
CH$_4$	10^2	He–Ne, 3.39	5×10^{-7}	Differential scheme with nonresonant cells	[5.15]
CO	1.5×10^2	CO, 4.75		Resonant with radial modes	[5.16]
NO	0.1	Spin–flip, 5.3		Nonresonant, rectangular cross section, six microphones	[5.17]
SF$_6$	10^{-2}	CO$_2$, 10.57		Resonant, H geometry	[5.18]

Besides its high sensitivity (up to 10^{-2}–1 PPB), the OA method is also characterised by such merits as a large dynamic range (up to 10^5), a small volume of the sample to be analysed (up to 0.5–3 cm^3 under normal conditions), a high speed of response (tens of seconds) and a high accuracy of measurement (up to 5–15%). In addition, the use of two-channel schemes with normalisation of the OA signal and calibration of OA cell sensitivity directly in the process of measurement by means of a stable acoustic radiator built into the chamber enables the accuracy of OA measurement to be increased by up to 3% [5.15]. The gas pressure in the OA cell is usually in the range 10–760 Torr. Below this pressure the sensitivity of the OA method begins to decrease significantly [5.19].

One of the main applications of the OA method may be to control the impurity of different gases including Xe, Kr, N$_2$, Ar and others, as well as to calibrate binary gas mixtures in the low-concentration range, up to 1 PPB.

(d) *Spectral methods for molecular identification*
The spectral methods for molecular identification are most effective in the presence of narrow isolated lines or a 'fine' structure in the absorption spectra

of the molecules to be detected and, of course, when this structure can be resolved by the methods of laser OA spectroscopy. This situation is specific for OA analysis of relatively simple molecules, like NO, NH_3, CO, C_2H_4, and their isotopic modifications. For example, the OA detection of $^{13}C^{16}O_2$ molecules mixed with $^{12}C^{16}O_2$ using the $^{13}C^{16}O_2$ laser instead of the $^{12}C^{16}O_2$ laser using the scheme with two spectrophones [5.20] makes it possible to increase the detection selectivity of $^{13}C^{16}O_2$ even without spectral scanning. If the OA spectrum is measured with a resolution of 0.05 cm^{-1} using a high-pressure tunable CO_2 laser, $^{15}NH_3$ molecules mixed with $^{14}NH_3$ and HDS mixed with H_2S in their natural concentrations, 0.3 and 0.045% respectively, may be selectively detected [5.4].

Spectral scanning is very useful for OA control of atmospheric pollution in the presence of a background from the molecules H_2O and CO_2 as well as from other impurities in the air. In [5.21], for example, the measurement of the OA spectrum of air at a pressure of 500 Torr in the spectral range 1815–25 cm^{-1} by a spin–flip laser made it possible to detect NO molecules in a concentration of up to 0.1 PPM in the presence of H_2O. The spectral identification of SF_6 molecules against a background of absorption of CO_2 and H_2O was realised in [5.22] at diffusion of SF_6 in room air using a CO_2 laser and a through-flow OA cell.

The presence of a 'fine' structure in the absorption spectra of some molecules makes it possible in some cases to use a relatively small range of spectrum for identification, e.g. from 0.1 to 10 cm^{-1}.

The identification of polyatomic molecules with slightly specific absorption spectra without a fine structure can be carried out, by analogy with classical absorption spectroscopy, with the use of the characteristic vibrations of the groups C–H, C–O, and others. For this purpose moderate resolution OA spectra in a wide spectral region or, at least, in separate ranges that can be realised by using lasers with a relatively wide tuning range, are needed. Several lasers of this type may be required. In [5.23], for example, it has been shown that it is possible to selectively detect nitroglycerine in the air in a concentration of up to 1–10 PPB, as well as dinitrotoluene and ethylene glycol dinitrate, by measuring the OA spectrum in the 6 μm region with a CO laser (NO_2 group) and in the 11 μm region with a $^{13}CO_2$ laser (O–H group).

Another approach to selective OA detection of polyatomic molecules may be based on the use of a fine structure in the spectra of sub-Doppler absorption of these molecules produced by the methods of saturation OA spectroscopy. In practice this approach has been supported by the results of work described in [5.24] in which the method of intermodulation OA spectroscopy of absorption saturation with a CO_2 laser was used to measure the sub-Doppler structure of CH_3OH molecules over the pressure range 20–100 mTorr with a sufficient sensitivity margin. Such a technique holds promise for selective analysis of hydrocarbon mixtures with the use of a tunable He–Ne laser near $\lambda = 3.39$ μm (see §1.5.2 and [1.90]).

Laser frequency modulation (FM) (for example by means of periodic displacement of one of the mirrors of the laser cavity) is most effective for suppressing a spectral nonselective background [5.25]. The selectivity in this case may be as high as 10^5 to 10^6.

To obtain information on the concentration of different types of molecules, when several absorption bands of these molecules overlap at the same time, one can use the usual technique of measurement of the absorption factors in the mixture for a number of laser lines with subsequent calculation of the concentrations to be sought from the given (standard) absorption cross sections. To apply this technique it is necessary to have *a priori* information on the qualitative composition of the mixture etc.

In some specific cases the selectivity of OA spectroscopy can be increased due to the difference between some physical (nonspectral) parameters of the molecules, such as saturation intensity, multiphoton absorption [5.26], nonradiative relaxation times [5.27], the dependence of the absorption cross section on temperature [5.28], or the sensitivity to electric field. For example, the modulation of the electric field in the OA cell as nonmodulated laser radiation runs through it makes it possible to increase the detection selectivity of molecules with a high Stark effect. In [5.29] use of this technique with a CO_2 laser at $\lambda = 9.07$ μm allowed an increase in the detection sensitivity of NH_3 molecules against the background of C_2H_4 of about two orders. In this case the window background was also suppressed.

A similar effect can be attained in detecting molecules active to the Zeeman effect as the magnetic field is modulated in the OA cell. The isotropic background from the windows can also be reduced by polarisation modulation of radiation with simultaneous application of a static magnetic field to paramagnetic molecules in the OA cell, for example NO molecules [5.30].

The method of OA Raman scattering spectroscopy realised as a medium is exposed to two-frequency radiation and the frequency difference coincides with the frequency of the Raman-active transition of the molecules is highly promising for selective analysis. The threshold of molecular detection by this method is still rather low—several PPM for the CH_4 and CO_2 molecules [5.31]. It is still widely applicable, however, since with one set-up comprising two dye lasers it permits the analysis of many molecules including such nonpolar molecules as H_2, N_2 and O_2.

The cooling of the gas to its liquefaction point and the recording of OA signals in the condensed phase make it possible to realise cryogenic OA spectroscopy [5.32]. Since the vibrational–rotational structure of molecules degenerates into separate bands, 5–10 cm^{-1} in width, this method holds much promise for selective spectral analysis of mixtures of such molecules as CCl_2F_2, C_2H_3Cl and so on. The concentration sensitivity of such a method is rather high, up to 10^{-1}–10^{-2} PPB.

5.2.3 *Opto-acoustic spectroscopy of liquids*

The direct detection of acoustic vibrations in a medium under analysis is the simplest and most sensitive method among different methods of detection of the OA effect in liquids [5.2, 5.4, 5.33]. In OA analysis the liquid is placed into a measuring cell composed of a closed chamber with optical windows and an acoustic transducer inside. OA signals in liquids can be detected most efficiently with piezoelectric transducers. This is due to the fact that such transducers have sufficient sensitivity for many wide-band applications (up to 10 MHz) and, more important, are in good acoustic agreement with liquid media.

The most widely used OA cell design is one in which the transducer is placed directly in the liquid under analysis and thus direct contact of the liquid with the transducer's surface is achieved. It is advisable in this case to use either a tiny cylindrical transducer with an operating end face or a hollow cylindrical transducer inside which laser radiation runs.

To avoid scattered light and possible environmental pollution the transducer should sometimes be placed outside the chamber with the liquid, say, on the outside of the chamber wall. To ensure a good acoustic transducer–wall contact, an immersion medium, for example in the form of a special lubrication, is introduced. The weak electric signals from the piezoelectric transducer can be amplified by using low-noise field-effect transitor preamplifiers.

The highest sensitivity of OA spectroscopy of liquids can be achieved under pulsed laser operation. With a laser pulse energy equal to 10^{-3} J and a repetition rate several tens of hertz, the sensitivity of the method makes it possible to measure absorption of about 10^{-6} cm^{-1} with an accumulation time of several seconds. With modulation of CW radiation the sensitivity ranges from 10^{-5} to 10^{-6} cm^{-1} with a radiation power of about 1 W.

The high sensitivity of the OA method enables it to be used for detecting some nonfluorescent impurities in extremely small concentrations. The results of some experiments in this field are listed in table 5.3. Compared with gases, it is rather difficult to eliminate the background absorption in solvents when analysing impurities in liquids ($\kappa_B = 10^{-4}$–10^{-6} cm^{-1}). This problem can be partially solved by using differential schemes, or FM operation. In [5.34], for example, discrete FM was applied by alternating Ar laser radiation at a frequency of 700 Hz between the two wavelengths by 488 and 514 nm. Such modulation was accomplished by spatial dispersion of the multifrequency laser radiation with a dispersing prism, and alternating mechanical modulation of the chosen pair of waves with their subsequent spatial realignment using the second prism. Such a scheme gave a decrease of the background signal in chloroform from 9×10^{-4} cm^{-1} to 2.2×10^{-5} cm^{-1} which corresponded to a detection threshold of β-carotene at the 514 nm line of about 0.08 ng cm^{-3} or 12 PPT with a radiation power of 0.7 W. Sensitivity of the same order (0.02 ng cm^{-3}) was obtained in the process of OA detection of cadmium in chloroform extracted from penicilline fungus [5.35]. The

Table 5.3 Concentration sensitivity of the OA method in analysis of impurities in liquids.

Detected component	Solvent	Laser and wavelength, λ (nm)	Limits of detection for $S/N = 1$		Reference
			Absorption (cm^{-1})	Concentration (ng/ml^{-1})	
β-carotene	Chloroform	Argon, 488 514.5	2.2×10^{-5}	0.08	[5.34]
Se	Chloroform	Argon, 488, 514.5	3.5×10^{-5}	15	[5.34]
Cd	Chloroform	Argon, 514		0.02	[5.35]
Bacterio-chlorophyll	Ethanol	N_2, 337	4×10^{-6}	1	[5.36]
Chlorophyll b	Ethanol	N_2, 337	4×10^{-6}	0.3	[5.36]
Cytochrome c	Water	N_2, 337	4×10^{-6}	30	[5.36]
Haematopor-phyrin	Ethanol	N_2, 337	4×10^{-6}	0.3	[5.36]
Vitamin B_{12}	Water	N_2, 337	4×10^{-6}	4	[5.36]
U(IV)	Water	Dye, 660	2×10^{-5}	$8 \times 10^{-7} mol l^{-1}$	[5.37]
U(VI)	Water	Dye, 414	2×10^{-5}	$10^{-6} mol l^{-1}$	[5.37]

measurements were performed with amplitude-modulated radiation of an Ar laser at a frequency of 185 Hz with a laser power of 0.7 W at a wavelength of 514 nm. The OA cell had a cylindrical piezotransducer. The solvent background was premeasured in the pure solvent and then the value obtained was electrically subtracted from the results obtained in measuring the samples. The calibrated characteristic turned out to be linear within three orders, approximately between 0.05 and 50 ng cm^{-3} Cd. On average, the sensitivy of such measurements is about two orders higher than that obtained with the conventional photometric technique. In [5.36] the detection thresholds are estimated for about 30 dyes and medicines with the use of the OA method and a pulsed N_2 laser with pulse energy of 1.3 mJ and pulse repetition frequency of 20 Hz. Some of these results are illustrated in table 5.3. The best limit—0.03 ng cm^{-3} or 30 PPT with $S/N = 1$—is obtained for protoporphyrin IX. These results show that the OA method is promising for medical and biological studies and particularly for research into the distribution of drugs in organisms.

One possible way of reducing the influence of background in such measurements is to use a differential scheme with two liquid cells, one of which is filled with a pure solvent and the second one with the solvent and the impurity to be detected. In [5.28], for example, where the OA cells were placed in parallel and the Ar laser radiation was directed into them alternately by an acousto-optic modulator, the background was reduced by a factor of 70. The same differential scheme with the OA cells in series and a pulsed dye layer was used in [5.37] to detect aqueous solutions of U(IV) and U(VI) in a concentration of up to 10^{-6} mol l^{-1}.

Analysis of more complex mixtures by OA spectroscopy is rather a difficult problem because the absorption bands of most impurities in liquids are rather wide at room temperature.

5.3   Laser opto-acoustic spectroscopy in gas chromatography

Chromatography is a method of separating complex mixtures into individual components based on the difference in equilibrium distribution of these components between two nonmixing phases: movable and fixed. In gas chromatography (GC) the movable phase is a gas, usually He, and the fixed phase is a liquid or a solid. The fixed phase is placed into a chromatographic column in the form of a long hollow tube made of metal, glass or quartz. The technique of gas-chromatographic analysis consists of pulsed introduction of a gas sample into the continuous flow of the carrier gas at the inlet of the column with subsequent detection of the separated components coming out of the column in series using a suitable detector. The recorded signal from the detector as a function of time is called the chromatogram. Individual compounds are usually identified by the time of their passage through the column, called the retention time. A gain in selectivity can be attained in this case by improving the separation of the compounds in the column by choosing the optimal fixed phase, the temperature in the column, the carrier gas rate, and so on. The limits of applicability of GC can be extended into the analysis of gases and volatile liquids provided their boiling points are no higher than 300–400 °C and their molecular weights no higher than 300. The theoretical questions of gas chromatography are covered in more detail in [5.39, 5.40], for example.

5.3.1   Features of detection in gas chromatography

Most gas-chromatographic detectors are nonselective, and the identification of substances in them is characterised by the retention time. This refers, in particular, to the widely used flame-ionisation (FI) detector that records the ionic current resulting from the combustion of organic substances in a hydrogen flame. Specific detectors (for example, electron-capture detectors) are highly selective to only a definite class of compounds (which contain phosphor or nitrogen) but are not selective within this class.

In the analysis of complex compounds with partially inseparable components identification by retention time alone is rather time-consuming and sometimes unreliable. Therefore in GC one also tries to use selective detectors which permit independent identification of substances by both retention time and some other parameters. A good illustration of this is chromato-mass spectrometry, when after the column the compounds pass into a mass spectrometer and are then analysed by their mass spectra. Independent identification of substances can be also carried out by their specific IR spectra.

The application of IR spectroscopy in GC began in the 1960s but remarkable progress has only recently been achieved. In the GC–IR spectrometer system there are three basic methods of analysis:

(i) capture of separate components in a trap with subsequent measurement of IR spectra under stationary conditions;

(ii) the stopping of the chromatographic process (the flow of carrier gas is interrupted) as the component to be analysed enters the absorption cell and

(iii) detection of the spectra directly in the flow of carrier gas after the column.

The third method is of most practical interest but places high requirements (increased sensitivity because of a small amount of substance, fast response and a small volume of detection to avoid chromatographic peak broadening) on the method of IR spectroscopy.

For illustration, the dispersion IR spectrometer used in [5.41] permitted the detection of spectra over the ranges 2.7–5 μm and 5–9 μm with a resolution of 20 cm^{-1} in 0.5 s. This sensitivity made it possible to analyse samples in amounts of 25–100 μg which, however, is not quite sufficient for solving most practical problems.

The advent of commercial IR Fourier spectrometers (FS) has considerably increased the potentialities of IR spectroscopy in GC. The advantage of these instruments is that they permit the measurement of spectra over a wide spectral region in a short time (up to 0.1 s) with a reasonable resolution (5–10 cm^{-1}) and sensitivity (100 ng and in some experiments up to 1 ng). The latter can be attained by using long (up to 50 cm) light-guide absorption cells of a small diameter (up to 0.5 mm) in combination with integration and smoothing of spectra by computers. Even though the FS–GC method is widely used in analysis of complex mixtures [5.42, 5.43], there is a practical demand for increasing the sensitivity of this method.

Further progress in the use of spectroscopy in GC is expected with the use of laser spectroscopy methods after the results of the first few experiments in this region. In [5.44] the method of spontaneous Raman scattering with excitation by an argon laser was used to detect substances directly in a free jet of carrier gas escaping from the open end of a capillary column. This method is characterised by high response, universality, a small volume of detection, but still has a low sensitivity of about 0.1–0.5 μg. In [5.45] the method of two-photon resonance ionisation by pulsed UV laser radiation with signal detection by a proportional ion counter was used for highly sensitive (0.3–50 pg) and selective detection of polycyclic aromatic hydrocarbons at the column outlet (see Chapter 7). Preliminary estimations for the applicability of the OA method in GC were performed in [5.46]. The detection thresholds for ethylene and

Table 5.4 Comparison of different gas-chromatographic detectors.

Detectors	Threshold absorption (cm^{-1})	Threshold concentration (g)	Dynamic range	Selectivity	Identification	Typical matter	Reference
Flame ionisation (FI)		$10^{-11}-10^{-12}$	10^7	No	No	Organic substances	[5.40]
Electron capturing		10^{-13}	10^2	High, invariable	No	Phosphor- and nitrogen-containing	[5.40]
Dispersion IR spectrometer	10^{-3}	$10^{-4}-10^{-5}$	10^3	Moderate, variable	Moderate	Absorbing in the IR spectrum	[5.41]
IR–Fourier spectrometers	$10^{-4}-10^{-5}$	$10^{-7}-10^{-8}$	10^4	Moderate, variable	High	Absorbing in the IR spectrum	[5.42, 5.43]
Mass spectrometry		$10^{-10}-10^{-11}$	10^4	High, variable	High	With different mass spectra of ions	[5.39]
Laser							
Spontaneous Raman scattering		10^{-7}		Moderate, variable	Moderate	Active in Raman scattering	[5.44]
Two-photon resonance ionisation		10^{-9}	10^4	High, variable	Moderate	With a low ionisation potential	[5.45]
Opto-acoustic	$10^{-7}-10^{-8}$	$10^{-10}-10^{-12}$	10^6	High, variable	Moderate	Absorbing in the IR spectrum	[5.46–5.48]

2-pentanon in the case of the CO_2 laser were 0.2 and 26 pg s^{-1} respectively. Table 5.4 gives results which enable one to compare the potentialities of different methods of detection in GC. It can be seen that the laser methods are the most sensitive of the spectral techniques.

5.3.2 Opto-acoustic spectrometer for gas chromatography

When an OA spectrometer operates as part of a chromatograph, a number of specific requirements are placed upon it. The most important requirements and the possible methods of fulfilling them are discussed below.

(a) General scheme of the spectrometer

The combined system 'gas chromatograph–laser opto-acoustic spectrometer' (GC–LOAS) is shown schematically in figure 5.2. Its principle of operation consists of the following. A gas sample is introduced into the flow of carrier gas (He) continuously flowing through the chromatographic column. At the outlet of this column the components of the mixture pass in the flow of He through the OA detector chamber. Intensity-modulated radiation of the corresponding laser is simultaneously passed through the same chamber. The absorption of the radiation by the components at any instant in the chamber generates acoustic vibrations, i.e. OA signals which are detected by a microphone. After the OA detector the flow of He carrying the components of the mixture goes into a nonselective detector FI and this allows a comparison of the potentialities of both detectors almost in real time.

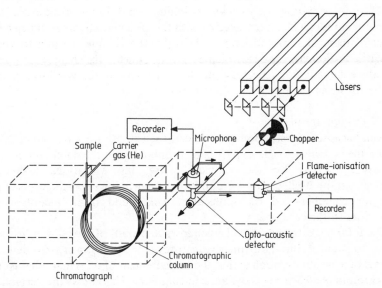

Figure 5.2 Scheme of a combined gas-chromatograph–laser opto-acoustic spectrometer (GC–LOAS) system [5.47].

A commercial version of the 'Biochrom-1' instrument with a built-in OA detector designed at the Zelinsky Institute of Organic chemistry, Academy of Sciences USSR, was used as a chromatograph. To widen the spectral range of the GC–LOAS system, several types of IR lasers were employed at the same time: a $^{12}C^{16}O_2$ laser (tuning range from 9.2 to 10.8 μm); a $^{13}C^{16}O_2$ laser (9.6–11.4 μm); a $^{12}C^{16}O$ laser (5.2–6 μm) and a He–Ne laser with $\lambda = 3.39$ μm.

(b)   Opto-acoustic detector design

The main requirements for an OA detector working as a chromatographic detector are:

(i) to ensure a continuous flow of gas through the chamber at a rate of up to 50 ml min^{-1};

(ii) a wide temperature range, up to 400 °C, in order to eliminate the condensation of the vapour of the compounds on the elements of the OA detector (the windows, the walls or the microphone);

(iii) a small chamber volume in order to eliminate chromatographic peak broadening;

(iv) high response;

(v) maintenance of the high sensitivity of conventional OA detectors.

Some of these requirements are inconsistent and can only be met in practice by compromise.

To eliminate chromatographic peak broadening the detector volume should not exceed the volume of the sample in the flow of carrier gas. In practice, in operating with packed columns about 3 mm in diameter, this requirement is met by an OA detector with the following chamber dimensions: inner diameter 0.5 cm and length 14 cm (i.e. the volume is equal to about 2.7 cm^3). To decrease the dead volume near the outlet in the chamber as well as to reduce the radiation reflected from the outlet window this window is placed at the Brewster angle to the optical axis.

(c)   Detection threshold

The main causes of restriction of OA detector sensitivity are discussed in §5.2.2. When the OA detector acts as part of a chromatograph the dominant effect is usually produced by the acoustic noise and vibrations from the chromatograph itself, the background from the windows and the walls of the detector chamber and the molecular background from uncontrollable impurities in the carrier gas.

The total background from the windows and walls of the OA detector was about 10^{-6} cm^{-1}. But the distinctive feature of the operation of an OA detector as part of a chromatograph is that the measurement of OA signals against the background of a base line is dynamic in character. This allows the permanent background to be compensated for quite easily in an electrical way, for example by displacing the zero line of the recorder. In such a situation the sensitivity restriction is determined not by the absolute value of the background but by its relative fluctuations related, for example, to the noncompensated fluctuations of laser radiation power. For the OA detector design under consideration the threshold of detection by absorption determined for propane and methanol was about $2 \times 10^{-7}\,cm^{-1}$. The chromatographic peaks of methanol measured under the same conditions with an OA detector and a FI detector given in figure 5.3 demonstrate a high sensitivity of the OA detector. The figure shows that the sensitivities of both the detectors are of the same order. It should be noted that the sensitivity of OA detection for some gases (ethylene, ammonia, etc) with a higher absorption coefficient than methanol, where $\kappa = 10.6\,cm^{-1}\,atm^{-1}$, is higher by one or two orders.

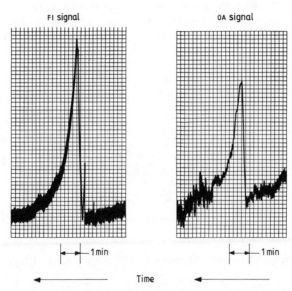

Figure 5.3 Comparison of the signals from the OA detector at the CO_2 laser line with $\lambda = 9.695\,\mu m$ and the FI detector with 20 ng of methanol introduced into the column. The conditions of measurement are: immobile phase 'Apiezon-L', $T_{col} = 100\,°$, the rate is $30\,cm^3\,min^{-1}$ [5.52].

According to the accepted classification of chromatographic detectors [5.40] the OA detector belongs to detectors of the concentration type. For most compounds its 'flow' threshold of detection equals 10^{-10} to $10^{-11}\,g\,cm^{-3}$.

(d) *Tunable lasers and basic ways of scanning*
One of the main requirements for lasers for gas-chromatographic measurements is the possibility of realising different methods of scanning and, primarily, 'fast' scanning which makes it possible to measure OA spectra of

individual compounds during their passage through the OA detector chamber. This requirement is fulfilled to the highest degree by the tunable laser scheme described in [5.50]. A modified version of this scheme with different scanning elements is presented in figure 5.4. In this scheme the spatial analysis of the generated spectrum is performed inside the laser cavity using a fixed diffraction grating operating under a nonautocollimation regime in combination with two additional mirrors. The spatial analysis of the entire generated spectrum is realised by an additional plane mirror. In such a scheme radiation with different wavelengths is reflected in the first order from the grating in the form of a fan of divergent monochromatic beams. The distance between the grating and the spherical mirror is equal to the focal length of this mirror. Under this condition the radiation falls on the end plane mirror as parallel beams in the same plane, and is reflected back in the same way. The wavelength required and the modes of scanning are preset by special diaphragms and the law of their motion in the plane near the end mirror.

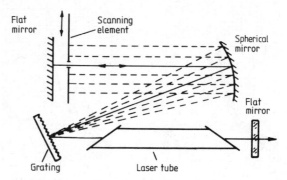

Figure 5.4 Scheme of CO_2 laser with rapid scanning of lasing lines [5.51].

The method of laser radiation scanning by wavelength defines the volume of spectral information obtained in the GC–LOAS system and thus affects the analytical potentialities of this system.

The simplest mode is successive measurement of OA chromatograms with different laser lines, i.e. first the chromatogram is measured at one line and then, after the chromatographic cycle is over, at another line, and so on. Under this mode of operation the OA detector acts as a chromatographic detector with varying selectivity that depends on the degree of coincidence of the laser line with the absorption bands of the compounds under analysis. In this case the sensitivity of the OA method may be maximum since it is possible to choose optimal values of laser radiation modulation frequency and OA signal integration time. The production of the required wavelength is ensured by choosing a stationary position of a single slit in the plane of scanning. The disadvantage of this mode is that a long time is needed to measure OA chromatograms at many laser lines (up to several hours) and the volume of the sample must be increased accordingly.

The highest information capacity in combination with rapidity can be attained in the mode of 'fast' scanning which in the simplest case is realised using a rotating disc with a spiral-form slit or a single, translationally moving slit (figure 5.4). These methods, however, are characterised by a high pulse period–to pulse duration ratio, especially when the generation spectrum is discrete in character, as in the case of a low-pressure CO_2 laser. They are also characterised by difficult programmed scanning in separate spectral intervals. The use of a rotating disc with concentric slits excludes the idle scanning time between separate lines or sections of the spectrum. The number of slits dictates the number of scanned lines; the distance between each slit and the centre of the disc defines the wavelength. The width of the slits affects the radiation monochromaticity and their angular length; with the rotation rate of the disc preset the generation time is set at one line. Besides an increased scanning rate, a high speed of response in this mode can also be provided by widening the electronic bandwidth, increasing the modulation frequency (up to 800 Hz) and using a synchronous detector with a small time constant (up to 10^{-3} s). However, these steps bring about a decrease in the detection threshold in this mode as compared to the measurement at one line (about an order).

If just one compound in a complex mixture is to be selectively detected, frequency modulation (FM) may be preferred. The advantage of this mode is that it is possible to effectively suppress both a spectral-nonselective instrumental background and a selective background from impurities. The latter is particularly important when the chromatographic resolution of peaks from different compounds is poor.

The mode of discrete FM can be realised using a rotating disc with two slits or a shutter mechanism. In the second case the required laser lines are isolated by two fixed slits which are alternately opened with shutters driven by an electromagnetic relay. In the case of a continuous generation spectrum it is possible to realise harmonic FM by means of a single vibrating slit or a rotating disc with a wavy slit geometry. It should be stressed that in the FM mode the OA signal is detected at the commutation frequency of the laser wavelengths.

When a rather small number of compounds is to be detected, time discrimination of OA signals at different wavelengths is advisable. In this mode the laser is switched successively as the chromatographic peak passes through the OA detector chamber at several wavelengths in combination with higher frequency amplitude modulation of the radiation. The processing of the information obtained consists of synchronous detection of OA signals at the modulation frequency with their subsequent recording by a chart recorder. The advantage of this mode is the rather simple recording of OA chromatograms at several wavelengths (usually no more than five) can be carried out at the same time. This mode can be realised using a raster with specified arrangement of the slits and with slower rotation of the disc.

The multiplex mode of operation comprises simultaneous transmission of several laser beams with different wavelengths modulated at different sonic frequencies through the OA detector chamber. The OA signals which carry information on absorption at different wavelengths are discriminated by the use of electronic filters situated at the OA detector outlet and tuned to suitable modulation frequencies. Such a scheme is a discrete version of Fourier spectroscopy and can be realised using several narrow-band lasers. In the case of wide-band lasers, for example, dye lasers, selective wavelength modulation can be achieved by transmitting wide-band laser radiation through a scanning interferometer as in classical Fourier spectrometers with nonlaser sources.

The same scheme can be applied to realise the multiplex mode with Hadamard transformation. In this mode in the time the peak passes through the OA detector its multiwave irradiation is performed with successive commutation of wavelengths by a mask with the Hadamard code. The main advantage of the two latter modes is that it is possible to obtain continuous information on OA signals at many laser lines. The scanning in this scheme can also be realised with individual shutters for each line (for example, electromagnetically driven shutters) which are switched in a definite order by a special program.

Most of the above-described modes of measurement have been subjected to experimental test in solving a number of tasks of practical importance. The application of the GC–LOAS system is considered in more detail in §5.3.4. Here these modes are compared only briefly (table 5.5). None of them is sufficiently universal. Therefore, it is advisable to apply different modes at the same time. The mode of fast scanning must be preferred in the case of spectral identification of individual components, particularly when the GC–LOAS system employs a microcomputer.

5.3.3 Detectors for capillary gas chromatography

Further progress in gas chromatography is connected with the application of highly effective capillary columns. The advantage of capillary columns over packed ones lies in their higher separation of the components in a mixture. But in capillary columns very narrow chromatographic peaks are formed. Their equivalent carrier gas volume may be no more than $0.1 \, \text{cm}^3$. Therefore, the advantages of capillary columns can be realised only with detectors having a small detection volume.

(a) Opto-acoustic detector

The basic problem resulting from the decrease of the OA detector chamber volume is that the background from the walls and the windows of the chamber increases. Study of an OA detector with a constant length of 15 mm and different

Table 5.5 Comparison of main scanning modes in the GC-LOAS combined system.

Parameters	Mode				
	Successive scanning	Multiplex	Fast scanning	With time separation	Frequency modulation
Sensitivity threshold (cm^{-1})	2×10^{-7}	6×10^{-7}	2×10^{-6}	2×10^{-7}	$\sim 10^{-7}$
Spectral range (μm)	3.39, 9.2–11.4	9.2–10.8, 3.39, 10.6–11.4	9.2–10.8	9.2–10.8	9.2–10.8
Selectivity	Moderate	Moderate	Moderate	Moderate	Moderate
Identification	High	Moderate	High	Moderate	Moderate
Rate of response	Low	High	High	Moderate	Moderate
Range application	High-sensitivity detection of microimpurities	Detection of the wanted compound in a complex mixture	Recording of the 'review' spectrum	Detection of the wanted compound in simple mixture	Selective detection of inseparable components
References	[5.47, 5.51–5.54]	[5.55]	[5.48, 5.51]	[5.54]	[5.54]

chamber cross sections has shown that the dependence of the background on the chamber diameter for $\lambda = 10.6\ \mu m$ begins to manifest itself from about 4 mm. When the diameter is decreased below 2 mm the growth of the background OA signal becomes rather strong, almost exponential. The studies performed concluded that the following parameters of OA detector look promising for capillary chromatography: a chamber length of 20 mm and internal diameter is 2 mm (i.e. a chamber volume of about 0.06 cm³).

An increase in background can be compensated for by using a differential scheme with two OA detectors. Through one of them the carrier gas flow is passed from the chromatographic column and through the other reference detector the 'pure' carrier gas is passed. The detection threshold in such a scheme may range from 10^{-7} to $10^{-8}\ cm^{-1}$.

A radical way of decreasing the detection volume consists of using a scheme with the flow irradiated in the direction perpendicular to its motion. In this case the window and wall background can be eliminated by using a free gas jet at the column outlet. With such a geometry the OA method is most effective with pulsed irradiation of the medium.

In this excitation scheme it seems highly promising to apply other methods of optocalorimetric spectroscopy and particularly the optothermal (OT) method (see Chapter 1).

(b) *Optothermal detector*

After the chromatograph column the carrier gas flow is directed by an additional capillary to a thermal transducer. The focused and modulated laser radiation irradiates the gas flow transversely to its motion. If the flow contains compounds absorbing radiation, a thermal distribution is formed in the direction of motion of the flow. In a first approximation this is similar to the distribution of the concentration of these compounds. The thermal transducer detects this thermal distribution, which is discontinuous due to the use of modulated laser radiation. The OT signal detection system is identical to that used in OA measurements.

The design of the OT detector is shown in figure 5.5. The chamber body and the thermal transducer holder are made of stainless steel. In the bottom of the body is fixed the output end of the connecting capillary. This acts as a nozzle which forms the gas jet falling on the thermal transducer. The distance between the transducer and the nozzle was 3 to 4 mm and that between the transducer and the laser beam waist 1.5 to 2 mm. This design has no windows and does not have to be gas-tight. A pyroelectric detector acted as a thermal transducer. This consisted of a thin film of pyroactive material between two aluminium electrodes deposited on a glass substrate [5.19]. Two adjacent pyroelectrics connected differentially were used to reduce the influence of the background OT signals caused by the scattered radiation falling on the pyroelectric surface. They were equally sensitive to scattered light but just one of them recorded the temperature distribution in the flow. For this purpose a thin NaCl plate was

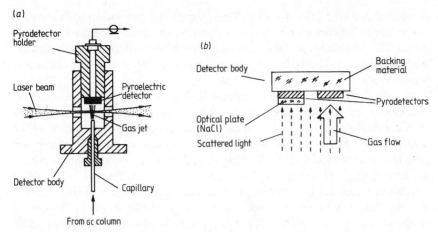

Figure 5.5 Designs of pyroelectric OT detectors of nondifferential (*a*) and differential (*b*) types [5.57].

stuck onto the surface of the second pyroelectric. This prevented thermal contact between the heated gas and the pyroelectric surface but was sufficiently transparent to IR radiation. Such a scheme permitted the suppression of the background signals from scattered radiation by about 10^2 to 10^3 times, up to the level of intrinsic electronic noise.

The OT detector was used to operate with both packed and capillary chromatographic columns. The conditions of operation with packed columns were as follows: fixed phase of 'Apiezon-L'; average rate of carrier gas 30 cm^3 min^{-1}; source of radiation a He–Ne laser; and modulation frequency, f, 70 Hz. The studies have shown that the OT signal amplitude depends slightly on the gas flow rate in the range from 5 to 50 cm^3 min^{-1} and, as the frequency varies between 30 and 300 Hz, the OT signal decreases as f^{-1}. Due to the mutually perpendicular geometry of the gas flow and the laser beam the detection volume was a minimum and did not affect the broadening of chromatographic peaks (in comparison with the FI detector). The absorption detection threshold in the OT detector was about 10^{-7} cm^{-1} for the radiation power of 1 W [5.56]. Like the OA detector, the OT detector is linear over a wide concentration range of compounds, at least 10^4.

In the region of comparatively high sample concentration there are some specific features in the use of the OT detector. Firstly, the value of gas heating, with constant absorbed power, begins to depend on the thermodynamic parameters (thermal conductivity, heat capacity) of the compounds being analysed, while at low concentrations it depends only on the parameters of the carrier gas. Secondly, even without laser radiation at the OA detector outlet, signals are formed whose origin is explained by a different rate of cooling of the carrier gas and the carrier gas with a compound having an excellent thermal conductivity at the column outlet. These signals are proportional to the time derivative of the concentration distribution of substance in chromatographic peaks. The potentialities of the OT detector combined with packed columns are demonstrated in [5.56] by the detection of a mixture of five saturated hydrocarbons.

In the detector with a capillary column a phase of the OV-101 type was used and the carrier gas rate was about 2 cm^3 min^{-1}. A carrier-gas divider is placed at the column inlet, the division ratio being 1:50. The parameters of the 'capillary' OT detector were studied using ethanol samples, the maximum absorption of which at the $^{12}C^{16}O_2$ laser line with $\lambda = 9.488 \mu$m is about 6.7 cm^{-1} atm^{-1}. The detection threshold of ethanol was around 10^{-9} g cm^{-3} which corresponded to an absorption detection threshold of 6×10^{-6} cm^{-1} (for the radiation power of 1 W). A comparison of the shapes of the signals from the OT and FI detectors shows that in the OT detector the chromatographic peaks of different compounds with a volume of up to 0.15 cm^3 are broadened by no more than a factor of 1.5.

The main advantage of flow OT detectors over OA detectors is increased spatial resolution with sensitivity sufficient for many applications. It is also important that OT detectors are less liable to acoustic noise and vibrations. Improvement of OT detectors lies in using more sensitive thermal transducers, for example differential bolometers, as well as in increasing the absorbed power due to multiple transmission of the radiation through the gas flow with the use of cylindrical optical systems. It is possible to increase the spectral resolution in OT spectrometers by decreasing the gas pressure in the zone of detection and also by applying methods of nonlinear laser spectroscopy. The technique of molecular beams formed at the chromatographic column outlet also holds much promise.

5.3.4 Some analytical applications of gas chromatography with opto-acoustic detection

In gas-chromatographic analysis of complex mixtures there are three typical situations:

(i) the components of the mixture can be fully separated in the chromatograph column;

(ii) the components can be partially separated;

(iii) the components cannot be separated in the column used.

In the first case the use of OA detection makes it possible to increase the reliability of identification of separate compounds by their IR absorption spectra. In the second case the chromatographic resolution can be improved significantly by choosing a proper spectral range. In the third case the OA detector makes it possible to solve the problem of quantitative analysis of inseparable compounds in the presence of *a priori* information on their absorption spectra. This problem cannot be solved with nonselective detectors. Here more detailed consideration is given below to possible applications of the GC–LOAS system.

(a) *Analysis of hydrocarbon mixtures*

Analysis of complex mixtures of hydrocarbons (HC) is an important task in petroleum chemistry and geochemistry, and the question of most importance here is what classes of HC the components of the mixture belong to. The presence of information about one or other class of HC in natural mixtures, for example, makes it possible to define the character, quality and prehistory of deposits and sediments. The application of conventional chromatographic or spectroscopic methods for such analysis presents difficulties for it is a complicated task to place each component into one or other class using their retention times in the superposition of their spectra. The results of analysis using a GC–LOAS system of a typical HC mixture consisting of 18 components are given in figure 5.6. The mixture comprised the following HC classes: paraffins, cycloparaffins, olefines, alkyl benzene and HC groups of naphthalene. The experiments were performed with successive scanning at the lasing lines of He–Ne, $^{12}C^{16}O_2$ and $^{13}C^{16}O_2$ lasers and fast scanning in the lasing range of the

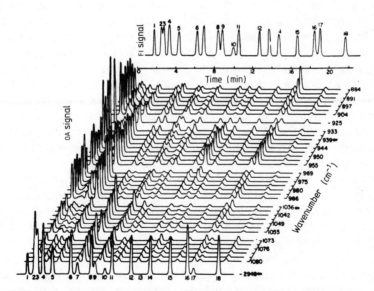

Figure 5.6   Analysis of a mixture of hydrocarbons. The top diagram shows a 'standard' chromatogram produced with a FI detector and the bottom an OA spectrochromatogram. The conditions of measurement are: immobile phase 'Apiezon-L', temperature programming from 70 to 245 °C at a rate of 8 °C min^{-1}, the carrier gas rate is 30 cm^3 min^{-1} [5.52].

$^{12}C^{16}O_2$ laser. The volume of the sample in the chromatograph column was equal to 0.3 μl.

The OA spectrochromatogram for the He–Ne laser line with $v = 2948$ cm^{-1} has very strong OA signals (the signal value at this line is reduced by a factor of five in the figure) for saturated HC (peaks 2, 3, 4, 6, 8, 11, 12, 14, 15, 16, 18), less intense signals for unsaturated HC (peaks 1, 5) and low intensity signals for aromatic HC (peaks 7, 9, 13, 17). In contrast, in the generation range of CO_2 lasers (880–1080 cm^{-1}) the strongest OA signals are characteristic of unsaturated (olefines) and aromatic (alkyl benzenes and HC groups of naphthalene) while the saturated HC have no strong absorption in this spectral region and produce weak OA signals. As can be seen from figure 5.6, the lasing range of CO_2 lasers is very useful for selective analysis of classes of unsaturated and aromatic HC since olefines (peaks 1, 5) have intense characteristic absorption bands in the 970–930 cm^{-1} region and aromatic HC (benzene homologues, peaks 7, 9) have absorption bands shifted towards higher frequencies, although less intense (1070–1030 cm^{-1}). As for the compounds of the naphthalene group (peaks 13, 17), they have an absorption band of rather low intensity in the region 960–930 cm^{-1} that, however, does not exclude the possibility of their selective definition. Thus, even though the lasers used have a limited spectral lasing range, the GC–LOAS system can be used in selective definition of classes of organic compounds in complex mixtures.

Fast scanning allows measurement of the OA spectra of individual compounds during a chromatographic cycle (§5.3.2). This mode was realised by the use of a rotating disc with 42 slots which corresponded to the following lines of the $^{12}C^{16}O_2$ laser: R(28)–R(10) and P(12)–P(30) of the 9.4 μm band and R(30)–R(12) and P(12)–P(34) of the 10.6 μm band of the $^{12}C^{16}O_2$ laser. The slots were placed so that the tuning from one line to another one, as well as from branch to branch, was effected without time losses. The CO_2 laser cavity had a diffraction grating, flat end mirrors and a spherical intermediate mirror with a radius of 1 m. With these parameters the linear width of the scanned generation spectrum on the plane of the flat mirror was about 90 mm, and the distance between separate lines was 1–1.5 mm.

Since in the analysis of complex mixtures a large amount of information is obtained with fast scanning, identification is only reliable with computers.

In routine analysis it is possible in principle to set up a library of OA spectra of individual compounds. It should be emphasised again that, as shown in practice, many compounds can be identified by analysing the specific features of OA spectra in a comparatively small spectral range of the lasers used.

Besides petroleum chemistry and photochemistry, the technique described is also promising for medical and biological studies, particularly in analysis of the microcomposition of exhaled air in the diagnosis of some diseases. It has been found that exhaled air contains about 150 compounds including ammonia, aldehydes, fatty acids, ketones, hydrocarbons, acetaldehyde, acetone, ethanol, etc. The concentrations of these compounds and their variations (the average level is 0.1–0.01 PPM) carry information about human biological processes.

(b) *Selective detection of individual compounds*

Due to tunable selectivity in the GC–LOAS system it is possible to provide a higher sensitivity to a given definite compound in a mixture. This is important, for example, for control of some impurities in the production of some substances.

In the case of poor chromatographic separation of compounds the selective detection of one of them is equivalent to an increase in chromatographic resolution. For illustration, figure 5.7(*a,b*) demonstrates the possibility of reliable detection of a 10% impurity of isopropyl alcohol in ethanol using an OA detector and the $^{12}C^{16}O_2$ laser where the ratio of absorption factors of isopropanol and ethanol is maximum (figure 5.7(*c,d*)). Under the same conditions of measurement with a FI detector it is almost impossible to distinguish these compounds (figure 5.7(*a*)).

The OA detector also gives information on the concentration of compounds in the case when they cannot be separated in the chromatographic column and leave it in the form of one peak. This was demonstrated in [5.47] by selective analysis of a mixture of two isomers, metaxylol and paraxylol, which could not be separated in the column. The measurement procedure in this case consisted of measuring the OA chromatogram at several wavelengths with subsequent solution of the equations for the concentrations provided that the standard spectra of these compounds are known.

When it is necessary to detect just one of the unseparated compounds, the optimal mode is discrete FM. In this mode a pair of laser lines is chosen on the basis of the following requirements: a maximum spectral contrast for the compound under detection; a minimum possible (i.e. equal to one) spectral contrast for the background compound; and a maximum signal-to-noise ratio at the chosen analytical line. The FM mode was realised in the scheme with spatial dispersion of spectrum generated by the CO_2 laser by separating two two wavelengths, λ_1 and λ_2, with the use of diaphragms and fast switching between them by an electromagnetic shutter. Using additional gates near the diaphragm it was possible to obtain independent control over the radiation power at every line. Figure 5.7(*e,f*) demonstrates the possibilities of such a mode as a 10% impurity of isopropanol is detected in ethanol. In the FM mode the spectral-selective background from the undesired compound was suppressed by a factor of 25–30. It can be seen that, as in the previous cases, it is difficult to separate the peaks of the compounds from each other in the AM mode with one line at $\lambda_1 = 10.182 \ \mu m$. But in the FM mode, on account of selective suppression of the ethanol signal ($\lambda_2 = 10.303 \ \mu m$), it is quite possible to detect the OA signal from isopropanol (figure 5.7(*f*)). Due to a significant increase in the signal-to-noise ratio the detection threshold of isopropanol in

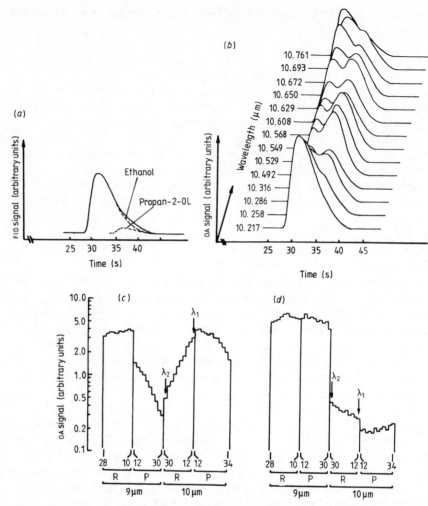

Figure 5.7 Spectrochromatographic analysis of a 10% admixture of isopropyl alcohol in ethanol with a FI detector (a) and an OA detector at several laser lines (b); individual OA spectra of 'pure' isopropyl alcohol (c) and ethanol (d) produced under 'fast' scanning; (*continued opposite*)

this case is improved at least by an order of magnitude as compared with the AM mode.

If rapid selective detection of individual compounds in more complex mixtures is required, simultaneous multiplex detection at several laser lines may prove useful (figure 5.8). This scheme allows simultaneous detection of three OA chromatograms at three wavelengths during one chromatographic cycle with the use of one OA detector. For this purpose the beams of three lasers

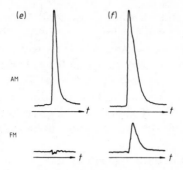

Figure 5.7 comparison of OA chromatosignal of pure ethanol (*e*) and a mixture of the said alcohols (*f*) under amplitude (AM) ($\lambda = 10.182 \mu m$) and frequency (FM) modulation of laser radiation ($\lambda_1 = 1.182 \mu m$, $\lambda_2 = 10.303 \mu m$) [5.47, 5.48].

modulated at different acoustic frequencies are passed through the OA detector at the same time. The OA signal from each laser is discriminated at the OA detector outlet with frequency-selective filters.

(c) *Conclusions*
It is advisable to define the place of OA laser spectroscopy among other analytical methods applied in gas chromatography (see table 5.4). Due to its high sensitivity the laser OA method stands between mass spectroscopy and Fourier spectroscopy, one of the best spectroscopic (nonlaser) methods. These

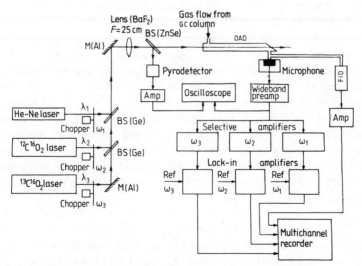

Figure 5.8 Scheme of multiplex OA detection [5.55].

methods can complement each other, for example, when the compounds have identical mass spectra but different IR spectra. The advantage of Fourier spectroscopy over laser OA spectroscopy is its wider spectral range which allows the identification of many compounds by characteristic bonds (C–H, O–H, etc). The decisive advantage of the OA method is its higher sensitivity, approximately from one to two orders. In addition the identification of individual compounds, after the library of their laser spectra is set up, can be realised with the characteristic 'fine' structure in the tuning range of the lasers used.

Further prospects for development of the GC–LOAS system consist of increasing its sensitivity up to 10^{-8}–10^{-9} cm^{-1}, widening the spectral range and automation of measurement. The most optimal field of application of this system is routine selective analysis of individual compounds in complex mixtures with requirements additional to the conventional chromatograph parameters.

5.4   Laser opto-acoustic spectroscopy in liquid chromatography

Gas chromatography is an effective method of separation of volatile substances. But it cannot be applied to substances with a high molecular weight (over 300) or to thermally and chemically unstable compounds. In such cases high-performance liquid chromatography is usually used (HPLC). The separation of substances in it occurs mainly at room temperatures in steel columns with a fine-grain sorbent through which a movable phase is pumped at a high pressure. The typical dimensions of the column are internal diameter 2–4 mm; length 15–30 cm; and rate of movable phase 1–4 ml min^{-1}. The procedure of HPLC is described in more detail in [5.59].

Progress in HPLC is strongly connected with the development of high-sensitivity detectors with a small detection volume (less than 1–10 μl). UV absorption detectors stocked with spectrum-scanned spectrophotometers are widely used in HPLC. It is also possible to use refractometric detectors, the potentialities of which became greater with the advent of laser sources. The sensitivity and selectivity in HPLC have been increased significantly using laser fluorescent detectors [5.59]. But these detectors are less effective when substances with a nonradiative relaxation channel are analysed, and the suppression of scattered light presents a serious problem. In such a situation fluorescent detectors can be complemented by OA detectors whose potential in HPLC is under study.

5.4.1   Opto-acoustic detectors for liquid chromatography

The laser OA detector consists of the following basic elements: a laser radiation source; an optical system providing the transfer of laser radiation into the detection space; an OA detector connected with the outlet of the liquid chromatograph column; a reference radiation detector designed for detecting the energy or power of laser radiation and a recording system. The type of recording system depends on the mode of laser operation.

Up to the present the potentialities of the OA detector as part of a liquid chromatograph have been investigated under both continuous laser operation with amplitude modulation [5.60] and under pulsed operation [5.61–5.65]. Figure 5.9 shows schematically a measuring set-up for pulsed laser operation. The radiation source (a pulsed N_2 laser) had the following parameters: $\lambda =$ 337.1 nm; pulse duration 5 ns; repetition frequency up to 10 Hz; and pulse energies up to 7×10^{-3} J. The laser radiation was focused with a quartz lens into the OA detector chamber. The signal from the acoustic (piezoelectric) transducer was amplified in a wide-band preamplifier with a bandwidth of 10 kHz to 1 MHz and then passed into the pulse gate-integrator. A starting electrical pulse from the reference radiation detector was fed to the second integrator inlet; the signal from the integrator outlet then entered the recorder.

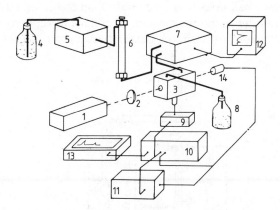

Figure 5.9 Opto-acoustic detection in liquid chromatography with a pulsed laser: 1, laser; 2, lens; 3, OA detector; 4, solvent pan; 5, pump; 6, column; 7, UV detector; 8, tank for liquid discharge; 9, preamplifier, 10, gate integrator; 11, generator; 12, 13, chart recorders [5.65].

When a CW laser is used, the radiation is premodulated before entering the OA detector, for example with the use of an acousto-optic modulator [5.60]. In this case the standard scheme of lock-in detection is used to treat the OA signal.

In combination with a liquid chromatograph the OA detector can operate in two main modes—with discontinuous and continuous flows of liquid. In the first case, as the component to be detected leaves the column, the chromatographic process is stopped or the solvent is recovered and then placed into the OA detector chamber. In this case the OA detector works under ordinary conditions and so ultimate sensitivity is possible. The operation with discontinuous flow, however, can be applied only in analysing a small number of components and it is characterised by a long measurement process.

In continuous flow operation the basic requirements placed on the OA detector are somewhat inconsistent: a high sensitivity and a small detection volume. The volume required depends on the type and dimensions of the column and in most cases must not exceed 5–10 μl.

There are two basic types of OA detectors used in liquid chromatography: closed chamber and open chamber detectors. A typical design for the first type is shown in figure 5.10. The detector was made of stainless steel. The optical channel was 5 mm long with an internal diameter of 1 mm, i.e. about 4 μm in volume. The liquid was introduced to and removed from the chamber through capillary tubes with internal diameter 0.5 mm. The windows were made of quartz (3 mm thick and 10 mm in diameter) and were packed by thin Teflon washers with clamping metal flanges. A strip of piezoceramic fixed on the side wall in parallel with the optical axis of the chamber acted as an acoustic transducer. The transducer was acoustically connected with the chamber by means of several small holes in the side wall. Thin nickel foil was introduced between the transducer and the medium to exclude direct entry of scattered light on the surface of the transducer. One of the contacts of the transducer was

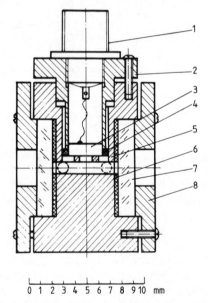

0 1 2 3 4 5 6 7 8 9 10 mm

Figure 5.10   Design of OA detector for liquid chromatography in the case of continuous flow of liquid through a spatially limited detection volume: 1, electric connector; 2, transducer body; 3, transducer (piezoceramic); 4, Teflon ring; 5, detector body; 6, Teflon washer; 7, quartz window; 8, flange [5.65].

earthed and the second terminated at a plug-and-socket unit. The same design of OA detector was used earlier in [5.60], but was large, the length being 11 mm, internal diameter 1.5 mm and the volume about 20 μl.

The limitations on the sensitivity of the OA detector operating as a part of a liquid chromatograph are caused mainly by the effect of turbulence of the flow, the formation of air bubbles, the background absorption in the solvent and the background OA signals of radiation absorption in the windows and the walls of the measuring chamber.

In the OA detector shown in figure 5.10 the detection threshold with respect to absorbed energy limited by the purely electronic noise of the preamplifier is about 10^{-8} J cm^{-1} with a repetition frequency of 20 Hz and integration time of several seconds [5.65]. When the N_2 laser pulse energy equals 7 mJ, such a sensitivity allows the measurement of absorption of around 10^{-6} cm^{-1}. The effect of flow turbulence can be reduced by careful elimination of dead volumes and by using a pump with minimum flow fluctuations. It is also quite possible to attain a positive effect by maximising the signal-to-noise ratio.

Under continuous operation the signal-to-noise ratio is maximised by increasing the modulation frequency and using acoustic resonances of the chamber since the vibration spectrum is concentrated mainly in the low-frequency region. In [5.60], for example, an almost complete suppression of fluctuation was achieved at a frequency of about 4 kHz.

Since the OA signals are dynamic in chromatographic measurements, the effect of the solvent background can be reduced by electrical compensation. In this case it is necessary to provide high stability of the zero line by normalising the OA signals to the energy level of laser pulses. Most solvents absorb in the visible and UV at the level of 10^{-3} to 10^{-4} cm^{-1} so that, with a typical zero line stability of 5%, the detection threshold in chromatographic OA detectors ranges from 5×10^{-5} to 5×10^{-6} cm^{-1}.

The effect of the background from the chamber walls in OA detectors with small-volume closed chambers is a serious problem. In the OA detector design shown in figure 5.10, where the laser beam had high divergence, the effect of the background from the metal walls showed itself at 10^{-4} cm^{-1}. This background was electrically reduced by more than an order of magnitude. However, it was difficult to keep the background OA signal level stable because of spatial fluctuations of the laser beam. In the case of highly directional radiation, that of the Ar laser for example, the wall background in a chamber 1.5 mm in diameter can be reduced at least down to 8×10^{-6} cm^{-1} with $S/N = 2$ [5.60]. It is also possible to reduce the background from the walls if they are made of optically transparent materials, for example, in the form of quartz capillaries connected to the column outlet [5.62, 5.64].

The problem of reduction of wall background can be solved by using a measuring scheme with a free flow of solvent at the column outlet which has been realised in fluorescent detection [5.59] and subsequently in OA measurements [5.61, 5.63]. In this scheme the flow goes freely out of the metal tube connected with the column and falls onto a nearby metal rod of the same diameter as the tube. The OA signals formed in the exposed volume of liquid are detected by a piezoelectric transducer joined to the lower end of the rod. The volume of a drop of liquid between the tube and the rod is usually equal to several microlitres. The detection volume, however, is defined by the dimensions of the focused laser beam and may be less than $1 \mu l$. A small detection volume allows such a scheme to be used with advantage in liquid chromatography with microcolumns.

Along with the OA detector, it is advisable to use the standard UV detector in many measurements which widens the scope of the HPLC–LOAS system and allows comparison of the characteristics of the both detectors under almost the same conditions. The detectors can be connected to the column outlet either in turn at different instants of time or in sequence. In the second case allowance must be made for the influence of the volume of the first detector on the broadening of chromatographic peaks in the second. It is also useful to apply other methods of detection including fluorescence and photoionisation methods [5.63].

5.4.2 Some analytical applications of the HPLC–LOAS system

Experiments in this field are still demonstrative rather than quantitative. In [5.60] the radiation source was a CW Ar laser with a power of 0.5 W and amplitude modulated at a frequency of about 4 kHz. The radiation was focused by a lens with $f' = 15$ cm into an OA detector with a closed chamber (parameters given in §5.4.1). This scheme was used to detect three isomers of chloro-4-(dimethylamino) azobenzenes: 2′Cl-DAAB, 3′Cl-DAAB and 4′Cl-DAAB. These isomers have similar absorption spectra in the visible range. Figure 5.11(a) shows the absorption spectrum of 2′Cl-DAAB. The coefficients of molar absorption in 2′Cl-DAAB at the UV detector wavelength (254 nm) and the OA detector wavelength (488 nm) are equal to 9600 and 7700 $mol^{-1} cm^{-1}$ respectively, i.e. they are comparable in a first approximation. The comparison of the OA and UV chromatograms (figure 5.11(b,c)) shows that the sensitivity of the OA detector is higher than that of the UV absorption detector by a factor of approximately 25. The detection threshold for 2′Cl-DAAB in the OA detector is about 0.2 ng with $S/N = 2$. The OA signal depends linearly on the concentration of 2′Cl-DAAB within four orders, from 6×10^{-8} to 5×10^{-4} M. The reproducibility of OA detector sensitivity was about 3%.

Similar results were obtained in [5.65] from detecting anthracene in hexane with a pulsed N_2 laser. It was found that with effective electric compensation for the wall background signal the OA detector sensitivity exceeds the UV detector sensitivity ($\lambda = 254$ nm) by about an order.

When the OA detector with a chamber volume of 4 μl was placed after the UV detector with chamber volume 20 μl, the chromatographic peak width in the OA detector was narrower by about 30% than in the UV detector. This points to the optimality of the OA detector volume and the small effect of the diffusion

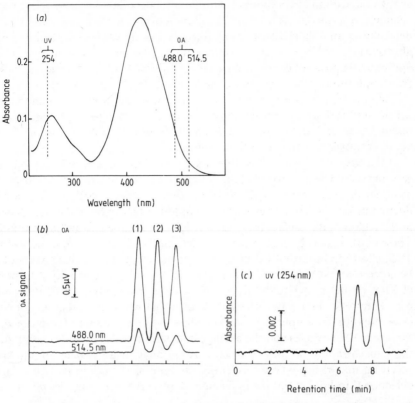

Figure 5.11   (*a*) The absorption spectrum of 2′Cl-DAAB in methanol, the concentration is 10^{-5} M, the optical path is 10 mm, the position of laser lines and the UV detector wavelength (broken lines).

(*b*) Opto-acoustic chromatograms of Cl-DAAB isomers measured at two wavelengths of Ar laser, (1) 2′Cl-DAAB; (2) 3′Cl-DAAB; (3) 4′Cl-DAAB;

(*c*) Ultraviolet chromatogram measured by a conventional absorption detector with the following parameters: length 10 mm, volume 8 μl, $\lambda = 254$ nm, absorption threshold 2×10^{-4} cm^{-1} (for $S/N = 2$). The flow rate is 1 ml min^{-1} [5.60].

processes in HPLC compared with GC. An excimer laser, like KrF, with $\lambda = 0.25$ μm was also used to detect substances with intense absorption bands near $\lambda = 0.25$ μm.

It is possible to widen the analytical scope of HPLC on the basis of simultaneous use of several types of laser detector. In [5.61, 5.63], for example, the OA detector was used together with fluorescence and photoionisation detectors. Such a combined system was designed on the basis of a throughflow cell. The OA and fluorescence detectors complement each other well when more sensitive detection of compounds with nonradiative and radiative relaxation channels, respectively, is needed. The photoionisation detector is optimal in detecting compounds with a comparatively low ionisation potential. The absorption detector is more universal but it is usually less sensitive. The approximate ratios of concentration detection thresholds in the fluorescence, photoionisation and uv detectors are 10^{-1}–10, 1–10^2 and 10–10^3 ng ml^{-1}.

The possibilities of operation of a liquid chromatograph in combination with a laser OA detector were demonstrated in [5.65] by detecting vitamin A in blood plasma. The detection threshold of vitamin A was about 0.3 pg ml^{-1} or about two orders better than the uv detector threshold.

In the case of OA detectors with a small detection volume sufficiently pure solvents should be used to avoid the potential effect of impurities in the form of solid suspended particles. This effect can manifest itself as increased fluctuations of the baseline caused by a random intersection of the laser beam in the region of its caustics by single particles due to their Brownian motion. These particles may be dust from the air, fragments of biomolecules, and so on. Their effect can be partially decreased by defocusing the laser beam. It should be noted that it is reasonable to apply this effect for high-sensitivity detection of such particles in liquids. Experiments on water samples with different concentrations of latex particles have shown that with strong focusing of N_2 laser radiation the amplitude of separate fluctuations contains information on the size of the particles and the average fluctuation rate gives information on the concentration of these particles. Additional measurement of the OA spectrum makes it possible to identify the type of absorbing particles. The detection threshold in this method comes to several tens of particles per cubic centimetre.

5.4.3 Ways of improving the HPLC–LOAS system

According to the results given above, the sensitivity of laser OA detectors can exceed that of conventional uv detectors by one or two orders of magnitude and there are potential reserves for further improvements of this parameter. These reserves lie in optimising the characteristics of OA detectors and decreasing the instrumental and solvent backgrounds by using differential schemes with two OA detectors as well as frequency modulation.

The analytical potentialities of OA detectors can be widened by increasing their spectral range with the use of different lasers or even a series of such lasers; for example, the second (0.53 μm) and fourth (0.265 μm) harmonics of the Nd laser, the second harmonic of ruby laser (0.347 μm), the first and second harmonics of dye lasers, excimer lasers and Ar and N_2 lasers. For spectral selection of separate compounds it is advisable to use the measuring modes described in §5.3.2 as applied to gas chromatography [5.66].

In analysing compounds with different physical properties the goal must be to design combined detectors making use of several analytical methods at the same time. However, there is a way of designing more universal types of OA detector. In fluorescence detectors, for example, nonfluorescent samples can be analysed by mixing them in the flow with special reagents which convert them to fluorescent compounds for detection. In much the same way, although with fewer technical problems, it is possible to develop effective methods for quenching the fluorescence in samples with a dominant channel of radiative relaxation to provide high-sensitivity detection of these samples with the help of OA detectors.

In comparison with widespread UV detectors the OA detector is simpler and more compact in design. But the introduction of laser sources into them may complicate them considerably and raise their cost. Therefore the optimal field of application of these detectors is routine high-sensitivity analysis of compounds when the detection threshold is inaccessible for UV detectors.

5.5 Opto-acoustic spectroscopy in thin-layer chromatography

Thin-layer chromatography (TLC) is in wide use, mainly for qualitative analysis of complex mixtures in chemistry, medicine and biology. This is explained by the simple equipment required, its high selectivity, efficiency and universality. The fundamental element in TLC is a plate of metal, glass or plastic material coated with a thin layer (0.15–0.75 mm) of adsorbent (very often silica gel). When the substance and the solvent move through the adsorbent under diffusion and capillary forces, they are subjected to spatial separation based on their different velocities of motion in the adsorbent layer. When the chromatographic process is over, zones with an increased concentration of substance are formed as two-dimensional spots at different distances from the place of introduction of the sample. The two-dimensional character of blurring of such chromatographic spots is explained by the possibility of the substances to diffuse not only along but partially across the gel. The procedure of TLC is described in more detail in [5.67].

It is somewhat difficult in TLC to obtain quantitative information on the concentration of separate substances. The spectrodensitometers and fluorimeters which have been developed recently do not always effectively solve this problem because of strong light scattering and absorption in an adsorbent of nonuniform thickness. Therefore, the use of new analytical methods in TLC including OA spectroscopy is of great interest.

5.5.1 Opto-acoustic detectors for thin-layer chromatography

Opto-acoustic spectroscopy of condensed media with indirect recording of acoustic vibrations in the gas in contact with the sample is of principal interest for TLC. In this version of OA spectroscopy the sample is placed into an acoustically closed chamber with a gas and a microphone (figure 5.12). Under modulated radiation the sample surface is periodically heated and cooled and part of heat is transmitted to the gas in contact. As a result, acoustic vibrations

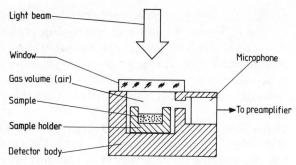

Figure 5.12   Design of OA detector with indirect detection of acoustic vibrations for analysis under stationary conditions [5.69].

recorded by conventional capacitor microphones are formed in the gas. The measuring scheme incorporates a light source, a modulator, an optical system, an OA cell and a reference channel with a photodetector. The modulation frequency range runs from tens to hundreds of hertz.

Due to direct recording of the energy absorbed in the sample this method turns out to be almost insensitive to scattered light which enables it to be used effectively in spectral analysis of highly scattered and highly absorbing substances in the form of different powders, biological objects, etc. An important characteristic feature of such OA spectroscopy is the dependence of the OA signal on both the optical and thermal properties of the sample. So, to produce an OA spectrum similar to the ordinary absorption spectrum, it is necessary to comply with a number of requirements the most basic of which is $l_T \ll l_\kappa$, where $l_\kappa = 1/\kappa$ is the optical absorption depth, $l_T = (2k/\omega)^{1/2}$ is the thermal diffusion length in the sample, κ is the absorption coefficient, k is the thermal diffusion coefficient and ω is the radiation modulation frequency. To eliminate the direct entry of scattered light to the microphone surface, the latter must be somewhat removed from the chamber with the sample and connected to it with a small-diameter channel. The OA detector sensitivity can be increased by minimising the chamber volume. In analysing highly absorbing substances the sensitivity of the OA method is quite sufficient for solving many practically important tasks when monochromatised light from powerful incoherent sources (for example, xenon lamps) with a power of 300 to 500 W is used.

Two basic designs of OA detectors with closed and open measuring chambers have been used in TLC. In the stationary design the sample specially prepared for analysis is placed inside the detector chamber. The sample may be a cut or drilled part of a chromatographic plate with compounds to be analysed or a small portion of the adsorbent with the compound extracted from the plate mechanically or by the use of vacuum devices. To increase the metrological characteristics the adsorbent is usually placed in a special holder as is shown in figure 5.12.

The second design of OA detector is intended for analysing compounds immediately on the chromatographic plate. This design is characterised by the presence of an open measuring chamber. During measurements the OA detector is located on the plate so that the zone under study could fall within the hole in the lower part of the measuring chamber. Acoustic coupling is achieved by pressing the lower polished surface of the detector onto the surface of the sample. If, for example, the OA detector weighs 330 g, this weight provides the necessary contact tightness [5.68]. It is very convenient to illuminate the zone under study with a flexible light guide.

5.5.2 Some applications of opto-acoustic spectroscopy in thin-layer chromatography

The possibilities of OA spectroscopy in combination with TLC were first demonstrated in [5.69] by an example of spectral identification of five compounds: p-nitroaniline, benzylidene-acetone, salicycladehyde, i-tetralone and fluorenone in the range 200 to 400 nm. A powerful xenon lamp, the radiation of which was monochromatised by a high-aperture monochromator, was used as a radiation source. In the experiments a plastic plate was used with a layer of silica gel 0.25 mm thick. The measuring technique consisted of cutting out appropriate zones of the plate which were then placed into a stationary OA cell (figure 5.12). The measured OA spectra of the compounds turned out to be identical with their ordinary absorption spectra produced in solutions. The detection threshold of benzylidene-acetone is about 0.1 μg over an area of 0.3 cm^3, which corresponds approximately to one monolayer of compound.

In [5.70, 5.71] the method of OA spectroscopy was applied to detect fluorescein on the silica gel surface of chromatographic plates made of aluminium and glass. In the case of the aluminium plate the sample with the zone to be analysed was cut out as a circle 20 mm in diameter and placed into the OA cell chamber. In the second case the zone under study, 9.5 mm in diameter, was first marked using a vacuum intake with a filter. Then the collected substance was placed into the OA detector chamber on a special holder (figure 5.12). As in the previous example, the sample was exposed to the monochromatised radiation of a xenon lamp. The detection threshold of fluorescein was 20 ng which corresponds to a relative concentration of 4.4 μg g^{-1}. The relative standard deviation in the middle of the dynamic range was 0.08 to 0.1. The main source of error was the replacement of the sample from the chromatographic plate into the OA detector chamber.

A similar procedure with vacuum intake was used in [5.72] to determine the concentration of synthetic dyes in food confectionary products. The content of tartrazine (absorption maximum at $\lambda = 425$ nm), carmoisine ($\lambda = 510$ nm) and brilliant blue fcf ($\lambda = 625$ nm) was measured in several food samples of different colours in the following concentration ranges: 3 to 40, 20 to 60 and 1 to 22 ng respectively. Such a technique of analysis is very useful for control over the quality of different food products.

oa detection and identification of compounds directly on the surface of the chromatographic plate is of particular interest. This was demonstrated in [5.69] with the use of an 'open' oa cell by quantitative analysis of pink Bengalese dye and palmitic acid. The measurement of pink Bengalese dye was performed at the wavelength of its absorption band maximum ($\lambda = 541$ nm) with compensation for the influence of the pure silica gel background. The calibrating characteristic was linear in the region of low concentrations (0.4 to 2 mg) and highly nonlinear in the region of high concentrations. The latter was most probably caused by the thermal saturation effect. Since the thermal diffusion length l_T is proportional to $\omega^{-1/2}$, it is necessary to increase the modulation frequency until the condition $l_T \ll l_\kappa$ is achieved in order to eliminate thermal saturation. By replacing an 'open' oa cell on the surface of a chromatographic plate it is possible to investigate the spatial distribution of the compound to be detected. A combination of spatial and spectral scanning permits two-dimensional oa spectromicroscopy which is very useful for identifying individual compounds when they are partially separated.

The results presented show that the use of oa spectroscopy is promising for the detection and identification of compounds in TLC. However, the first experiments have only just been carried out in this direction, and further efforts must be made to improve the technique and the equipment. In all the experiments, for example, powerful nonlaser radiation sources were used because of a relatively high absorption factor of the compounds analysed (of the order of 1 to 10 cm^{-1}). But the use of laser sources, dye lasers in particular, will make it possible to considerably widen the scope of the oa method in terms of an increase in its sensitivity. This is of particular importance in operating at high modulation frequencies when the sensitivity of the method is significantly reduced. A high signal-to-noise ratio will also allow more effective subtraction of background from the adsorbent itself. It is also advisable in this case to use radiation frequency modulation as well as differential measuring schemes with the two oa cells. One of them contains the zone of a plate with a compound and the other one has just a plate with a pure solvent. A modification of this scheme is spatial modulation of the radiation beam by its alternate entry into the zone with a compound and then into the zone with a pure adsorbent.

In the application described above the use of piezoelectric recording of absorbed energy, for example, by joining the piezoceramics to the plate surface is not effective due to strong attenuation of acoustic waves in a viscous adsorbent. But in the case of throughflow TLC, an analogue of column liquid chromatography with flat geometry, the piezoelectric recording may be very useful. In this version the laser beam irradiates a permanent zone on the plate where a flow of liquid moves on the surface and the piezoelement is joined to the lower surface of the plate near the irradiation zone.

5.6 Conclusions

The results presented in this chapter point to the significant potential of OA spectroscopy in different chromatographic methods of analysis of complex multicomponent mixtures. As applied to gas chromatography, the methods of OA spectroscopy enable the selectivity of analysis to be significantly increased at a sufficiently high sensitivity. In liquid chromatography the main advantage of the OA method is a significant increase of the detection threshold as compared with standard UV detectors. The application of OA spectroscopy in TLC allows the development of reliable methods of quantitative measurement in this type of chromatography which is to date still a qualitative analytical method. On the other hand, the use of chromatographic methods in OA spectroscopy simplifies the problem of selective sampling and makes it possible to measure, with a high accuracy, the OA spectra of many individual compounds. Under standard conditions this is a complicated problem due to the influence of impurities.

The possibilities for development of these methods lie in further modification of the techniques described as well as in their combination with other promising methods for selective analysis of complex mixtures. Strong possibilities are offered where a flow of a gas or liquid is cooled at the chromatographic column outlet and individual compounds are subjected to OA detection in the condensed cryogenic phase. The possibilities of the selection methods described in §5.2 have not yet been completely exhausted. Taking into account the progress achieved in experimental technique, the primary task is now to develop and design commercial OA detectors for operation together with different types of chromatograph.

In conclusion the author would like to express his gratitude to Professor V S Letokhov for his help and fruitful discussions.

References

5.1   Rosencwaig A 1980 *Photoacoustics and Photoacoustic Spectroscopy* (New York: Wiley)

5.2   Tam A S 1983 Photoacoustic spectroscopy: spectroscopy and other applications in *Ultrasensitive Laser Spectroscopy* ed D S Kliger (New York: Academic) pp. 2–101

5.3   Zharov V P 1982 Optoacoustic method in laser spectroscopy in *New Spectroscopy Methods* (Nauka: Novosibirsk) pp. 126–203 (in Russian)

5.4   Zharov V P and Letokhov V S 1984 *Laser Optoacoustic Spectroscopy* (Moscow: Nauka) (in Russian) (translation published in 1986 as Springer Series in Optical Sciences vol. 37)

5.5   Gerlach R and Amer N M 1980 *Appl. Phys.* **23** 319

5.6   Dewey C F 1974 *Opt. Eng.* **13** 483

5.7   Bagratashvili V N, Knyazev I N, Letokhov V S and Lobko V V 1976 *Opt. Commun.* **18** 525
5.8   Bruce C W *et al* 1976 *Appl. Opt.* **15** 2970
5.9   Bagratashvili V N, Zharov V P and Lobko V V 1978 *Kvant. Electron.* **5** 637 (in Russian)
5.10 Zharov V P and Montanari S G 1984 *Zh. Prikl. Spectrosc.* **41** 401 (in Russian)
5.11 Deaton T E, Depatie P A and Walker T W 1975 *Appl. Phys. Lett.* **26** 300
5.12 Gomenyuk A S *et al* 1974 *Kvant. Elektron.* **1** 1805 (in Russian)
5.13 Koch K P and Lahmann W 1978 *Appl. Phys. Lett.* **32** 289
5.14 Angus A M, Marinero E E and Colles M J 1975 *Opt. Commun.* **14** 223
5.15 Zharov V P, Gomenyuk A S, P'aset'zki V B and Tokhtuev E G 1983 *Zh. Prikl. Spectrosc.* **39** 1029 (in Russian)
5.16 Gerlach R and Amer N M 1978 *Appl. Phys. Lett.* **32** 228
5.17 Patel C K N and Kerl R J 1977 *Appl. Phys. Lett.* **30** 578
5.18 Nodov E 1978 *Appl. Opt.* **17** 1110
5.19 Dorozhkin L M, Zharov V P, Makarov G N and Puretzky A A 1980 *Pis'ma Zh. Techn. Fiz.* **6** 979 (in Russian)
5.20 Gomenyuk A S, Zharov V P, Letokhov V S and Raybov E A 1976 *Kvant. Elektron.* **3** 369 (in Russian)
5.21 Patel C K N and Kreuzer L B 1971 *Science* **173** 45
5.22 Gomenyuk A G, Zharov V P, P'aset'zki V B and Shaidurov V O 1981 *Zh. Prikhl. Spectrosc.* **35** 1112 (in Russian)
5.23 Claspy P C, Pao V-H, Kwong S and Nodov E 1976 *Appl. Opt.* **15** 1506
5.24 Di Lieto A, Minguzzi P and Tonelli M 1979 *Opt. Commun.* **31** 25
5.25 German K R and Gornal W S 1981 *J. Opt. Soc. Am.* **71** 1452
5.26 Shtepa V I, Vereshchagina L N, Osmanov R R, Putilin F N and Zharov V P 1982 *J. Photoacoustics* **1** 181
5.27 Zharov V P and Montanari S G 1981 *Opt. Spectrosc.* **51** 124 (in Russian)
5.28 Zharov V P, Letokhov V S and Ryabov E A 1977 *Appl. Phys.* **12** 15
5.29 Compillo A J, Lin H-B and Dodge C J 1980 *Opt. Lett.* **5** 424
5.30 Bridges T J and Burkhardt E G 1977 *Opt. Commun.* **22** 248
5.31 Siebert D R, West C A and Barrett J J 1980 *Appl. Opt.* **19** 53
5.32 Brueck S R J, Kildal H and Belanger L J 1980 *Opt. Commun.* **34** 199
5.33 Patel C K N and Tam A C 1981 *Rev. Mod. Phys.* **53** 517
5.34 Lahmann W, Ludewig H J and Welling H 1977 *Anal. Chem.* **49** 549
5.35 Oda S, Sawada T and Kamada H 1978 *Anal. Chem.* **50** 865
5.36 Voigtman E, Jurgensen A and Winefordner J D 1981 *Anal. Chem.* **53** 1442
5.37 Schrepp W, Stumpe R, Kim J I and Walther H 1983 *Appl. Phys.* B **32** 207
5.38 Sawada T and Oda S 1981 *Anal. Chem.* **53** 539
5.39 Eltre L S (ed) 1969 *Encyclopedia of Industrial Chemical Analysis* (Port Chester, NY)
5.40 Brazhnicov V V 1974 *Differential Detectors for Gas Chromatography* (Moscow: Nauka) (in Russian)
5.41 Dolin S A and Kruegle V A 1967 *Appl. Opt.* **6** 276
5.42 Ericson M D 1979 *Appl. Spectrosc. Rev.* **15** 261
5.43 Smith S L, Garlok S E and Adams G E 1983 *Appl. Spectrosc.* **37** 192
5.44 D'Orazio M 1979 *Appl. Spectrosc.* **33** 278
5.45 Klimcak C M and Wessel J E 1980 *Anal. Chem.* **52** 1233

5.46 Kruezer L B 1978 *Anal. Chem.* **50** 597A

5.47 Zharov V P, Letokhov V S and Montanari S G 1983 *Laser Chem.* **1** 163

5.48 Zharov V P, Gavrilov V V, Litvin E F and Montanari S G 1985 *Zh. Prikl. Spectrosc.* **42** 506 (in Russian)

5.49 Welti P 1970 *Infrared Vapor Spectra* (New York: Heyden)

5.50 Yakobi Yu A 1981 *Kvant. Elektron.* **8** 555 (in Russian)

5.51 Zharov V P, Letokhov V S, Montanari S G and Tumanova L M 1983 in *Abstr. III Int. Meet. on Photoacoustic and Photothermal Spectroscopy, Paris, France, 1983* report 7.14

5.52 Zharov V P, Letokhov V S, Montanari S G and Tumanova L M 1983 *Dokl. Acad. Nauk SSSR* **269** 1079 (in Russian)

5.53 Zharov V P, Montanari S G and Tumanova L M 1984 *Zh. Anal. Khim.* **39** 551 (in Russian)

5.54 Zharov V P, Letokhov V S, Montanari S G and Tumanova L M 1984 *J. Mol. Struct.* **115** 261

5.55 Zharov V P and Montanari S G 1982–3 *J. Photoacoustics* **1** 355

5.56 Zharov V P, Montanari S G and Letokhov V S *J. Phys., Paris* **44** C6 Suppl. 10 573

5.57 Zharov V P and Montanari S G 1985 *Laser Chem.* **5** 133

5.58 Zharov V P and Montanari S G 1983–4 *J. Photoacoustics* **1** 445

5.59 Froehlich P and Wehry E L 1981 Fluorescence detection in liquid and gas chromatography in *Modern Fluorescence Spectroscopy* ed E L Wehry Chap. 2 (New York: Plenum) p. 35

5.60 Oda S and Sawada T 1981 *Anal. Chem.* **53** 539

5.61 Voigtman E, Jurgensen A and Winefordner J D 1981 *Anal. Chem.* **53** 1921

5.62 Lai E P C, Su S V, Voigtman E and Winefordner J D 1982 *Chromatographia* **15** 645

5.63 Voigtman E and Winefordner J D 1982 *J. Liq. Chromatogr.* **5** 2113

5.64 Voigtman E and Winefordner J D 1983 *J. Liq. Chromatogr.* **6** 1275

5.65 Gomenyuk A S *et al* 1984 in *Travaux III Symp. Franco-Sovietique en Instrumentation, Aussois* p. 156

5.66 Zharov V P 1984 *Zh. Anal. Khim.* **39** 780 (in Russian)

5.67 Berezkin V G and Bochkov A S 1980 *Qualitative Thin-Layer Chromatography* (Moscow: Nauka) (in Russian)

5.68 Fishman V A and Barda J 1981 *Anal. Chem.* **51** 102

5.69 Rosencwaig A and Nall S S 1975 *Anal. Chem.* **47** 548

5.70 Castleden S L and Kirkbright G F 1979 in *Abstr. I Int. Meet. on Photoacoustic Spectroscpy, Iowa, USA, 1979* paper FB3-1

5.71 Castleden S L, Elliott C M, Kirkbright G F and Spillane D E M 1979 *Anal. Chem.* **51** 2152

5.72 Ashworth C M, Castleden S L, Kirkbright G F and Spillane D E M 1982 *J. Photoacoustics* **1** 152

6 Laser Fluorescence Analysis of Organic Molecules in Solid Solutions

R I Personov

6.1 Introduction

The fluorescence method is distinguished from other analytical methods by its extremely high sensitivity. In principle under optimal conditions the limit of detection for this method is one atom or one molecule (see Chapter 1). Many complex organic compounds which are of both considerable scientific and practical interest possess intensive luminescence (fluorescence and phosphorescence). These include aromatic hydrocarbons, heterocyclic compounds, porphyrins including chlorophyll and its nearest analogues, dyes and many others. For dozens of years luminescence of organic compounds has been used as an important source of information about properties and structure of complex molecules. It has also been used for spectrochemical analysis of different natural and industrial products.

In the case of one-component solutions rather low detection limits are achieved easily enough. For example, investigating fluorescence of aqueous solutions of rhodamine 6 G (excited by argon laser radiation) the dye is easily detected at a concentration of 10^{-13} mol l^{-1} (in a working volume of solution 10^{-12} litre with the number of the detected molecules about 10^4 [6.1]). At this stage, as the authors note, there are great reserves for a further considerable reduction of the detection limit. Even this example shows that the sensitivity necessary for practical analytical purposes can be achieved rather easily.

The problem of selectivity improvement appears to be more important and difficult so far as the analysis of complex organic molecules goes. In fact, in the

272

majority of practical cases, it is necessary to analyse multicomponent mixtures. Here considerable difficulties emerge, which are connected with an insufficient selectivity due to the large width and the overlapping of spectral bands of separate components. At room temperature the absorption and luminescence spectra of organic compounds usually contain one or several bands whose width is hundreds or thousands of wavenumbers. In this respect searching for conditions and means for obtaining more informative fine-structure spectra of complex molecules is of paramount importance.

A well known method due to Shpol'skii [6.2] is one of the important methods for obtaining quasi-line spectra. It was established that the spectra of many organic molecules in an appropriate crystalline matrix selected from the short-chain n-paraffins at low temperature consist of narrow bands—quasi-lines with a width of 2–20 cm^{-1}. At present several hundreds of compounds which give quasi-line spectra in n-paraffin matrices are known (see e.g. [6.3–6.5]). The above method has had wide recognition and is successfully employed both in investigating molecular spectroscopy problems and in spectrochemical analysis [6.6–6.8].

However, the presence of quasi-line spectra for a number of molecules in specially selected n-paraffin matrices does not eliminate the general problem of the origin of broad-band molecular spectra and searching for means for obtaining fine-structure spectra. As a matter of fact, many compounds have low or practically no solubility in n-paraffins. Many molecules that can be introduced into n-paraffin matrices even under such conditions possess broad-band spectra. Taking into account the great variety of organic compounds and the wide choice of different solvents, it can be said that in the majority of cases the spectra of solutions of complex molecules consist of broad bands even at helium temperatures.

In the last decade considerable progress has been made in understanding the spectral broadening nature of organic compounds. An important advance along these lines was the establishment in our laboratory of the fact that in many cases, at sufficiently low temperatures, the spectra of solutions are broadened, mainly inhomogeneously, and possess a concealed line structure [6.9, 6.10]. This circumstance served as a starting point for the development of a new trend—laser fine-structure selective spectroscopy of complex molecules in frozen solutions—often called 'site selection spectroscopy'. By now methods have been worked out which eliminate inhomogeneous broadening and reveal line structure in the fluorescence, phosphorescence and absorption spectra of molecules in arbitrary solutions by means of selective laser excitation (see reviews [6.11, 6.12]). All the principal features and characteristics of line spectra obtained by these procedures have been studied. New vast opportunities were thus created for fine-structure spectroscopic investigation of complex molecules and their interaction in various media, as well as for development of new selective techniques of spectrochemical analysis for complex organic mixtures.

It is worth noting that inhomogeneous broadening is also quite significant in the quasi-line spectra of molecules in *n*-paraffin matrices. Consequently in this case selective laser excitation also proves to be rather effective. It allows simplification of the so-called 'multiplet structure' (connected with several non-equivalent ways of introducing the molecules under investigation into the crystallised matrix) and, upon eliminating the inhomogeneous broadening, permits narrowing of the spectral lines by one or two orders.

Employing cooled supersonic molecular beams [6.13, 6.14] is another new and interesting opportunity for obtaining fine-structure spectra of complex molecules. However, this prospective method for investigating energy structure and relaxation processes in complex molecules (see, for example, [6.15, 6.16]) has not yet been widely employed in spectrochemical analysis of multicomponent mixtures. Serious complications here will probably be connected with the introduction of a complex sample under study into the beam while preserving the necessary conditions of its purity and cooling.

The present chapter deals with the main principles and techniques of selective spectroscopy of complex organic molecules in solid solutions, and certain applications of these methods are considered. While paying most attention to fluorescence spectra we shall also briefly discuss some related problems of phosphorescence and absorption spectra.

Before considering and discussing experimental results, we shall recall certain general principles of the theory which will be required later.

6.2   General theoretical principles

Below we will consider three types of electron–vibronic spectra of complex organic molecules in solution: absorption, fluorescence and phosphorescence spectra (see figure 6.1). Molecules in solution are similar in many aspects to impurity centres in crystals. For this reason we shall make use of the main concepts of the impurity centre theory to analyse the peculiarities of the molecular spectra (see, for example, [6.17–6.19]). In accordance with the conditions commonly realised in experimental investigations we shall assume that the electronic excited states of the solvent are considerably higher than those of the molecules—i.e. the 'impurities' being investigated. This means that in the spectral region which interests us only the electrons of the impurities interact with the light. We shall also consider the concentration of the impurity molecules to be low enough to neglect interaction between them.

The nature of the impurity molecule spectrum is determined by electron–vibrational interactions of two types: interaction of the molecule's electrons with its intramolecular vibrations (vibronic coupling) and interaction with the intermolecular vibrations of the solution (electron–phonon coupling). Vibronic coupling leads to the presence of a series of vibronic bands in the

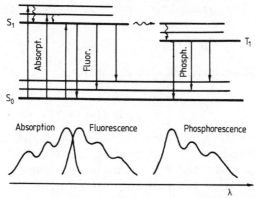

Figure 6.1   Diagram of electron–vibrational transitions in a molecule.

molecular spectrum, along with the band in the region of a purely electronic transition. The nature and shape of each vibronic band (including a purely electronic one) are determined by electron–phonon coupling. In principle, each vibronic band may consist of two parts (figure 6.2): a narrow zero-phonon line (ZPL), corresponding to transitions in the impurity molecule without any change in the number of phonons in the matrix (an optical analogy of the resonance γ-line in the Mössbauer effect); and a relatively broad phonon wing (PW) due to the phototransitions in the molecule with the simultaneous creation or annihilation of matrix phonons. What particularly should be observed in the spectrum (ZPL, PW or ZPL + PW) depends upon the strength of the electron–phonon coupling and upon the temperature.

Let us consider more thoroughly the structure of a spectral band which corresponds to one vibronic transition. In the region of other transitions the picture will be qualitatively repeated. Without going into theoretical detail (see [6.17–6.19] and also review [6.11]), we shall demonstrate here certain principles.

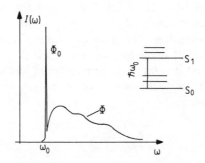

Figure 6.2   The general nature of the spectral band of an impurity centre (one vibronic transition).

The profile $I(\omega)$ of a spectral band (which we shall take as normalised, i.e. $\int_{-\infty}^{+\infty} I(\omega)\, d\omega = 1$) can be represented as follows:

$$I(\omega, T) = \Phi_0(\omega, T) + \Phi(\omega, T). \tag{6.1}$$

The first term defines a zpl of Lorentzian shape:

$$\Phi_0(\omega, T) = \frac{1}{\pi}\, \frac{e^{-f(T)}\, \Gamma(T)}{[\omega - \Omega(T)]^2 + \Gamma^2(T)} \tag{6.2}$$

where $\Gamma(T)$ and $\Omega(T)$ express the temperature broadening and shift of the zpl, determined by the electron–phonon coupling. In case of $T \to 0$ the quantities $\Gamma(T)$ and $\Omega(T)$ go to zero and the zpl turns into a δ-shaped peak or, strictly speaking, into a narrow line possessing natural width. The temperature-dependent function $f(T)$ is determined by the impurity crystal phonon spectrum and by the electron–phonon coupling character.

The second term in (6.1) describes the profile of a broad pw and can be expressed as:

$$\Phi(\omega, T) = \sum_{m=1}^{\infty} \Phi_m(\omega, T)$$

$$\int_{-\infty}^{+\infty} \Phi_m(\omega, T)\, d\omega = \Phi_m(T) = \frac{f^m(T)}{m!}\, e^{-f(T)}. \tag{6.3}$$

With a sufficiently good approximation (in the case of the so-called line electron–phonon coupling when the zpl temperature broadening and shift are not present and the zpl has natural width) quantities $\Phi_m(T)$ have a simple physical meaning: they represent contributions to the observed pw of the transitions with creation or annihilation of m phonons of the matrix. In the same approximation the function $f(T)$ has a relatively simple form:

$$f(T) = \int_0^{\infty} f_0(v)\left[\frac{2}{\exp(hv/kT) - 1} + 1\right] dv \tag{6.4}$$

where $f_0(v) = 6N\xi^2(v)\rho(v)$ is the so-called weighted density of phonon states, $\rho(v)$ is the phonon state density of the impurity crystal; $\xi^2(v)$ is the electron–phonon coupling function—the average square displacement of the equilibrium positions of oscillators of frequency $v = v_q$ upon an electron transition in the impurity and N is the number of molecules in the crystal. Finally, for the sake of experimental data analysis we introduce a very convenient parameter α:

$$\alpha = \frac{I_{ZPL}}{I_{ZPL} + I_{PW}} = e^{-f(T)} \tag{6.5}$$

where $I_{ZPL} = \int_{-\infty}^{+\infty} \Phi_0(\omega, T)\, d\omega$ and $I_{PW} = \int_{-\infty}^{+\infty} \Phi(\omega, T)\, d\omega$ are integrated intensities of the zpl and the pw, respectively. The quantity α indicates the share of the ZPL intensity in the integrated intensity of the whole spectral band, and may serve as a characteristic of the strength of the electron–phonon coupling. The parameter α (by analogy with the theory of neutron and x-ray scattering) is usually called the Debye–Waller factor.

Taking into consideration the above relations we can come to a number of very important conclusions. It is seen from (6.2), (6.4) and (6.5) that the stronger the electron–phonon coupling the lower the ZPL intensity. At a given strength of electron–phonon coupling the ZPL intensity and the α factor go down very quickly with increasing temperature. Here the ZPL intensity is transferred to the PW, for the integrated intensity of the whole ZPL + PW band is independent of the temperature. In the temperature region where $kT \gg h\nu_{max}$ (ν_{max} is the maximum phonon frequency with which the impurity molecule interacts) the ZPL intensity drops exponentially with the increase in the temperature.

Comparing (6.3) and (6.5) it becomes evident that a single-valued relation exists between $\Phi_m(T)$ and α. Knowing α we can easily find the contribution of the m-phonon transitions to the PW (and vice versa). The required calculations were carried out in [6.21] and their results are demonstrated in figure 6.3. From this figure it follows that while for $\alpha = 0.9$, for example, the PW is practically determined by one-phonon transitions, for $\alpha = 0.1$ then a substantial contribution to the PW comes from two-, three- and four-phonon transitions. Since the Debye frequency (ν_{max}) of molecular crystals usually lies in the range of 40 to 100 cm^{-1} (see, for example, [6.20, 6.21]), the PW can occupy a spectral range from several dozens to several hundreds of wavenumbers.

Thus, a temperature increase results in a reduction in α, a weakening of the ZPL and a relative strengthening and broadening of the PW. Therefore, in order to reveal narrow ZPL, sufficiently low temperatures should be used.

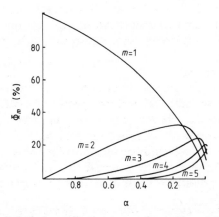

Figure 6.3   Relative contribution to PW of transitions with the creation of various number of phonons as a function of α [6.21].

6.3   Principles of selective spectroscopy of molecules in matrices

6.3.1   *Introduction*

From the viewpoint of the aforesaid concepts about the character of spectral bands of the impurity centres, there are two possible extreme cases that concern the origin of the broad bands in the spectra of organic molecules in solutions.

(1) Strong electron–phonon coupling takes place between the molecules and the solvent. In this case the ZPL intensity can be very low even at low temperatures (see equations (6.4) and (6.5)), and wide bands which are observed are well developed PW. These PW are due to both one- and multiphonon transitions. The broadening of the spectral band of a molecular ensemble is connected with the spectrum broadening of each molecule and is called homogeneous.

(2) The electron–phonon coupling is weak, and the spectra of all molecules consist of narrow ZPL. But, in this case, the molecules under investigation are in different local conditions in the field of a solvent. This leads to statistical scattering in the positions of their electronic levels and to relative displacement of their spectra on the frequency scale. Here each broad band observed in the spectrum is the envelope of a great number of ZPL and the band broadening is inhomogeneous†.

As a result of the experiments conducted in our laboratory employing fluorescence excitation by means of a monochromatic laser, it was established that at sufficiently low temperatures the decisive role, in many cases, is played by the inhomogeneous broadening [6.9, 6.10, 6.23, 6.24]. This was subsequently confirmed in the works of other authors [6.25–6.28]. Later papers developed selective methods of eliminating inhomogeneous broadening and of revealing line structure in fluorescence, phosphorescence and absorption spectra. Main features of this structure were also investigated (see reviews [6.11, 6.12] and references therein). Let us consider each of the above three types of spectra, paying special attention to fluorescence spectra.

6.3.2   *Fluorescence spectra*

At sufficiently low temperatures and with not too strong electron–phonon coupling the absorption and fluorescence spectra of molecules in a solid solution consist of narrow vibronic ZPL, accompanied by broad PW (at the longwave side of the ZPL in emission and the shortwave side in absorption). However, differences in local conditions for dissolved molecules in the solid solution (due to the differences, for example, in the mutual orientation of impurity and solvent molecules) result in various displacements of their

---

† It should be noted, that a number of other causes of the broadening of molecular spectra have been discussed in the literature (see [6.22]). However, these have nothing in common with the spectra considered below.

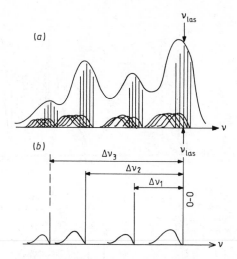

Figure 6.4 Diagram of the formation of the inhomogeneously broadened spectrum. (*a*) spectrum of the whole set of molecules; (*b*) spectrum of one-type centres upon selective excitation.

electronic levels and in statistical scattering of their spectra on a frequency scale (figure 6.4(*a*)). This scattering is usually expressed in hundreds of wavenumbers. In this case, generally speaking, the local fields shift various electronic levels differently. Therefore the band displacements corresponding to different electronic transitions may not be correlated†. Upon summation of a large number of the aforesaid spectra a broad-band spectrum is observed (figure 6.4). For the purpose of eliminating the said inhomogeneous broadening in emission spectra and of revealing its line structure we can use monochromatic laser excitation with a frequency in the region of the purely electronic or the lowest vibronic transitions. In this case mainly those molecules that possess their absorption ZPL at the laser frequency will undergo selective excitation. Then in the emission spectrum narrow lines belonging only to those molecules will appear. Some examples which illustrate this are given in figure 6.5. It can be seen that when monochromatic excitation is resorted to, narrow ZPL are displayed in the spectra instead of a few broad bands. These ZPL are accompanied on the longwave side by broad PW. Such line-structure spectra can be obtained for various compounds both in the neutral and in the ionic forms, and in a great variety of solvents (glassy and crystalline, polar and nonpolar). These data unambiguously demonstrate the

† The above local differences can also result in a certain scattering of frequency values of normal vibrations in impurity molecules, but it is usually very small ($\lesssim 1$–$2 \ cm^{-1}$); and the position of the whole spectrum is determined by a purely electronic level position.

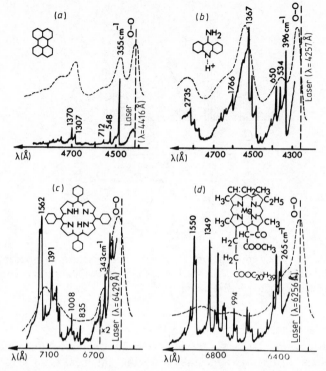

Figure 6.5 Fine-structure fluorescence spectra of solutions of organic compounds upon selective laser excitation (full curves) at $T = 4.2$ K: (a) perylene in ethanol; (b) protonated form of 9-aminoacridine in ethanol with HCl addition; (c) tetraphenylporphyrin in polystyrene; (d) protochlorophyll in ether. Broken lines indicate the spectra upon ordinary UV excitation by the light of the mercury discharge lamp.

essential role played by inhomogeneous broadening of the spectra and the possibility of its elimination by selective excitation.

Let us now consider the principal characteristics of the fluorescence line spectra obtained by this method.

(a) General nature of spectral bands

Strictly speaking, even by monochromatic laser excitation one cannot completely eliminate inhomogeneous broadening in fluorescence spectra. First and foremost we consider the significant role played by the so-called 'nonresonantly' excited centres. Thanks to the latter the longwave wings, observed close to the ZPL, are not simply phonon ones but have a more complicated origin. They include the true PW of the resonantly excited centres, as well as the fluorescence bands (ZPL + PW) of the nonresonantly excited centres, i.e. excited not through the ZPL but through the PW in absorption. To elucidate this point consider figure 6.6, showing the spectrum pattern in the O–O transition region. The absorption bands of the centres, laser-excited through the PW are shown by broken lines in the upper part of the figure. Since the intensity at a PW maximum is usually considerably less than that at a ZPL maximum, the fluorescence of each nonresonantly excited centre is relatively weak. But since the width of the PW is much greater than that of the ZPL, the number of nonresonantly excited centres is large. Together they make an appreciable contribution to the integrated intensity of the longwave wing close to the fluorescence line. (This contribution corresponds to the hatched area in the lower part of figure 6.6.) This leads to an apparent increase in the PW and an apparent reduction in the Debye–Waller factor as compared with the true values. Let us consider this in more detail.

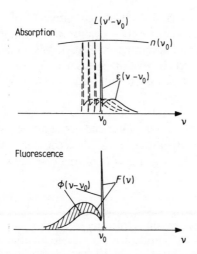

Figure 6.6 Diagram of the formation of a longwave wing near a ZPL in a selectively excited fluorescence spectrum.

Let $\varepsilon(v' - v_0)$ and $\phi(v - v_0)$ be the true spectral curves of the O–O band in absorption and fluorescence spectra belonging to one type of centre with the ZPL maximum at frequency v_0. We assume that the shape of the spectra for all types of centres is the same, and that they differ only in the value of v_0. Let $L(v' - v_1)$ be a laser line profile (where v_1 is the frequency of the maximum), and $n(v_0)$ be the centre distribution function along the O–O transition frequency. Then the spectral curve for the O–O band in the fluorescence spectrum is expressed in the form of convolution:

$$F(v) = \int_0^\infty \int_0^\infty dv' \, dv \, \phi(v - v_0)\varepsilon(v' - v_0)n(v_0)L(v' - v_1). \qquad (6.6)$$

When the laser linewidth $L(v' - v_l)$ is much narrower than the ZPL width and the laser line can be considered as a δ function and, in addition, the width of the centre distribution function $n(v_0)$ is considerably larger than the PW width (which is valid only for an infinitely large inhomogeneous broadening), then, instead of (6.6) we can write:

$$F(v) \simeq \int_0^\infty dv_0 \, \phi(v - v_0)\varepsilon(v_1 - v_0). \qquad (6.7)$$

Making use of (6.7), it is possible to establish the relation between the true factor α and the 'observed Debye–Waller factor' α_1, which can be defined, similar to (6.5), as follows:

$$\alpha_1 = \frac{I_{ZPL}}{I_{ZPL} + I_W} \qquad (6.8)$$

where the observed I_W is the integrated intensity of the observed wing. Presenting each of the functions $\phi(v - v_0)$ and $\varepsilon(v_1 - v_0)$ as the sum of the curves for the ZPL and the PW and assuming the Debye–Waller factors to be the same for both fluorescence and absorption bands (i.e. $\alpha_{fluor} \simeq \alpha_{absorp} = \alpha$), on the basis of (6.7), (6.8) and (6.5) it is not difficult (see [6.11]) to obtain

$$\alpha_1 \simeq \alpha^2. \qquad (6.9)$$

The relations (6.7) and (6.9) are obtained in the rather rough approximation $n(v_0) = $ constant. It is evident that in real-valued systems the distribution $n(v_0)$ width is not infinite. Making use of (6.6) one can show that in general the observed factor α_1 is found between

$$\alpha^2 < \alpha_1 < \alpha \qquad (6.10)$$

and depends on the excitation frequency (see below).

From the relations (6.9) and (6.10) it is evident that the fluorescence of nonresonantly excited centres considerably 'spoils' the line pattern of the spectrum, reducing the observed Debye–Waller factor α_1 as compared with α in the homogeneous spectrum. The α variations due to the effects of some factors lead to the sharper α_1 variations.

To give a concrete example of how α variation affects the appearance of the spectrum figure 6.7 demonstrates the spectra of the same dye (thioindigo) in three different solvents upon selective excitation in the O–O transition region (in the case of *n*-paraffin as solvent the excitation wavelength was taken somewhat shorter than in the other two cases due to the general displacement of the spectrum in this solvent). Rough approximate estimates, based on measurements in these spectra using (6.9), yield the values $\alpha = 0.3, 0.15$ and 0.07 respectively for the three above cases. It follows from figure 6.7 that a further small reduction in α would lead to the complete disappearance of the ZPL in the spectrum and to the appearance of a structureless band.

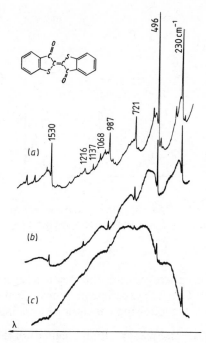

Figure 6.7 Fluorescence spectra of thioindigo solutions upon laser excitation ($T = 4.2$ K) [6.59]: (*a*) in paraffin oil ($\lambda_{laser} = 5550.6$ Å); (*b*) in ethanol ($\lambda_{laser} = 5621.5$ Å); (*c*) in chloroform ($\lambda_{laser} = 5621.5$ Å).

From the aforesaid it follows that when the line structure in the spectrum is absent, even at low temperatures and with corresponding monochromatic excitation, the reason should be primarily sought in a large magnitude of the electron–phonon coupling. The latter could in principle be weakened by selecting an appropriate matrix.

(b) *Temperature dependence of spectra*
Following from (6.4) and (6.5) the temperature increase results in a drastic ZPL weakening. This ZPL weakening in fluorescence spectra upon selective excitation is revealed particularly strongly due to the relations (6.9) and (6.10). In many cases the ZPL practically disappear from the spectrum at a temperature between 40 and 60 K [6.9, 6.10]. This is illustrated by a typical example shown in figure 6.8 taken from an earlier paper [6.9]. Thus it is evident that in order to reveal the fine structure of a spectrum a sufficiently low cooling of the solution is necessary along with selective excitation.

(c) *Dependence of fluorescence spectra on the laser excitation wavelength*
In the above examples (figure 6.5 and 6.7) the monochromatic excitation of fluorescence was made in the region of a purely electronic transition. The

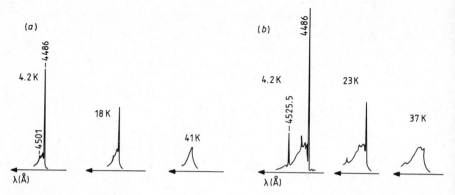

Figure 6.8   Vibronic band (0–355 cm^{-1}) in the fluorescence spectrum of perylene in ethanol (*a*) and in *n*-undecane (*b*) at various temperatures ($\lambda_{\text{laser}} = 4415.6$ Å in both cases) [6.9].

question which naturally arises concerns the spectra's dependence on the laser excitation wavelength. This was experimentally investigated in a number of papers [6.27–6.32] and the following conclusions were made.

(1) Variation of the laser frequency within the limits of the inhomogeneous broadening in the O–O transition region causes a corresponding displacement of the fluorescence spectrum preserving frequency intervals between vibronic ZPL. However, if the width of the function $n(\nu_0)$ of the centre distribution along the O–O transition frequency is close to or less than the PW width, the variation of the exciting line frequency in the limits of a longwave absorption band can result in a considerable variation of relative contributions into fluorescence of the resonantly and nonresonantly excited centres†. In fact, in the case discussed the longwave part of the broad inhomogeneous absorption band is mainly formed from the ZPL, and the shortwave one mainly from the PW. Therefore upon excitation in the shortwave region of the O–O absorption band the selectivity goes down and the spectrum structure becomes less sharp.

(2) Upon a considerable increase in the laser frequency, when the excitation is performed in the region of high vibronic transitions of the excited electronic state (exceeding the O–O transition frequency by 3000 to 5000 cm^{-1} and more), the fluorescence spectrum is smeared. In fact, it does not differ from the spectra obtained with broad-band source excitation. It is not difficult to qualitatively understand this smearing of the spectrum upon a large increase in the laser frequency. As a matter of fact, high vibronic levels in the region of overtones and combination tones and the corresponding ZPL are considerably broader than the purely electronic ones. Moreover, there is a significant increase in the density of states in this region. Hence, one can speak of an

† This is demonstrated by simple pattern calculations [6.29].

appreciable overlap of spectral bands belonging to various centres in the region of high vibronic states in the absorption spectrum. In this region the laser excites practically all types of centres [6.30], and inhomogeneous broadening is not eliminated.

(3) More important and interesting is the fluorescence spectrum dependence on v_{laser} excitation to not very high vibronic levels (up to 2000 or 3000 cm^{-1}), when the spectrum is still of the line type. Under these conditions a detailed structure of the spectrum strongly depends on the wavelength of the laser excitation [6.30]. If the excitation is performed in the O–O transition region the fluorescence spectrum consists of single vibronic lines. On the contrary, with the excitation to the vibronic levels, the fluorescence spectrum becomes more complicated and consists of groups of lines—'multiplets'. The number and the arrangement of the lines in a multiplet depend on v_{laser}. This is illustrated by figure 6.9, which shows the O–O transition region in the fluorescence spectrum of porphin in polystyrene upon selective excitation in the region of various vibronic transitions. In each case, it is evident that the 'O–O multiplet' contains from three to six lines and all these multiplets are quite different from one another. The most intense components of these multiplets are subsequently repeated at vibronic transition frequencies throughout the whole fluorescence spectrum.

The occurrence of the said multiplets can be explained by employing the simple diagram in figure 6.10. Assume that in the excited state S_1 the molecules

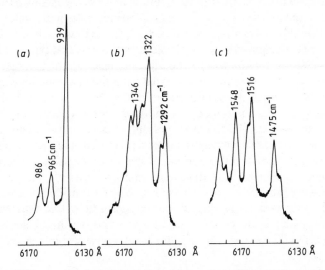

Figure 6.9 'O–O multiplets' in the fluorescence spectrum of porphin in polystyrene upon laser excitation to various vibrational sublevels of the excited electronic state ($T = 4.2$ K) [6.34]: (a) $\lambda_{laser} = 5807$ Å; (b) $\lambda_{laser} = 5688$ Å; (c) $\lambda_{laser} = 5634$ Å.

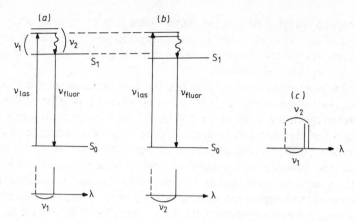

Figure 6.10 Diagram of the 'O–O multiplets' formation in a fluorescence spectrum.

under investigation have two close vibronic levels corresponding to the excitation of vibrations with frequencies v_1 and v_2. It is assumed later that for the dissolved molecules the laser frequency corresponds to the excitation to a vibronic level with the frequency v_1. After vibronic relaxation to the S_1 state these molecules emit a quantum of fluorescence. The O–O line in the fluorescence spectrum of these molecules is separated from the laser line by a distance v_1 (figure 6.10(a)). If the magnitude of inhomogeneous broadening is larger than the difference $v_2 - v_1$ there will be molecules in the solution whose energy of a purely electronic transition S_1–S_0 is less than the energy of the first type just by the magnitude $v_2 - v_1$. Molecules of the second type will be excited by the same laser radiation to the vibronic level v_2. In their fluorescence spectrum the O–O line will be separated from the laser one by a distance v_2 (figure 6.10(b)). Since the molecules of both types are present in the solvent simultaneously, a doublet emerges in the fluorescence spectrum (figure 6.10(c)). If, in the excited state, the dissolved molecules have not two but several closely spaced vibronic levels neighbouring the laser frequency, a corresponding, more complex 'multiplet' emerges in the O–O transition region of the fluorescence spectrum. The distances from the laser line to the separate components of this multiplet correspond to the vibrational frequencies of a molecule in the excited state S_1. By varying the exciting line frequency we can 'probe' various regions of the vibronic states and obtain all the vibrational frequencies in the excited state which prove to be active in a given electronic transition.

The above mechanism explains the 'multiplet structure' dependence on the laser excitation frequency and its independence of the solvent. The validity of the above 'multiplet' interpretation has been experimentally confirmed on a number of molecules, for which the vibration frequencies in the excited electronic state were determined by independent means [6.30]. Later this method was employed to obtain vibration frequencies of molecules in the excited electronic state (and symmetry, when measuring the polarisation degree, as well) from the fluorescence spectra [6.33–6.35]. So far as the vibrations of molecules in the ground state are concerned, it is evident that they are determined by the vibronic structure of the fluorescence spectrum.

It should be noted that if an absorption spectrum contains sufficiently intense PW, with relatively narrow maxima, a supplementary structure, due to these PW, may appear in the multiplets. However, as a rule the PW do not contain such narrow peaks.

The above data lead to the conclusion that from the viewpoint of analytical applications, when it is more important to obtain simple spectra with the intense narrow ZPL, it is better to perform selective excitation in the longwave part of a purely electronic absorption band or of the lowest vibronic transition bands of compounds under investigation.

(d) Zero-phonon linewidth

Measurements of the ZPL width in the fluorescence spectra of organic molecules (at 4.2 K) upon excitation in the O–O transition region by narrow laser lines ($\Delta v \leqslant 0.1\,\mathrm{cm}^{-1}$) usually yield values within the range of 1 to $10\,\mathrm{cm}^{-1}$. All these ZPL correspond to vibronic transitions and their width is determined to a large degree by the relaxation times of the vibrational excitations of the molecules under investigation. This width cannot be appreciably reduced. The corresponding times, estimated from the width of the vibronic ZPL, turn out to be of the order of 10^{-11} or 10^{-12} s.

At a very low temperature, when the ZPL temperature broadening due to the electron–phonon coupling is negligibly small, the homogeneous width of the purely electronic ZPL is significantly less than the homogeneous width of vibronic lines, and is determined by a lifetime of the electronic state. For the excited electronic state S_1 this lifetime equals 10^{-8} or 10^{-9} s, and the broadening magnitude, related to it, lies in the range of 5×10^{-4} to $5 \times 10^{-3}\,\mathrm{cm}^{-1}$. However, this width cannot be determined from the fluorescence line spectra. Upon excitation in the O–O band the observation is hindered by scattered laser radiation. Upon excitation in the vibronic absorption band the O–O fluorescence line will be inhomogeneously broadened. In the latter case, the profile of the O–O line in the fluorescence spectrum is a convolution of a comparatively wide profile belonging to the vibronic absorption line with the O–O fluorescence line of a separate type of centre (see relation (6.7)). In this case, therefore, the O–O linewidth in the fluorescence spectrum cannot be less than the homogeneous width of the vibronic absorption line.

The homogeneous width of the purely electronic line can be determined from experiments on 'burning' narrow holes in an inhomogeneously broadened absorption band (see §6.3.4(b)). The appropriate measurements indicate that the homogeneous width of purely electronic lines is in fact very small. It is particularly small in the case of crystalline matrices, where at 2–4 K it ranges from 10^{-2} to 10^{-3} cm^{-1} [6.36–6.38].

It is worth noting that for various systems of impurity matrix the purely electronic ZPL width and its temperature dependence can vary widely. In the case of glassy solutions this width (at 2–4 K) is, as a rule, one or two orders larger than in the case of crystals. Along with this, due to a substantial role of low-frequency phonons, glassy solutions are characterised by considerable (in comparison with crystalline matrices) temperature variations of the ZPL width in the low-temperature region. To illustrate the above figure 6.11 gives the data for homogeneous widths of the O–O line of porphin in different matrices, taken from [6.40]. It is evident that in a certain number of amorphous matrices the ZPL width can become two or three times smaller if the temperature falls from 4 to 2 K.

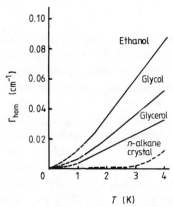

Figure 6.11　Temperature dependence of the homogeneous width of the purely electronic absorption line of porphin in various organic glasses (the broken line indicates similar data for a crystalline matrix) [6.40].

In some cases, the extrapolation to $T \to 0$ of the observed temperature dependence of the purely electronic ZPL width leads to theoretical values ranging between 10^{-3} and 3×10^{-4} cm^{-1} which are determined by the natural lifetime of the excited electronic state [6.38–6.41].

It can be concluded from the above, so far as analytical applications go, that in a number of cases it is advisable to select an appropriate solvent in order to obtain sharper spectra; and in some extreme situations, perhaps, to further decrease the temperature (below 4 K).

6.3.3　Phosphorescence spectra

Many organic compounds possess phosphorescence as well as fluorescence. Selective excitation enables us also to eliminate the inhomogeneous broadening and to reveal the line structure in phosphorescence spectra. However, to achieve this, monochromatic excitation should be employed directly in the $T_1 \leftarrow S_0$ transition region, forbidden with respect to spin. Population of the triplet state through the $S_1 \leadsto T_1$ conversion after excitation in the $S_1 \leftarrow S_0$ transition region (as usually done) is out of place here. Therefore it is rather difficult to perform such experiments, for the absorption coefficient in the $T_1 \leftarrow S_0$ transition region is 10^6 or 10^7 times less than in the case of the $S_1 \leftarrow S_0$ transition. By using sufficiently intense laser radiation, special phosphoroscopes and a sensitive registration system of weak spectra, these difficulties can be overcome and fine structure of the phosphorescence spectra can be revealed [6.42, 6.43]. Figure 6.12(*a,b*) shows us two such examples.

It should be noted that in a phosphorescence spectrum upon selective $T_1 \leftarrow S_0$ excitation, not only vibronic but also purely electronic resonance lines can be rather easily recorded at the laser frequency, due to the long lifetime of the triplet state. These lines are so narrow that in a number of cases with a high spectral resolution it is possible to observe the so-called zero-field splitting, determined by a spin–spin magnetic dipole interaction. The splitting, usually investigated using ESR methods, was observed for the first time (directly in the optical spectrum of the complex molecule) on coronene where it was $\simeq 0.09 \text{ cm}^{-1}$ (see figure 6.12(*d*) of [6.44]).

6.3.4 Absorption spectra

The above methods of selective excitation enable us to obtain line spectra in emission. It is of no less interest to reveal an analogous fine structure in absorption spectra. But it is obvious that for absorption spectra other techniques have to be employed.

(a) Excitation spectra

One of the possibilities for revealing line structure in inhomogeneously broadened absorption spectra concerns the recording of excitation spectra for narrow spectral parts of a fluorescence band. As is well known an excitation spectrum is obtained by scanning the wavelength of the exciting light and measuring the resulting variation in fluorescence intensity in a fixed (usually wide) spectral range. When the solution is sufficiently dilute and there is no dependence of the fluorescence quantum yield on the wavelength of the exciting light, the excitation spectrum exactly reproduces the absorption spectrum. Under conditions of inhomogeneous broadening it is possible to separate a narrow fluorescence spectral range which is determined mainly by one type of centre. Then the corresponding excitation spectrum will mainly reproduce the absorption of these centres, and it should display a fine structure. This can be observed experimentally (see, for example, [6.23, 6.45]).

The above method can be employed for investigating absorption spectra of luminescent molecules. In the last decade, however, another more universal method has been developed which reveals a fine structure in the absorption

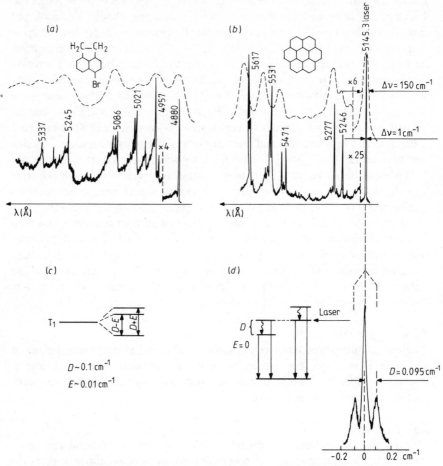

Figure 6.12   Fine-structure phosphorescence spectra upon laser $T_1 \leftarrow S_0$ excitation ($T = 4.2$ K) [6.44]: (*a*) 5-bromacenaphthene in butyl bromide ($\lambda_{\text{laser}} = 4880$ Å); (*b*) coronene in butyl bromide ($\lambda_{\text{laser}} = 5145$ Å); (*c*) diagram of zero-field spin sublevels; (*d*) zero-field splitting of the O–O line of coronene upon high-resolution recording. The broken curves in (*a*) and (*b*) indicate phosphorescence spectra of the same samples upon conventional $S_1 \leftarrow S_0$ excitation by the light of the mercury lamp. Spectrum (*a*) is taken from [6.93].

spectra. It is applicable, in principle, to both luminescent and nonluminescent molecules.

(b)   *'Hole-burning' spectra*
Fine structure in the inhomogeneously broadened absorption spectra can be revealed with the so-called 'hole-burning effect' [6.46, 6.47]. The essence of this

method is as follows. By means of monochromatic laser radiation it is possible to selectively modify ('burn') the centres of predominantly one type in an inhomogeneous system. The specific mechanisms of this 'burning' can be quite different. They can include a photochemical reaction, ionisation of selectively excited molecules, reorientation of impurity molecules in a matrix upon irradiation, transfer of molecules to a long-lived metastable state, etc. Among various types of burning there is a rather universal type observed in many photochemically stable molecules in glassy matrices. This type of burning is evidently connected with variation in the mutual orientation of the dissolved molecules and the close matrix molecules and is often called nonphotochemical. Howefer, independently of a specific burning mechanism the absorption spectrum of selectively burnt molecules differs from the initial one. Therefore, after burning, narrow holes appear in the broad-band absorption spectrum at the vibronic transition frequencies of the burnt centres. The difference in the absorption (transmission) spectra of the sample before and after hole burning determines the ensemble of all holes in their pure form. This difference spectrum—the 'hole burning spectrum'—carries direct information on the absorption line spectrum of molecules without inhomogeneous broadening [6.48].

As an example, two hole-burning spectra are shown in figure 6.13†. These spectra are recorded by means of a special two-channel spectral set-up. The latter is capable of registering changes in the optical density ΔD of the order of 10^{-3} or 10^{-4}. In these spectra, together with the O–O line observed at the laser frequency in the longwave absorption band, there are also weaker vibronic absorption lines. As expected, the narrowest lines (holes) in hole burning (absorption) spectra correspond to the purely electronic transitions. As an example, a narrow hole, burnt in a broad longwave absorption band of chlorin by a single-frequency laser, is demonstrated in figure 6.14.

The 'hole-burning' method proved to be exceedingly useful for a number of fine spectroscopic investigations of complex molecules in condensed media. Among these investigations are: the investigations of the homogeneous width of purely electronic and vibronic lines and, similarly, the investigation of electronic and vibrational relaxation processes; the investigation of features and properties of glassy media; the investigation of the Stark and the Zeeman effects on complex molecules; the investigation of photochemical reaction velocities in a solid medium, etc. The consideration of all these problems is beyond the scope of the present chapter, and references to original papers can be found in reviews [6.11, 6.12, 6.49–6.51]. Although hole-burning spectra have not been used in analytical practice, one can be sure that in future they

---

† The first case (perylene) is a typical example of nonphotochemical hole-burning. The second case (porphin) is an example of photochemical hole-burning due to phototautomerisation (i.e. movement of two protons in the centre of a porphin ring).

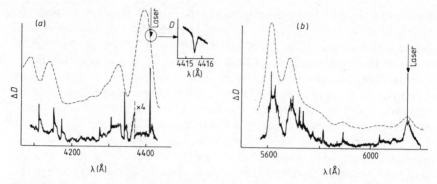

Figure 6.13 Broad-band absorption spectra (indicated by broken curves) and fine-structure hole-burning spectra (indicated by full curves) at $T=4.2$ K: (*a*) perylene in ethanol (burning time $t=5$ min, irradiation power density $P=5$ mW cm^{-2}); (*b*) porphin in polystyrene ($t=1$ min, $P=5$ mW cm^{-2}). At the top right there is a hole in the absorption spectrum at laser frequency. (Spectra obtained by Al'shits and Kharlamov.)

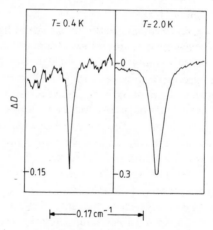

Figure 6.14 Hole in the absorption band of chlorine in polyvinylbutyral, burnt by a single-frequency laser ($\lambda_{\text{laser}}=6335$ Å) [6.41].

will be applied in selective spectrochemical analysis of complex impurities, for instance for selective determination of nonfluorescent compounds.

6.3.5 Selective fluorescence excitation of impurity molecular crystals with quasi-line spectra

As was noted earlier (see §6.1) certain impurity molecular crystals (e.g. impurity *n*-paraffin crystals) possess spectra with a quasi-line structure. For such systems, contrary to solutions with broad spectra, the problem of the nature of the investigated bands does not arise for the ZPL are present in the spectrum upon normal excitation. In these cases the width of the narrowest purely electronic ZPL is usually (at 4 K) 1 to 5 cm^{-1} (vibronic lines are several times broader). This width is of two or three orders of magnitude greater than the natural one, and is determined by inhomogeneous broadening due to imperfection of the crystals (defects, dislocations, etc). For instance, the latter is proved by the ZPL width dependence on the means of sample preparation, on the velocity of the crystal cooling, etc†.

Besides this inhomogeneity, we often come across another type of inhomogeneity which reveals itself through the presence of a specific 'multiplet structure' of spectra due to several discrete ways of the impurity molecule entering into a matrix lattice. In this case the spectra of each type of centre have a quasi-line character, but they are displaced along the frequency scale, with respect to each other, by dozens of wavenumbers. The observed pattern has a form of 'multiplets' (possessing a common length of 100–200 cm^{-1} and sometimes having from five to ten components) repeated throughout the spectrum. It can be said that in these cases the distribution density of impurity molecules $n(v_0)$ along the O–O transition frequency v_0 (in contrast with glassy media, where it has a form of a smooth bell-shaped function) has relatively narrow peaks at the component frequency of the 'O–O multiplet' [6.54].

It is evident, from the above, that selective excitation in the case of impurity crystals with quasi-line spectra can be useful both for investigating the ZPL homogeneous width and for simplifying 'the multiplet structure' of the spectrum. Line narrowing in quasi-line fluorescence spectra upon narrow laser line excitation inside the inhomogeneous ZPL absorption profile was observed long ago (see, for example, [6.55–6.57]). However, the vibronic lines (the homogeneous width of which is usually no less than 1 cm^{-1}; see §6.3.2(d)) are not as narrow as the purely electronic ZPL should be. But it is rather difficult to observe a purely electronic narrow resonance line in the fluorescence spectrum, due to the reasons given in §6.3.2(d). To investigate its homogeneous width, therefore, it is more convenient to employ the hole-burning technique, which, as a matter of fact, is usually done [6.36–6.38, 6.50]. From the viewpoint of applying selective excitation in the case of fluorescence spectral analysis of crystalline objects with quasi-line spectra, the possibility of selective excitation of separate chemical components and simplifying multiplet structure (and not a weak narrowing of the vibronic ZPL) is of paramount importance (see §6.5).

† The present chapter deals with the spectroscopy of complex organic molecules. However, it should be stressed that inhomogeneous ZPL broadening is a common property of all impurity crystals. The narrowing of the inhomogeneously broadened ZPL upon monochromatic excitation was first observed on the R line of a chromium ion in ruby [6.52]. Selective excitation is widely employed in the investigation of ion spectra in crystals and glasses (see, for example, the review [6.53]).

6.4   Experimental technique

For the fluorescence analysis of complex molecules in solid solutions employing selective excitation two peculiarities of laser radiation are particularly important: *monochromaticity* and, in a number of cases, *spatial coherence*, which enables us to focus the radiation on to a small area while analysing rather small samples. In this case, as in other fluorescent analysis methods, there are certain technical problems. These include: providing a smooth tuning of the exciting radiation in a wide spectral range; recording weak light signals; eliminating scattered laser radiation; automation of experimental data processing and of directing the experiment, etc. These items have been thoroughly considered in a large number of papers and are not the concern of the present chapter (see, for example, the review [6.58] and references therein). In this section we shall briefly review those experimental details which are specific for the method under consideration.

6.4.1   Samples. Fluorescence excitation and its recording
The investigated solid solutions can be both glassy and mono- or poly-crystalline (snowlike) samples. The solubility of many organic compounds largely depends on the temperature. Upon cooling the dissolved substance can form aggregates, precipitate, etc. Therefore it is advisable to use dilute $(C \sim 10^{-5}-10^{-6} \, \text{g ml}^{-1})$ solutions, especially in the case of quantitative analysis. Cuvettes for solutions can be made of quartz, glass or some other material which withstands deep cooling. Depending on the specific experimental task it is convenient to use optical helium cryostats of various constructions.

For the purpose of selective fluorescence excitation one can use both conventional and tunable lasers. However, due to their wide tuning range, relative cheapness and accessibility, it is more convenient nowadays to employ tunable dye lasers (usually with another pumping laser with nontunable frequency) with pulse frequency ranging from 10 to 100 Hz. In this case since (see §6.3.2(d)) the homogeneous width of vibronic lines is of several wavenumbers, it is better to use a laser with a linewidth of $1-3 \, \text{cm}^{-1}$†. In the majority of fluorescence measurements considered here it is quite sufficient to have a laser with an average power of $1-10 \, \text{mW}$.

To record fine-structure fluorescence spectra, spectrometers with a resolution of $1-3 \, \text{cm}^{-1}$ are required. Upon fluorescence excitation by laser pulses it is necessary to measure the PM signal using a boxcar integrator.

---

† Laser line narrowing (with a fixed power) down to lower magnitudes is not only useless (from the viewpoint of obtaining narrower spectral lines) but in many cases quite harmful, for the number of excited centres and luminescence intensity is reduced. The effect of the observed fluorescence ZPL weakening due to the selective hole-burning processes is also increased (see §§6.4.2 and 6.5.5).

6.4.2 Synchronic restoration of burnt-out centres

As has already been noted (§6.3.4(b)), in many cases a reversible selective hole-burning of luminescence centres takes place upon laser excitation. This results in a decrease of the ZPL fluorescence intensity in the course of spectrum recording, so that the recorded spectrum does not give the true intensity distribution. In certain cases owing to speedy hole-burning it is impossible to record the narrow ZPL in a spectrum. The degree of distorting influence of such hole-burning greatly depends upon the properties of the investigated compounds, upon the matrix character, experimental conditions, and, in the first place, upon the laser radiation power on a sample and the recording time of the spectrum. As was established in [6.9, 6.10, 6.46] and in a great number of further works (see reviews [6.11, 6.49] and references therein) many photochemically stable molecules in glassy matrices are susceptible to the so-called 'nonphotochemical' hole-burning. This rather universal phenomenon is now connected with the presence of two-well potentials of the impurity molecule interaction with the environment in glasses (see reviews [6.19, 6.49]). Without going into detail we shall point out that the distorting influence of this phenomenon is quite noticeable at a radiation power density of 10 mW cm^{-2}. With such excitation powers and an exciting linewidth of 1–3 cm^{-1} the ZPL intensity in fluorescence spectra can decrease by dozens of per cent during several minutes. In the case of photochemical hole-burning a similar reduction of the line intensity (upon radiation) takes place even at considerably lower (by approximately two orders) densities of the radiation power. However, since the hole-burning processes can be reversed and the ZPL intensity restored after the irradiation of a sample by ultraviolet or 'white' light, to compensate the hole-burning effect, when recording the fluorescence spectrum one can use a pulsed lighting of the sample between exciting laser pulse spacings†.

6.4.3 Spectral set-up

As an example we shall briefly describe a relatively simple spectral set-up, which operates in our laboratory for obtaining and investigating fluorescence spectra upon selective laser excitation. It is shown schematically in figure 6.15. In this case a sample fluorescence is excited by a dye laser which is pumped by a pulsed nitrogen laser. The dye laser has the following characteristics: a pulse energy of 10^{-4} J, a pulse duration of 5 ns, a line generation width of 1–3 cm^{-1}, a pulse frequency of 10 to 20 Hz, and an average power about 1 mW. The fluorescence spectrum is recorded by a double diffraction spectrometer DFS-24 whose scanning system is operated by a computer (D3-28). A photomultiplier pulse signal is measured by a gating integrator, is digitised and then entered into the computer's memory. According to a special program

---

† In certain cases along with the reversible hole-burning we can observe irreversible hole-burning (see, for example, [6.60]). Then the said 'synchronic restoration' by the pulsed lighting is quite useless.

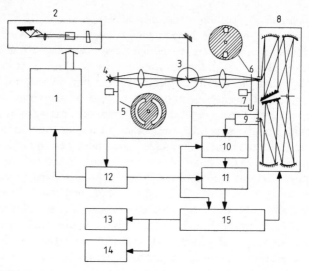

Figure 6.15 Diagram of experimental set-up for obtaining fluorescence spectra upon selective laser excitation. (1) nitrogen pumping laser, (2) tunable dye laser, (3) optical helium cryostat with sample, (4) white light source, (5), (6) choppers, (7) photohead, (8) spectrometer, (9) photomultiplier, (10) boxcar integrator, (11) analogue–digital converter, (12) pulse generator, (13) display, (14) graph plotter, (15) computer.

multiple scanning of the desired spectral fraction is performed and the signal accumulated to achieve the required signal-to-noise ratio.

A peculiar feature of the set-up is its system of synchronous restoration of the burnt centres with intense white light. An ordinary phosphoroscope is made of two continuously rotating choppers (5 and 6 in figure 6.15). Chopper 5 passes the intense restoring light of a lamp onto the sample during those intervals when chopper 6 obstructs the spectrometer slit. The paired pulse generator operated on chopper 6 by a photon-coupled pair, turns on the laser at the moment when the restoring light is cut off and the recording system is opened. The same generator synchronises the recording electronics with the pulsed laser.

The above system of synchronous restoration was widely used in our laboratory for investigating spectra of metal-less porphyrins [6.34, 6.61–6.63]. With such porphyrins (having two hydrogen atoms in the centre of the porphin ring which are easily displaced to neighbouring nitrogen atoms) the hole-burning is performed extremely easily, due to the proton tautomerisation. In these cases, in order to observe the narrow ZPL in a fluorescence spectrum, application of the restoration system proved to be necessary. Figure 6.16 illustrates this by demonstrating the recordings of the protoporphyrine fluorescence spectra made with and without restoration. In

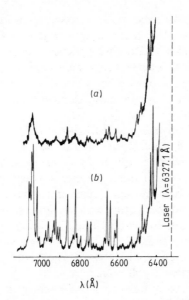

Figure 6.16 Fluorescence spectrum of protoporphyrine in *n*-octane ($T = 4.2$ K) upon laser excitation ($\lambda_{laser} = 6327$ Å, $\Delta\lambda = 1$ Å, $P = 0.1$ mW cm^{-2}): (*a*) without 'restoration'; (*b*) employing white light in the course of spectrum recording.

this case the ZPL in the spectrum are being weakened by one order over several minutes. By using lighting the restoration enables one to record more intensive ZPL in the spectrum (figure 6.16(*b*)).

6.5 Analytical applications

Fine-structure spectra obtained with the selective techniques discussed offer vast new opportunities for fluorescence analysis. Of special importance is the possibility of using various solvents. Free choice of solvent turns out to be particularly necessary in investigating molecular ions [6.64, 6.65], solvates [6.66], charge transfer complexes etc, when the preparation itself of such systems requires special conditions. In the ordinary case of neutral isolated molecules it enables one to use the most appropriate solvent and to obtain samples of high optical quality. In a number of cases given below, it is possible to carry out the analysis of a raw product without any solvents.

Thanks to the line narrowness and high peak intensity, it is rather easy to analyse compounds in a complex mixture at low concentration employing fine-structure spectra. These spectra prove to be very helpful in identifying not only compounds possessing similar structure but also various isomers of the same compound.

Potential possibilities for selective laser excitation in analysing mixtures of complex compounds at low temperatures began to be discussed while investigating solutions of 'a' and 'b' chlorophylls in reference [6.32]. However, up to now, most analytical papers where the above technique is employed deal with the investigation of polycyclic aromatic hydrocarbons and their derivatives. Let us consider certain characteristic instances.

6.5.1   Identification of polycyclic hydrocarbons in glassy solutions

Polycyclic aromatic hydrocarbons (PAH) and heterocyclic aromatic compounds are widespread in nature. They can be found in the pyrolysis products of organic substances and in the environment itself (in soil, natural waters, etc). Many of them are characterised by strong carcinogenic and mutagenic properties, and these properties can appreciably depend upon the isomeric structure of the compound [6.67]. The development of highly selective and sensitive methods of PAH analysis is extremely important. This concerns both the technological problems of producing fuel, oils, coal, etc, and the control of environmental pollution.

Many of the PAH in polycrystalline n-paraffin matrices at low temperatures have quasi-line spectra [6.2–6.5] which are used for analytical purposes [6.6–6.8]. However, there are certain difficulties in applying this method. Here we enumerate some of them: low solubility of certain PAH in n-paraffins; the necessity of using various n-paraffins to identify different PAH; strong light scattering in snowlike frozen solutions; the presence of the specific multiplet structure of a spectrum which depends on concentration and sample cooling velocity etc. To avoid these difficulties it was proposed in references [6.68–6.70] to use organic glasses as solvents. This enables one to increase selectivity and also to speed up the identification procedures. The PAH spectral bands in these cases are broadened inhomogeneously at low temperatures (the bandwidth is usually 200–300 cm^{-1}). Upon selective laser excitation this broadening is eliminated and the fluorescence spectra acquire a rich line structure.

Of the tested solvents the authors [6.68–6.70] chose aqueous glycerol mixtures: glycerol and water in proportions of 1:1 [6.68], and a glycerol–water–dimethylsulphoxide mixture (1:1:1) [6.69]. Dimethylsulphoxide, being an ideal PAH solvent, does not permit the isolated microcrystals to be formed in the solution. Besides, other solvents can be added if necessary. Such additions, made in small portions, do not affect the optical properties of the samples obtained [6.70]. The indicated mixtures are characterised by a number of advantages which are very important in analytical practice. Even upon a relatively quick (15–20 s) cooling down to the helium boiling point these mixtures form a homogeneous transparent glass and do not crack. This enables one to focus the laser radiation well and to collect effectively the fluorescence radiation, having sharply reduced the scattered light level. The absence of scattering permits the recording of the fluorescence spectra regions near the laser line. Since the mixtures discussed contain about 50% water the techniques being developed appear to be rather promising for direct identification of PAH in water (in sewage, natural basins, etc). In the papers mentioned the standard fine-structure fluorescence spectra of pyrene, anthracene, phenantrene, azulene, benzopyrene, fluorantene and of a number of methyl derivatives of these hycrocarbons were obtained (about 20 PAH in all) upon laser excitation in the lowest vibrational sublevel region of the S_1 state.

In experimenting on solutions which contain only one PAH a detection limit of 2×10^{-11} g cm^{-3} was reached (with a signal-to-noise ratio of 5:1) [6.69]. Taking into account the geometry of the experiment and the fact that the laser selectively excites only a small fraction of the molecules present in the solution, one can realise that the number of molecules detected in these experiments approaches 2.5×10^6. Moreover, the authors stress that their experimental conditions are far from optimal (the sensitivity is determined by the dark noise of the photomultiplier which can be reduced by cooling; light scattering due to bubbles in boiling helium which can be eliminated by lowering the temperature beyond the λ point, etc). Thus, the detection limit can be reduced by one order or more.

The investigations showed that the concentration dependence of the ZPL intensity in the spectra of anthracene and pyrene is linear in a broad concentration range of $10^{-9} - 10^{-6}$ g cm^{-3} [6.70]. The high optical quality of the samples and the corresponding good reproducibility of results grants the possibility of quantitative analysis using a constant calibration plot without resorting to the inner standard or the standard additions method.

Experiments with the 14 PAH artificial mixtures in water–glycerol solutions demonstrated that all the components can be reliably identified.

The method described above, with respect to the analysis of natural products, can be exemplified by the coal extract spectra shown in figure 6.17, taken from [6.69]. The only operation in preparing this sample for the PAH content analysis consisted of diluting the extract (1:10 000) in the following solvent mixture: glycerol 44%, water 36%, methyltetrahydrophuran 4% and ethanol 16%. Upon suitable monochromatic excitation narrow lines belonging to anthracene, 1-methylpyrene and pyrene are clearly revealed in the spectra. This solution was thoroughly analysed by the authors for the presence of seven PAH. The concentrations obtained were between the levels of 10^{-7} and 10^{-9} g cm^{-3}.

6.5.2 Direct analysis of raw products

In certain cases selective excitation enables one to directly analyse raw products without any preliminary treatment of a sample and without employing any solvents. But in doing so it is necessary to fulfil a number of requirements, especially when performing quantitative analysis. They are as follows: rather low optical density of the samples in the spectral domain investigated, which excludes the inner filter effect, and sufficiently low

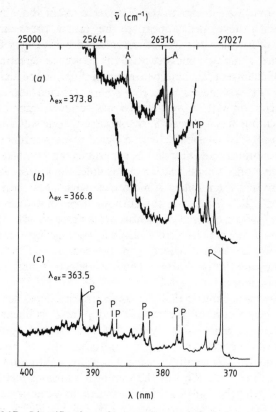

Figure 6.17 Identification of aromatic compounds in coal extract by fluorescence spectra upon selective excitation ($T = 4.2$ K). In the spectrum (*a*) anthracene lines (A) are visible, in (*b*) 1-methylpyrene lines (MP) are visible, and in (*c*) pyrene lines (P) are visible [6.69].

concentrations of the compounds to be identified, which excludes any possibility of energy transfer between them. As an example of a raw product analysis let us consider an experimental identification of certain PAH in motor gasoline performed in our laboratory [6.59, 6.71]. Figure 6.18(*a*) demonstrates a part of the gasoline spectra in the visible region upon conventional UV excitation by the light of a mercury lamp. Although several broad maxima are present in the spectrum it is rather difficult to identify any individual PAH. Quite a different picture is observed in the spectrum when the gasoline fluorescence is excited by a tunable dye laser. In this case one can observe highly resolved spectra with dozens of lines of 1–3 cm^{-1} width. The structure of the spectra depends on the excitation frequency. The number of different spectra is determined by the tuning step of the laser and can be rather large. Figure 6.18(*b,c,d*) shows three such spectra.

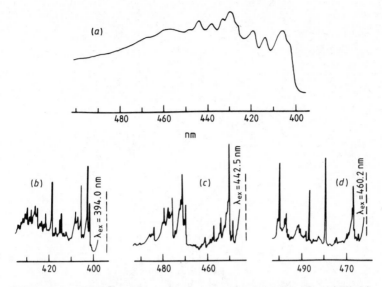

Figure 6.18 Sections of fluorescence spectrum of gasoline ($T = 4.2$ K): (*a*) conventional excitation ($\lambda_{Hg} = 365$ nm); (*b*), (*c*) and (*d*) monochromatic laser excitation [6.71].

Preliminary experiments are important in preparing the PAH content analysis. There one chooses laser frequencies which effectively excite the compound under investigation, and its reference spectrum is established. Figure 6.19 shows the reference spectrum of a carcinogenic hydrocarbon (3,4 benzopyrene) and a corresponding section of the gasoline spectra upon the same excitation. The measurements showed that more than 20 lines in both spectra coincide (within the limits of experimental error of about 1 Å). Another example illustrating the possibility of direct raw product analysis for PAH content is given in figure 6.20. This figure shows a fluorescence spectrum of solid paraffin upon conventional excitation, and a fine-structure spectral fragment of the sample in the fluorescence domain of perylene, obtained with suitable laser excitation. In the latter, the main lines of the fine-structure perylene spectrum are clearly revealed.

In order to perform quantitative spectral analysis with selective laser excitation one can make use of various techniques widely known in analytical practice. In the above example of gasoline the 3,4-benzopyrene content that was found by means of the standard additions method was about 10^{-7} g cm^{-3}. The perylene content was 5×10^{-8} g cm^{-3}. The limit of detection in such experiments depends upon the composition of gasoline being investigated and its 'background' luminescence and can be estimated as 10^{-10}–10^{-11} g cm^{-3}.

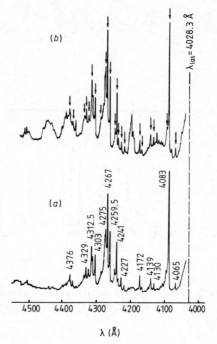

Figure 6.19 Identification of 3,4-benzopyrene in gasoline by the fluorescence spectrum upon selective laser excitation ($T = 4.2$ K): (*a*) reference spectrum of 3,4-benzopyrene in gasoline ($C = 10^{-4}$ g ml^{-1}), (*b*) part of the gasoline spectrum, in which the 3,4-benzopyrene lines are indicated by arrows [6.71].

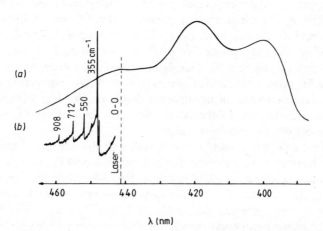

Figure 6.20 Fluorescence spectrum of commercial solid paraffin at 4.2 K: (*a*) conventional uv excitation, (*b*) selective laser excitation [6.71].

6.5.3 Fluorescence analysis of molecules in n-paraffin matrices upon selective excitation

As was noted in §6.3.5 for the objects possessing quasi-line spectra in Shpol'skii's *n*-paraffin matrices, monochromatic excitation in separate absorption quasi-lines enables one to selectively excite the fluorescence of separate compounds in a complex mixture. Though a polycrystalline snowlike nature of solutions results in the above complications (§6.5.1), in principle it is nevertheless possible to achieve a much higher degree of selectivity than in the case of glassy solutions. It has been demonstrated [6.72, 6.73] that some of the principal difficulties can be overcome. For example, nonreproducibility of the intensity distribution in multiplets, the analytical line intensity dependence on the solution cooling velocity, etc. This can be done by employing sufficiently dilute solvents and by properly controlling the sample cooling procedure. In the case of 3,4-benzopyrene in *n*-heptane (when two more hydrocarbons of similar structure were present) it was also established that its analytical line intensity changes linearly when the concentration changes by three orders (in the range of 10^{-10}–10^{-7} g cm^{-3}) [6.73]. The technique was used for PAH identification (pyrene, methylpyrene and benzopyrenes) in some samples of coal liquids and shale oil without any preliminary separation of the PAH fractions by means of chromatography [6.72]. In these experiments 0.1 g of liquid fuel was diluted 10^3 times by *n*-octane, and the fluorescence spectrum was recorded ($T = 15$ K) with tunable dye laser excitation. Thus direct analysis of the original product was performed.

6.5.4 Method of matrix isolation

Preparation of samples by evaporating the investigated liquid or solid substance, then mixing its vapours with a gas matrix and precipitation of this mixture onto a cooling base is what is conventionally called the method of matrix isolation [6.74, 6.75]. Two gases, argon and nitrogen, are most often used as a matrix. The method of matrix isolation is more advanced compared with the use of solutions, for it has no problem connected with the solubility of investigated substances and aggregate formation (especially in case of polar compounds). When we apply the above method the fluorescence spectra of the samples obtained with conventional excitation are characterised (at $T = 10$–15 K) by a considerable inhomogeneous broadening (the bandwidth in the PAH spectra is about 100 cm^{-1}). This broadening can be eliminated by means of selective laser excitation.

In many experiments inert gases are used as a gas 'solvent'. However, any other substance can be used as a matrix; it only requires a high enough vapour pressure. In this respect the experiments where *n*-paraffin vapours precipitated onto a cooling base served as a matrix are of a considerable interest [6.76, 6.77]. In these matrices the PAH spectra with conventional excitation have much narrower bands than in inert gases. The application of a tunable laser enables one to selectively excite individual PAH in a complex mixture. In some cases it becomes possible to analyse the samples for PAH content without any preliminary fractionation [6.77].

The limit of detection in the matrix isolation method with laser fluorescence excitation is in the picogram range. This has been established with individual PAH spectra on model samples [6.76].

6.5.5 Some general remarks

The versions of low-temperature laser fluorescence analysis discussed above differ mainly in sample preparation technique and the character of the matrix in which the molecules under investigation are introduced. Each of these versions has its merits and disadvantages that have already been indicated.

Among the analytical techniques considered above laser fluorescence analysis of solid glassy solutions is the most interesting. The simplicity of sample preparation, the possibility of obtaining such solutions for various objects and compounds (both polar and nonpolar), the absence of light scattering and reliable reproduction of results—all these merits make this technique quite universal. However, upon selective laser excitation in this case a rather small fraction of molecules under analysis is excited inside an inhomogeneously broadened band. This turns out to be a principal limitation of this method. For example, if the width of the whole inhomogeneous absorption band is 200–300 cm^{-1}, and the width of the laser line is 2–3 cm^{-1}, only a hundredth of the molecules in the compound under study will manifest fluorescence. This is equivalent to a situation in which we analyse a solution having a concentration 100 times less than its real concentration. This phenomenon, in principle, can lead to sensitivity reduction. However, as has already been outlined, in the analysis of organic compounds the problem of sensitivity is not so complicated as that of selectivity. Besides, as will be pointed out later, in the case of glassy solutions having inhomogeneously broadened spectra one can achieve very low limits of detection upon selective excitation, which are no worse than for objects having quasi-line spectra†. In performing low-temperature analysis of objects having inhomogeneously broadened spectra, when one employs selective excitation in order to obtain low detection limits and high determination accuracy, it is important to correctly select the spectral width and the laser line intensity and to use a highly sensitive fluorescence detection system. Let us consider this in more detail and make some estimates.

Let the fluorescence of the investigated compound (at 4.2 K) be excited by a pulsed tunable laser in the O–O transition region and fluorescence recording be performed on one of the vibronic lines. We shall take for the estimation the following values of the main parameters: width of the whole inhomogeneously broadened absorption band, 200 cm^{-1}, homogeneous width of a purely electronic absorption ZPL–O, 1 cm^{-1}, and vibronic ZPL width, upon which the

† The lowest limit of PAH detection in one-component solutions—5×10^{-13} g cm^{-3}—was achieved for 3,4-benzopyrene in *n*-octane [6.78].

recording is performed, 3 cm^{-1}. It is evident that the variations of the exciting laser linewidth in the region $\Delta\nu_{\text{laser}} < 3$ cm^{-1} will produce practically no effect upon the vibronic linewidth being recorded. Then the following question arises: how large should the optimal spectral width of the exciting irradiation be in this case at fixed power?

To answer this question one should take into account that in our case even with relatively low laser power, it is very easy to achieve the saturation conditions. Let us show that this is the case. Under stationary conditions, according to relation (1.3), the following intensity is required: $I_{\text{sat}} = 1/2\sigma\tau$ photon/s cm^2, where σ is the absorption cross section and τ is a lifetime of the excited state. A characteristic value of the excitation molar coefficient in the absorption band maximum at room temperature is $\varepsilon = 10^{4.5}$ for many PAH [6.79], that corresponds to the absorption cross section $\sigma \simeq 10^{-16}$ cm^2. Upon temperature decrease the broad absorption band 'collapses' into a narrow ZPL, and the absorption cross section at its maximum becomes approximately 10^3 times larger, i.e. at helium temperatures $\sigma \simeq 10^{-13}$ cm^2. Then, having $\tau = 5$ ns, one obtains: $I_{\text{sat}} = 10^{21}$ photon/s cm^2. Using a laser with $\lambda = 4000$ Å and pulse power of 10 kW (which with a pulse duration of 10 ns and pulse frequency of 10 pulse/s corresponds to an average power of 1 mW) and focusing onto an area of 0.1 cm^2, the laser radiation power density at the sample is $I = 2 \times 10^{23}$ photon/s cm^2. This is 200 times greater than that required for saturation†. Under saturation conditions the fluorescence signal will be almost linearly dependent on the laser linewidth, as the latter determines the number of centres excited in the inhomogeneous system. In this case it is much more convenient to use a line 2–3 cm^{-1} wide for excitation, because the fluorescence intensity will be approximately 20 to 30 times greater as compared to excitation by a line $\Delta\nu \lesssim 0.1$ cm^{-1} wide. If the homogeneous absorption ZPL width is less than in the case discussed (e.g. about 10^{-2}–10^{-3} cm^{-1} [6.80]), then upon fluorescence excitation with too narrow a laser line (10^{-2} cm^{-1}) one can loose a factor of 100 or more in the fluorescence signal intensity. Besides, the use of very narrow laser lines for excitation, i.e. the creation of great spectral power density, results in an increasing role for interfering hole-burning effects (see §§6.3.4, 6.4.2 and 6.4.3).

There can be serious errors if one performs the analysis regardless of the above circumstances. It might be that this (or some other experimental conditions which were not properly considered) enabled the authors [6.81] to obtain abnormally high detection limits for the studied PAH in glassy solutions upon selective laser excitation. Thus, for instance, the limit of detection for perylene in a glycerol–ethanol–water mixture was established as 10^{-7} g cm^{-3}. In reality this limit must be many orders lower. This can be exemplified by a vibronic line recording (0–355 cm^{-1}) or perylene in ethanol (figure 6.21(a)) with a concentration of 1.5×10^{-12} g cm^{-3} performed in our laboratory

---

† Strictly speaking this estimate is valid for $\tau_{\text{pulse}} \gg \tau$ but it is acceptable for our case.

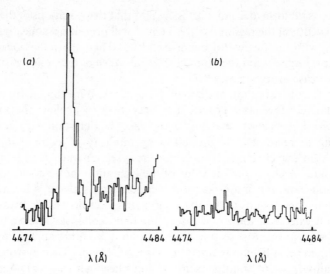

Figure 6.21　Photoelectric recording of the vibronic line $(0\text{--}355\text{ cm}^{-1})$ in the fluorescence spectrum of perylene in ethanol $(T = 4.2\text{ K})$: (*a*) perylene solution with concentration $C = 1.5 \times 10^{-12}\text{ g cm}^{-3}$; (*b*) chromatographically pure ethanol. Excitation: $\lambda_{\text{laser}} = 4408\text{ Å}$, laser linewidth $\Delta v = 1\text{ cm}^{-1}$, pulse duration $= 5\text{ ns}$, pulse energy density $= 0.4\text{ mJ cm}^{-2}$. The spectra are recorded upon 10-fold scanning (with 10 pulses at the point upon each scanning). (Spectra obtained by Romanovskii and Kulikov.)

employing the set-up described in §6.4.3. For the sake of comparison, figure 6.21(*b*) demonstrates the recording of the same spectral section when a cell contained a pure solvent only. It is clearly seen that at a concentration of $10^{-12}\text{ g cm}^{-3}$ the perylene line signal is several times stronger than the noise signal. Note that the concentration mentioned is not the limit of detection which is at least one order of magnitude lower. There is a possibility of improving the signal-to-noise ratio by means of increasing the time of fluorescence signal accumulation. In our experiments with model perylene and 3,4-benzopyrene solutions towards the lower concentration domain the restricting limit was expressed through the presence of these compounds at the level of about $10^{-13}\text{ g cm}^{-3}$ even in very pure solvents. From the viewpoint of technical opportunities in fluorescence recording we have here an appreciable reserve of sensitivity.

6.6　Some other applications of selective methods

Fine-structure spectroscopy of organic molecules in solid solutions upon selective laser excitation offers new possibilities, not only for the spectrochemical analysis of complex organic products but also for investigation of certain problems in the domain of molecular spectroscopy and spectroscopy of solids. A detailed treatment of these problems is not the subject of the present chapter. Therefore, we shall restrict ourselves by briefly enumerating certain applications of selective methods in the precise determination of molecular characteristics.

6.6.1 Establishment of energy level system

It is obvious that fine-structure spectra, upon selective excitation, permit us to determine the electronic and vibrational energy levels of a molecule with great accuracy. It is also possible to determine their symmetry by polarisation measurements [6.34, 6.43, 6.61, 6.82]. By now many aromatic hydrocarbons, heterocyclic compounds, porphyrins, phthalocyanines and some dyes have been investigated using selective spectroscopy methods. Among the investigated objects a number of biologically active porphyrins are also present, for example, the 'a' and 'b' chlorophylls and their close analogues [6.32, 6.35, 6.45, 6.66, 6.83]. We can also name various isomers of ethio-, copro-, meso- and protoporphyrin and their ionic forms [6.63, 6.84] as well as some others.

6.6.2 Homogeneous width of zpl and its temperature dependence

The homogeneous width of purely electronic and vibronic ZPL and their temperature dependence yield very important data regarding intramolecular relaxation processes and electron–phonon coupling properties in a solid solution. However, detailed and correct measurements of the homogeneous ZPL width can only be made upon elimination of the inhomogeneous width. The latter can be achieved by means of selective laser excitation. For the last five to seven years a great number of detailed investigations have been performed concerning the ZPL homogeneous width in the spectra of complex organic molecules, both in crystalline and various amorphous matrices. In this respect we should especially point out the method of stable hole-burning in the absorption spectra or fluorescence excitation spectra bands, which has proved to be very convenient. By means of this technique theoretical width limits, corresponding to the radiation limit (about 10^{-3} cm^{-1}) were obtained (which has already been discussed in §6.3.2). This was done for the purely electronic ZPL of large organic molecules for the first time. A more detailed description of the problem and original references can be found in the reviews [6.11, 6.19, 6.49, 6.50].

6.6.3 Determination of photochemical reaction rate

In molecular systems, photochemical processes proceed with the participation of excited electronic states. If such photochemical reaction is accomplished rapidly enough, the homogeneous width of the corresponding excited level is determined by the rate of the process. Thus, it is possible to obtain information on the photochemical reaction rate from the homogeneous linewidth. In the recent years a number of papers have been published [6.85–6.88] where the method of hole-burning in inhomogeneously broadened absorption bands was used to determine photochemical reaction rates in the solid state (at low temperatures). By means of the hole width, which is appropriately related to the homogeneous ZPL absorption width, the rate of photochemical reaction was established. Thus, for instance, in [6.86], employing the above technique, it was established that the proton phototransfer in glassy solutions of 1,4-dihydroxyanthraquinone at 2 K is performed in 5–10 ps.

Of considerable interest is the application of the above method in the investigation of the primary processes of photosynthesis in biological systems. Recently, the hole-burning technique has been employed in determining the rate of electron phototransfer in the reaction centres of algae (in the photosystem 2 of *Chlamydomonas*) [6.88]. Figure 6.22 demonstrates a hole burnt in the absorption band of an original photoelectron donor (that is of a special kind of chlorophyll 'a'—P-680). The rate of the electron phototransfer determined with this hole width is 3–4 ps.

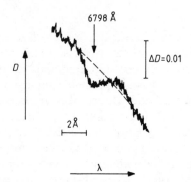

Figure 6.22 Hole in the absorption band of *Chlamydomonas reinhardii* 420/7 after light irradiation with $\lambda = 6798$ Å with power density of 5 mkW cm^{-2} over 40 min; $T = 4.2$ K [6.88].

The examples given above demonstrate a new and interesting way for the investigation of rapid processes without resorting to the picosecond technique.

6.6.4   The Stark and Zeeman effects

Line structure of the spectra obtained with selective methods enables one to investigate favourably the effect of the external electric and magnetic fields known as the Stark and the Zeeman effects.

From the Stark effect measurements, in particular, one can determine dipole moments and polarisability of molecules in the excited electronic states.

Since the Stark shifts in accessible fields yield values of several wavenumbers or fractions of $1\,cm^{-1}$, it becomes difficult to perform their measurements in solutions of complex organic molecules with broad-band spectra. At present, one can point out highly sensitive methods of Stark spectroscopy that are being successfully developed. In these methods holes are burnt in the absorption or excitation spectra [6.41, 6.89–6.91]. Due to a narrow hole width $(10^{-1}$–$10^{-3}\,cm^{-1})$, variations of its profile are easily recorded even with the addition of relatively small external electric fields. Along with this, one can speak of an approach to the development of new methods in investigating internal local fields in various solid materials.

Selective methods appreciably broaden the number of objects accessible for Zeeman experiments. It is evident that the possibility of direct observation of the line zero-field splitting (see §6.3.3) in the phosphorescence spectrum upon laser $T_1 \leftarrow S_0$ excitation also means the possibility of Zeeman effect investigations in magnetic fields (starting from $H = 0$). Some papers [6.92–6.94] have been devoted to the development of corresponding methods and to the analysis of specific features of the Zeeman effect for systems with inhomogeneously broadened phosphorescence spectra when this broadening is eliminated upon selective $T_1 \leftarrow S_0$ excitation. In particular, it is demonstrated that there are interesting possibilities for determining the radiative lifetime of spin sublevels belonging to the triplet state, and for determining spin–lattice relaxation constants.

Some works [6.95–6.97] have dealt with investigations of the square-law Zeeman effect on the S_1–S_0 transitions of some porphyrins (porphin, chlorine and their isomers) in the n-paraffin monocrystals by means of the hole-burning method. In these investigations line displacements of 10^{-1}–$10^{-2}\,cm^{-1}$ were recorded (in fields up to 80–90 kG), and subtle differences in the properties of impurity centres, differently oriented in a crystalline matrix lattice, were revealed.

6.7 Conclusions

Only certain applications of fine-structure spectra upon selective excitation have been considered above. It is clear that not all the possibilities of selective methods in spectroscopy of complex organic molecules have yet been discovered. Especially promising are the applications of site-selection spectroscopy to detailed research of various intra- and intermolecular interactions and to a spectrochemical analysis of many organic products. Further development of the methods for selective laser spectroscopy of complex molecules and their wide practical employment will undoubtedly give many important and interesting results.

The author expresses his sincere gratitude to L A Bykovskaya for her assistance in preparing the present review.

References

6.1   Dovichi N J, Marton J C, Jett J H and Keller R A 1983 *Science* **219** 845
6.2   Shpol'skii E V 1960 *Usp. Fiz. Nauk.* **71** 215 (in Russian)
6.3   Teplitskaya T A, Alekseeva T A and Val'dman M M 1978 *Atlas of Quasi-line Luminescence Spectra of Aromatic Molecules* Moscow University (in Russian)
6.4   Strokach N S, Shigorin D N and Scheglova N A 1982 *Electron–Vibrational Spectra of Polyatomic Molecules* Chap. 3 (Moscow: Nauka) (in Russian)
6.5   Kolotyrkin Ya M and Shigorin D N (ed) 1984 *Electron–Vibrational Spectra of Aromatic Compounds with Heteroatoms* Chap. 3–6 (Moscow: Nauka) (in Russian)
6.6   Teplitskaya T A 1971 *Quasi-line Luminescence Spectra as Investigation Method of Complex Natural Organic Compounds* Moscow University (in Russian)
6.7   Shabad L M 1973 *On Carcinogenic Circulation in the Environment* (Moscow: Medizina) (in Russian)
6.8   Alekseeva T A and Teplitskaya T A 1981 *Spectral Fluorometric Method of Aromatic Hydrocarbon Analysis in Natural and Technogenic Media* (Leningrad: Hydrometeo) (in Russian)
6.9   Personov R I, Al'shits E I and Bykovskaya L A 1972 *Pis'ma v Zh. Eksp. i Teor Fiz.* **15** 609 (in Russian); *Opt. Commun.* **6** 169
6.10   Personov R I, Al'shits E I, Bykovskaya L A and Kharlamov B M 1973 *Zh. Eksp. i Teor. Fiz.* **65** 1825 (in Russian)
6.11   Personov R I 1983 Site selection spectroscopy of complex molecules in solutions and its applications in *Spectroscopy and Excitation Dynamics of Condensed Molecular Systems* ed V M Agranovich and R M Hochstrasser Chap. 10 (Amsterdam: North Holland)
6.12   Personov R I 1983 *Spectrochim. Acta* **38** B 1533
6.13   Levy D H, Wharton L Jnr and Smalley R E 1977 in *Chemical and Biochemical Applications of Laser* vol. II ed C B Moore Chap. 1 (New York: Academic)
6.14   Levy D H 1980 *Ann. Rev. Phys. Chem.* **31** 197
6.15   Fitch S H, Hayman C A and Levy D H 1981 *J. Chem. Phys.* **74** 6612
6.16   Smalley R E 1982 *J. Phys. Chem.* **86** 3504
6.17   Maradudin A A 1966 *Theoretical and Experimental Aspects of the Effects of Point Defects and Disorder on the Vibrations of Crystals* (New York: Academic)
6.18   Rebane K K 1968 *Elementary Theory of the Vibrational Structure of Spectra of the Impurity Centers of Crystals* (Moscow: Nauka)
6.19   Osad'ko I S 1983 Theory of light absorption and emission by organic impurity centers in *Spectroscopy and Excitation Dynamics of Condensed Molecular Systems* ed V M Agranovich and R M Hochstrasser Chap. 8 (Amsterdam: North Holland)
6.20   Personov R I, Osad'ko I S, Al'shits E I and Godyaev E D 1971 *Fiz. Tverd. Tela* **13** 2653 (in Russian)
6.21   Osad'ko I S, Al'shits E I and Personov R I 1974 *Fiz. Tverd. Tela* **16** 1974 (in Russian)
6.22   Byrne J P and Ross I G 1971 *Aust. J. Chem.* **24** 1107
6.23   Personov R I and Kharlamov B M 1973 *Opt. Commun.* **7** 417
6.24   Bykovskaya L A, Personov R I and Kharlamov B M 1974 *Chem. Phys. Lett.* **27** 80
6.25   Avarmaa R A 1974 *Isv. Akad. Nauk Est. SSR, Fiz.-Matem. Ser.* **23** 93 (in Russian)
6.26   Eberly J H, McColgin W C, Kawaoka K and Marchetti A P 1974 *Nature* **251** 215

6.27 Cunningham K, Morris J M, Fünfschilling J and Williams D F 1975 *Chem. Phys. Lett.* **32** 581

6.28 Abram I I, Auerbach R A, Kohler B E and Stevenson J M 1975 *J. Chem. Phys.* **63** 2473

6.29 Kikas J V 1978 *Chem. Phys. Lett.* **57** 511

6.30 Al'shits E I, Personov R I, Pyndyk A M and Stogov V I 1975 *Opt. Spektrosk.* **39** 274 (in Russian); *Chem. Phys. Lett.* **33** 85

6.31 Rebane K K, Avarmaa R A and Gorokhovski A A 1975 *Izv. Akad. Nauk Est. SSR, Fiz. Ser.* **39** 1793

6.32 Fünfschilling J and Williams D F 1977 *Photochem. Photobiol.* **26** 109

6.33 Avarmaa R A and Mauring K Kh 1978 *Zh. Priklad. Spektrosk.* **28** 658 (in Russian)

6.34 Bykovskaya L A, Personov R I and Romanovskii Yu V 1979 *Zh. Priklad. Spektrosk.* **31** 910 (in Russian)

6.35 Bykovskaya L A, Litvin F F, Personov R I and Romanovskii Yu V 1980 *Biofiz.* **25** 13 (in Russian)

6.36 de Vries H and Wiersma D A 1976 *Phys. Rev. Lett.* **36** 91

6.37 Gorokhovski A A, Kaarli R and Rebane L A 1976 *Opt. Commun.* **16** 282

6.38 Voelker S, Macfarlane R M, Genak A Z, Trommsdorff H P and van der Waals J H 1977 *J. Chem. Phys.* **67** 1759

6.39 Gorokhovski A A, Kikas J V, Pal'm V V and Rebane L A 1981 *Fiz. Tverd. Tela* **23** 1040 (in Russian)

6.40 Thijssen H P H, Dicker A I M and Völker S 1982 *Chem. Phys. Lett.* **92** 7

6.41 Burkhalter F A, Suter G W, Wild U P, Samoilenko V D, Razumova N V and Personov R I 1983 *Chem. Phys. Lett.* **94** 483

6.42 Al'shits E I, Personov R I and Kharlamov B M 1976 *Opt. Spektrosk.* **41** 803 (in Russian); *Chem. Phys. Lett.* **40** 116

6.43 Brenner K, Ruziewicz Z, Suter G and Wild U P 1981 *Chem. Phys.* **59** 157

6.44 Al'shits E I, Personov R I and Kharlamov B M 1977 *Pis'ma v Zh. Eksp. Teor. Fiz.* **26** 751 (in Russian)

6.45 Burkhalter F A and Wild U P 1982 *Chem. Phys.* **66** 327

6.46 Kharlamov B M, Personov R I and Bykovskaya L A 1974 *Opt. Commun.* **12** 191

6.47 Gorokhovski A A, Kaarli R K and Rebane L A 1974 *Pis'ma v Zh. Eksp. Teor. Fiz.* **20** 474 (in Russian)

6.48 Kharlamov B M, Bykovskaya L A and Personov R I 1977 *Opt. Spektrosk.* **42** 755 (in Russian); *Chem. Phys. Lett.* **50** 407

6.49 Small G J 1983 Persistent nonphotochemical hole burning and the dephasing of impurity electronic transitions in organic glasses in *Spectroscopy and Excitation Dynamics of Condensed Molecular Systems* ed V M Agranovich and R M Hochstrasser Chap. 9 (Amsterdam: North Holland)

6.50 Rebane L A, Gorokhovski A A and Kikas J V 1982 *Appl. Phys.* B **29** 235

6.51 Personov R I 1984 *Vest. Akad. Nauk SSSR* N4 49 (in Russian)

6.52 Szabo A 1970 *Phys. Rev. Lett.* **25** 924; 1971 *Phys. Rev. Lett.* **27** 323

6.53 Avouris P, Campion A and El-Sayed M A 1977 *Proc. Soc. Photo-Opt. Inst. Eng.* **113** 57

6.54 Tamm T B, Kikas J V and Sirk A E 1976 *Zh. Prikl. Spektrosk.* **24** 315 (in Russian)

6.55 Korotaev O N and Personov R I 1974 *Opt. Spektrosk.* **37** 886 (in Russian)

6.56 Abram I I, Auerbach R A, Birge R R, Kohler B E and Stevenson J M 1974 *J. Chem. Phys.* **61** 3875

6.57 Marchetti A P, McColgin W C and Eberly J H 1975 *Phys. Rev. Lett.* **35** 387

6.58 Harris T D and Lytle F E 1983 Analytical applications of laser absorption and emission spectroscopy in *Ultrasensitive Laser Spectroscopy* ed D S Kliger (New York: Academic)

6.59 Al'shits E I, Bykovskaya L A, Personov R I, Romanovskii Yu V and Kharlamov B M 1980 *J. Mol. Struct.* **60** 219

6.60 Cuellar E and Castro G 1981 *Chem. Phys.* **54** 217

6.61 Bykovskaya L A, Gradyushko A T, Personov R I, Romanovskii Yu V, Solov'ev K N, Starukhin A S and Shul'ga A M 1980 *Izv. Akad. Nauk SSSR, Fiz. Ser.* **44** 822 (in Russian)

6.62 Bykovskaya L A, Litvin F F, Personov R I and Romanovskii Yu V 1980 *Biofiz.* **25** 13 (in Russian)

6.63 Romanovskii Yu V, Bykovskaya L A and Personov R I 1981 *Biofiz.* **26** 621 (in Russian)

6.64 Romanovskii Yu V 1981 *Opt. Spektrosk.* **50** 388 (in Russian)

6.65 Chiang I, Hayes J M and Small G J 1982 *Anal. Chem.* **54** 315

6.66 Romanovskii Yu V 1982 *Izv. Akad. Nauk Est. SSR, Fiz.-Matem. Ser.* **31** 139 (in Russian)

6.67 Lea M L, Novotny M and Bartle K D 1976 *Anal. Chem.* **48** 405

6.68 Brown J C, Edelson M C and Small G J 1978 *Anal. Chem.* **50** 1394

6.69 Brown J C, Duncanson J A Jnr and Small G J 1980 *Anal. Chem.* **52** 1711

6.70 Brown J C, Hayes J M, Warren J A and Small G J 1981 New laser-based methodologies for the determination of organic pollutants via fluorescence in *Lasers in Chemical Analysis* ed G M Nieftje, T C Travis and F E Lytle (New York: Humana)

6.71 Bykovskaya L A, Personov R I and Romanovskii Yu V 1981 *Anal. Chim. Acta* **125** 1

6.72 Yang Y, D'Silva A P, Fassel V A and Iles M 1980 *Anal. Chem.* **52** 1350

6.73 Yang Y, D'Silva A P and Fassel V A 1981 *Anal. Chem.* **53** 894

6.74 Meyer B 1971 *Low Temperature Spectroscopy* (New York: Elsevier)

6.75 Wehry E L and Mamantov G 1981 in *Modern Fluorescence Spectroscopy* vol. 4 ed E L Wehry (New York: Plenum)

6.76 Maple J R, Wehry E L and Mamantov G 1980 *Anal. Chem.* **52** 920

6.77 Perry M B, Wehry E L and Mamantov G 1983 *Anal. Chem.* **55** 1893

6.78 Tamm T B 1981 *Izv. Akad. Nauk Est. SSR, Chem.* **30** 44 (in Russian)

6.79 Klar E 1964 *Polycyclic Hydrocarbons* vol. 1, 2 (New York: Academic)

6.80 Thijssen H P H, Van Den Berg R E and Völker S 1983 *Chem. Phys. Lett.* **103** 23

6.81 Bolton D and Winefordner J D 1983 *Talanta* **30** 713

6.82 Bykovskaya L A, Gradyushko A T, Personov R I, Romanovskii Yu V, Solov'ev K N, Starukhin A S and Shul'ga A M 1978 *Zh. Prikl. Spektrosk.* **29** 1088 (in Russian)

6.83 Avarmaa R A, Tamkivi R, Kiisler S and Nymm V 1980 *Izv. Akad. Nauk Est. SSR, Fiz.-Matem. Ser.* **29** 39 (in Russian)

6.84 Sapozhnikov M N, Shubin A L and Rakhovskii V I 1983 *Khim. Fiz.* N3 351 (in Russian)

6.85 Maslov V G 1978 *Opt. Spektrosk.* **45** 824 (in Russian)

6.86 Drissler F, Graff F and Haarer D 1980 *J. Chem. Phys.* **72** 4996

6.87 Friedrich J and Haarer D 1980 *Chem. Phys. Lett.* **74** 503

6.88 Maslov V G and Chunaev A S 1982 *Mol. Biol.* **16** 604
6.89 Marchetti A P, Scozzafava M and Young R H 1977 *Chem. Phys. Lett.* **51** 424
6.90 Samoilenko V D, Razumova N V and Personov R I 1982 *Opt. Spektrosk.* **52** 580
6.91 Bogner U, Schätz P, Seel R and Maier M 1983 *Chem. Phys. Lett.* **102** 267
6.92 Kharlamov B M, Al'shits E I, Personov R I, Nizhankovsky V I and Nazin V G 1978 *Opt. Commun.* **24** 199
6.93 Kharlamov B M, Al'shits E I and Personov R I 1983 *Opt. Commun.* **44** 149
6.94 Kharlamov B M, Al'shits E I and Personov R I 1984 *Zh. Eksp. Teor. Fiz.* **87** 750 (in Russian)
6.95 Dicker A I, Noort M, Völker S and van der Waals J H 1980 *Chem. Phys. Lett.* **73** 1
6.96 Dicker A I, Noort M, Thijssen H P H, Völker S and van der Waals J H 1981 *Chem. Phys. Lett.* **78** 212
6.97 Dicker A I M, Dobkowski J, Noort M, Völker S and van der Waals J H 1982 *Chem. Phys. Lett.* **88** 135

7 Laser Photoionisation Spectroscopy and Mass Spectrometry of Molecules

V S Antonov and V S Letokhov

7.1 Introduction

The term 'multiphoton ionisation' (MPI) or 'resonant multiphoton ionisation' (RMPI) is often used to define the ionisation of molecules resulting from absorption of several photons. Depending on specific physical conditions, such ionisation can be realised in different ways (see [7.1–7.4]). Figure 7.1 shows some basic schemes of resonant molecular photoionisation. With all these schemes, photoionisation involves the photoexcitation of intermediate molecular quantum states, and so it is, in principle, a selective process.

Two-step photoionisation via intermediate excited electronic states (figure 7.1(a)) is the simplest type of selective stepwise molecular photoionisation. As the electronic states are excited, the ionisation energy decreases by several electron volts, and so it is possible to photoionise only the excited molecules by proper choice of ionising photon energy, $\hbar\omega_2$ (figure 7.1(a)).

Since most molecules have intense electron-absorption bands in the UV spectral range $\lambda < 200$–300 nm, which is rather difficult to cover for tunable lasers, molecular ionisation through intermediate multiphoton resonances has gained much recognition. The essence of the method consists in using a tunable dye laser with strong focusing (up to the diffraction limit). In this case the radiation intensity in the focus may be higher than 10^{10} W cm^{-2}, which provides for effective multiphoton absorption and ionisation of molecules. As the radiation wavelength is varied, resonance absorption occurs at the double

314

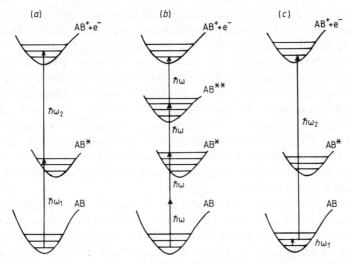

Figure 7.1 Schemes of resonant laser molecular photoionisation: (*a*) via intermediate electronic-excited states; (*b*) via an intermediate two-photon resonance; (*c*) via vibrational-excited states.

or triple laser frequency (figure 7.1(*b*)) with subsequent one- or two-photon molecular ionisation, thus forming a multiphoton ionisation spectrum. With this method almost any molecule can be ionised by means of a tunable laser radiation via intermediate multiphoton resonances.

For polyatomic molecules, vibrational spectra are more characteristic than electronic ones, and so the two-step photoionisation scheme involving intermediate vibrationally excited states (figure 7.1(*c*)) must be more selective than that involving electronic states. However, because of the fact that the change in the ionisation energy of the excited molecules is small (0.1–0.3 eV), unexcited molecules will also be ionised with this scheme, which makes the scheme more difficult to realise experimentally. This difficulty can be overcome by multiple-photon vibrational excitation or successive vibrational and electronic excitation of molecules.

According to the laser radiation intensity there are three characteristic modes of molecular ionisation.

(1) Resonant two (or more)-step photoionisation in weak laser fields through real quantum molecular states (the saturation parameter of optical transitions $G < 1$, [7.5]). In this case we have a pure n-step process that finishes as the molecule reaches the ionisation limit. The power dependence of the photoion yield on radiation intensity is defined by the number of absorbed photons, and the ion yield itself depends essentially on the properties of intermediate excited molecular states.

(2) Resonant stepwise photoionisation in strong laser fields (the saturation

parameter $G \gtrsim 1$). This case is of particular interest for analytical applications as the molecular photoionisation efficiency approaches 100%. But, when reaching the first ionisation potential the molecular ions, as a rule, go on absorbing the radiation and this leads to their subsequent fragmentation.

(3) Multiphoton ionisation. The radiation intensity is sufficient to ionise the molecules via intermediate multiphoton resonances. The subsequent resonant single-photon optical transitions are close to saturation. The molecules are subjected to strong fragmentation, up to their atomisation. The fragmentation under powerful laser radiation first observed in experiments on multiphoton ionisation has become the subject of many investigations as far as the physics of the process and the determination of the optimal conditions for molecular detection are concerned.

From the analytical standpoint, the interest in laser photoionisation methods is due to the fact that the resonant, and hence selective, interaction of molecules with photons gives rise to ions with a high probability. These ions can then be easily detected with an efficiency approximately 100% and with simultaneous measurement of their mass. As far as optical schemes and laser radiation intensities are concerned, almost all laser photoionisation methods can be applied for analytical purposes depending on specific tasks. Such a variety of approaches widens the resources of investigators but, on the other hand, it complicates the designing of universal instruments for analytical laboratories. Therefore, even though the laser photoionisation methods have proved to be very effective for solving a wide range of tasks, their application at present has not gone beyond research laboratories. In this connection much consideration will be given in the present chapter to basic physical processes taking place during resonant laser photoionisation of polyatomic molecules, defining the field of analytical applications and the choice of an optimal technique.

7.2 Principles of resonant multiphoton ionisation of molecules

First of all it should be noted that the essential feature of all the experiments on resonant molecular photoionisation is the use of powerful laser radiation which makes it possible to obtain a high ionisation efficiency in a multistep or multiphoton process. For example, in the case of two-step photoionisation via an intermediate electronic-excited state, to saturate the optical transition at the first step, the flux of the first laser pulse $\phi_1(\omega_1)$ must satisfy the condition

$$\phi_1(\omega_1) \gtrsim \phi_{sat}^{(1)} = \hbar\omega_1/2\sigma_{exc} \tag{7.1}$$

where $\phi_{sat}^{(1)}$ is the saturation energy and σ_{exc} is the resonant excitation cross section. This relation is valid in the optimum case when the exciting pulse duration is shorter than the excited state relaxation time. Considerable depletion of the excited electronic state due to stimulated transitions to the ionisation state can be attained when the flux ϕ_2 of the laser pulse with the frequency ω_2 is

$$\phi_2(\omega_2) \gtrsim \phi_{\text{sat}}^{(2)} = \hbar\omega_2/\sigma_i \tag{7.2}$$

where σ_i is the cross section of the transition from an excited state to the ionisation continuum. For characteristic cross sections of molecular excitation and ionisation (10^{-17} to 10^{-18} cm^2) and lifetimes of intermediate excited states ($\tau_{\text{exc}} \simeq 10^{-8}$ to 10^{-9} s) the saturation laser fluxes and intensities will accordingly be 0.1 to 1 J cm^{-2} and 10^7 to 10^9 W cm^{-2}. In the case of multiphoton ionisation the characteristic cross sections of two- and three-photon transitions will respectively range from 10^{-30} to 10^{-36} cm^4 W^{-1} and 10^{-43} to 10^{-49} cm^6 W^{-2} [7.6]. So, to reach a reasonable ionisation efficiency it is necessary that the radiation intensity should be $I > 10^{10}$ to 10^{11} W cm^{-2}.

Other essential requirements placed upon the laser radiation sources in experiments on multistep photoionisation are related to the specific spectral features of molecules. The electronic absorption bands for most molecules lie in the ultraviolet spectral range and the excitation of electronic singlet states needs photon energies $\hbar\omega \gtrsim 3\text{–}5$ eV. The ionisation potential for polyatomic molecules usually lies in the range of 8 to 12 eV, so photons with energy $\hbar\omega = 4\text{–}8$ eV are required for effective one-step photoionisation of molecules from electronic-excited states.

Considerable progress in quantum electronics in the designing of uv and vuv lasers has provided experimenters with some powerful laser radiation sources on excimer molecules XeCl ($\hbar\omega = 4$ eV), KrF ($\hbar\omega = 5$ eV) and ArF ($\hbar\omega = 6.4$ eV) [7.7] as well as with the vuv laser on the F_2 molecule ($\hbar\omega = 8$ eV) [7.8] in addition to the harmonics of the Nd–YAG laser. In experiments with tunable radiation the first and second harmonics of dye lasers ($\lambda = 210$ to 600 nm) pumped by nitrogen, excimer or Nd–YAG lasers are used.

7.2.1 Stepwise molecular photoionisation

Far from saturation the ion yield of two-step photoionisation via an intermediate electronic state in the absence of excited-state relaxation is defined by a simple expression

$$N_i = N_{\text{exc}}\sigma_i(\omega_2)\phi_2/\hbar\omega_2 = N_0\sigma_{\text{exc}}(\omega_1)\sigma_i(\omega_2)\phi_1\phi_2/\hbar\omega_1\hbar\omega_2 \tag{7.3}$$

where N_i is the number of ions formed in one pulse, N_{exc} is the number of excited molecules, N_0 is the number of molecules in the irradiated volume, $\sigma_{\text{exc}}(\omega_1)$ is the molecular excitation cross section, $\sigma_i(\omega_2)$ is the cross section of molecular photoionisation from the excited state and ϕ_1, ϕ_2 are the laser fluxes at the first and second steps. The dependence $\sigma_{\text{exc}} = \sigma_{\text{exc}}(\omega_1)$ defines the molecular excitation spectrum. The ionisation cross section of excited molecules $\sigma_i(\omega_1, \omega_2)$ depends both on the specific excited electronic–vibrational level and on the excess of the energy of two photons $\hbar(\omega_1 + \omega_2)$ over the ionisation limit of the molecule.

In the case of direct single-photon ionisation, that is a transition from the ground state of a molecule to the ion state, the dependence of the photoionisation cross section on the radiation wavelength near the ionisation threshold is approximated by a step function which reflects the Franck–Condon factors for the transitions to the vibrational states of the ion [7.9]. Far above the ionisation threshold the stepwise character of the dependence is smoothed out due to the overlapping of transitions to different vibrational ion states and the ionisation efficiency gradually reaches a plateau in the region of vertical Franck–Condon transitions. All this is valid for two-step photoionisation too, but in this case the molecular transition to the ion state takes place from excited electronic-vibrational states and so account must be taken of the Franck–Condon factors between these states.

For example, in the case of benzene, toluene and naphthalene molecules a delocalised π electron that slightly changes the parameters of chemical bonds in molecules detaches in the process of ionisation. As a result, transitions without change in vibrational state $\Delta v = 0$ [7.10–7.12] have the largest probabilities. Figure 7.2 presents photoionisation spectra $\sigma_i(\omega_2)$ from different vibrational levels of the excited electronic state S_1 of the naphthalene molecule [7.12]. The spectrum in figure 7.2(a) corresponds to the O–O vibrational transition and has the form of a well pronounced step dependence. The position of the step with $\lambda_2 = 297.2$ nm is defined by the adiabatic ionisation potential of the molecule I_{ad}. The ionisation transitions from the excited vibrational levels of the S_1 state (figure 7.2(b–d)) are characterised by a shift of the ionisation threshold to the longwave side by the value of the vibrational excitation but the photoionisation signal is increased most near $\lambda_2 = 297.2$ nm which corresponds to transitions without change in the vibrational state $\Delta v = 0$. Thus, in this case the value of the vertical ionisation potential I_{vert} from an excited state, i.e. the energy needed for a vertical Franck–Condon transition with a maximum probability, is equal to

$$I_{vert} = I_{ad} + E_{vib} \qquad (7.4)$$

where E_{vib} is the vibrational excitation energy in an intermediate electronic state.

Neglecting the transient region of increase in stepwise photoionisation yield $I_{ad} < \hbar(\omega_1 + \omega_2) < I_{vert}$ one can approximately assume that $\sigma_i(\omega_2) = $ constant. In this case the two-step photoionisation spectrum in both single-frequency and two-frequency laser fields coincides with the molecular excitation spectrum at the first step $\sigma_{exc}(\omega_1)$. This was proved in experiments for a number of molecules [7.13–7.16].

As in single-photon ionisation, in the two-step process the discrete states of a neutral molecule can be excited over the ionisation limit. Then in a nonradiative process these states can decay into the ionisation continuum (i.e. be autoionised) [7.9]. For many molecules, single-photon autoionisation states are well known but in the process of two-step photoionisation other

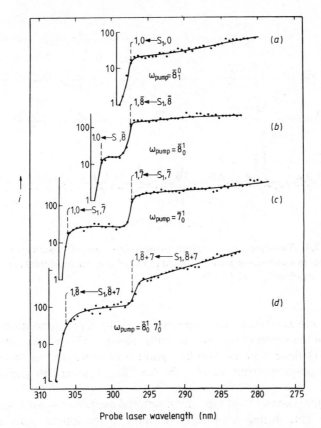

Figure 7.2 Photoionisation spectra of naphthalene produced with the radiation wavelength scanned at the second step of molecular photoionisation. The type of vibrational level in the intermediate S_1 electronic state is given for each spectrum [7.12].

states can be excited. In an experiment with the diazabicyclooctane molecule in a supersonic jet [7.17] it was possible to observe the autoionisation structure in the two-step photoionisation spectrum. Figure 7.3 shows the ionisation spectrum with a fixed wavelength at the first step tuned to the transition $S_0 \rightarrow S_1$ exciting the 1007 cm^{-1} vibration (20_0^1) and scanning the radiation wavelength at the second step. An autoionisation structure arises when the total photon energy is higher than 58 030 cm^{-1} (figure 7.3) that complies with the adiabatic ionisation potential $I_{ad} = 58\,032$ cm^{-1}. A sharp step at the energy $\hbar(\omega_1 + \omega_2) = 59\,030$ cm^{-1} corresponds to the vertical ionisation potential complying with the condition $\Delta v = 0$. In the spectrum (figure 7.3) one can observe three Rydberg series ($n = 11$–35) with quantum defects of 0.41, 0.23 and 0.05. The autoionisation structure in this experiment

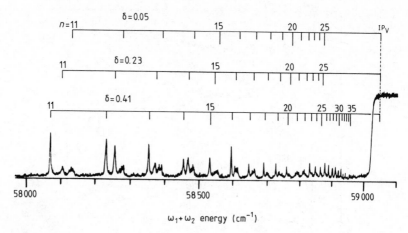

Figure 7.3 Two-frequency photoionisation spectrum of diazabicyclooctane molecules in a supersonic jet. ω_1 is tuned to the vibrational level 1007 cm^{-1} (20_0^1) in the transition $S_1 \leftarrow S_0$ [7.17].

was very distinct due to the suppression of the direct ionisation channel because of the action of the selection rule $\Delta v = 0$ [7.17]. In some other cases [7.10, 7.11] there were no autoionisation peaks observed in the two-step photoionisation spectrum despite the fact that they exist in single-photon transitions.

The conservation of the type of vibrational excitation in an intermediate electronic state during a vertical transition to the ionised state can also manifest itself in a change of molecular ion fragmentation as compared to single-photon ionisation. This question was studied in [7.18, 7.19] for the case of two-step photoionisation of benzaldehyde molecules. In the experiment [7.18] the energy of two laser photons ($\hbar\omega_1 + \hbar\omega_2 = 3.7 + 7.7 = 11.4$ eV) exceeded not only the molecular ionisation potential $I = 9.5$ eV but also the appearance potential of a fragment ion with detachment of a hydrogen atom. So there was a peak of fragment C_6H_5CO ions observed in the mass spectrum (figure 7.4). An interesting feature of the two-step photoionisation mass spectrum is that the yield of fragment ions in this case is much higher than for single-photon ionisation, the energy being the same (figure 7.4). This can be explained by the fact that the ion vibrational energy distribution over different modes may differ materially for one- and two-step photoionisation at incomplete stochastisation of vibrational excitation. The absorption of a photon with its energy of 3.7 eV by a benzaldehyde molecule corresponds to an electronic transition of the $n \rightarrow \pi^*$ type, i.e. to the transfer of an electron from the nonbonding orbital of the oxygen atom to the antibonding molecular π^* orbital. The electron configuration is rearranged mainly in the aldehyde CHO group. Since the antibonding molecular orbital is filled up, the bond

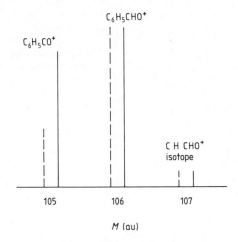

Figure 7.4 Photoionisation mass spectrum of the benzaldehyde molecule at two-step (full line) and one-step (broken line) photoionisation with the same photon energy, 11.4 eV [7.18].

parameters in the aldehyde group, particularly the C–O bond, change greatly. Because of this, it is mainly the vibrations in the region of the aldehyde group that are excited at electronic transitions. When a second photon is absorbed, ionisation occurs from a molecular state with a large store of vibrational energy (about 0.6 eV) localised in the aldehyde group which promotes the formation of an ion with high vibrational excitation in the aldehyde group and increasing probability of detachment of an aldehyde hydrogen atom (figure 7.5) [7.19].

Two-step photoionisation of polyatomic molecules is usually considered in a noncoherent way in the rate equation approximation [7.20, 7.21] because the phase memory of a system in an intermediate state is lost very quickly. Unlike atoms, electronic-excited molecules may undergo different photophysical and photochemical transformations which affect the dynamics of resonant stepwise photoionisation. There are two types of such processes bringing about a change in ion yield as the excited states are probed at different instants of time. The first type is connected with excited state deactivation. In this case the photoexcitation energy is spent on photon emission or the breaking of chemical bonds. In this case the ionising photon energy may turn out not to be sufficient for photoionisation [7.22].

In the second case the electronic–vibrational energy is redistributed inside the molecule (vibrational relaxation, intersystem crossing, internal conversion, isomerisation) which causes the vertical potential of ionisation, I_{vert}, to vary. Therefore, when choosing the photoionisation energy $I_{ad} < E_{phot} < I_{vert}$ the intermolecular relaxation dynamics can be determined by

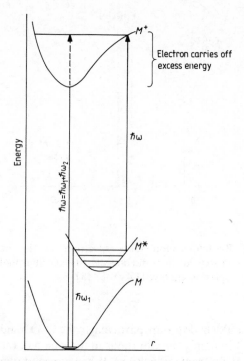

Figure 7.5 Franck–Condon transitions at one- and two-step photoionisation. The excess of photon energy ΔE over the molecular ionisation potential at one-photon ionisation is carried away by an electron, at two-step photoionisation it converts to the vibrational energy of the ion [7.19].

measuring the time dependence of the stepwise photoionisation yield [7.23–7.25].

When excited to the electronic state S_1, the toluene molecule gives an example of complex excitation evolution that causes the photoionisation yield to change in time by both the first and the second mechanisms. Figure 7.6 shows the two-step photoionisation yield for the toluene molecule measured in a supersonic jet as a function of the delay time between the exciting and ionising laser pulses [7.26]. The fast component in the observed two-exponential decay (figure 7.6) is related to two processes—radiative deactivation of the singlet S_1 state and $S_1 \rightarrow T_1$ intersystem crossing. The slow component is related to the conversion of the triplet state to the ground electronic state $T_1 \rightarrow S_0$.

Let us consider the two-step photoionisation dynamics of toluene in more detail. The first photon with energy $\hbar\omega_1 = 4.6$ eV excites the molecules to the lower vibrational level of the S_1 state (figure 7.7). The radiative decay of the S_1 state causes the photoionisation signal to fall until a fraction of the excited

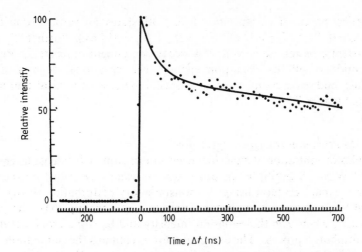

Figure 7.6 Photoionisation signal in toluene as a function of the delay time between laser pulses. The fast component corresponds to the decay of the S_1 state, the slow one corresponds to the decay of T_1 [7.26].

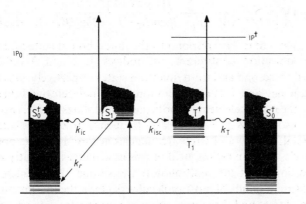

Figure 7.7 Scheme of two-step photoionisation of the toluene molecule: $S_0^\dagger$ denotes the high-lying vibrational levels of the ground electronic state S_0, S_1 is the first excited singlet state, $T^\dagger$ is the excited vibrational level of the triplet electronic state T_1, IP and IP† are the vertical potentials of ionisation from singlet and triplet states [7.26].

molecules reaches a comparatively long-lived triplet state T_1 in a competitive process of intersystem crossing. Both S_1 and the vibrational-excited T_1 state ($E_{vib}^T = \Delta E_{S-T} = 1.1$ eV, where E_{vib}^T is the vibrational excitation energy of the T_1 state and ΔE_{S-T} is the difference of electronic energies between S_1 and T_1 states) can be ionised with approximately the same efficiency since, as an ArF laser is used at the second step ($\hbar\omega_2 = 6.4$ eV) the total energy of laser photons exceeds the vertical potentials of ionisation from both states (for vertical transitions with $\Delta v = 0$ $I_{vert}^S \simeq I_{ad} = 8.82$ eV, $I_{vert}^T = I_{vert}^S + E_{vib}^T = 9.92$ eV). Further $T_1 \rightarrow S_0$ intersystem crossing leads to a slow decay component in figure 7.6 due to distribution of all the excitation energy over vibrational degrees of the molecule, and, as a result, there is a drastic increase of vertical ionisation potential (figure 7.7).

7.2.2 Multiphoton resonance spectroscopy

Multiphoton ionisation of molecules occurs in an ultimately intense laser field, $I > 10^{10}$ W cm^{-2}. Multiphoton processes are usually described by the first non-zero expansion term using the nonstationary perturbation theory. The general expression for n-photon ionisation probability has the form $W^{(n)} = \sigma_n I^n$, where I is the radiation intensity and σ_n is the cross section of the multiphoton process. The expression for σ_n contains the summation with respect to all molecular states including the continuum. For example, in the case $n = 4$, that is typical of multiphoton molecular ionisation, the expression for σ_4 has the form

$$\sigma_4(\omega) \sim \left| \sum_1 \sum_2 \sum_3 \frac{\langle f|V|3\rangle\langle 3|V|2\rangle\langle 1|V|i\rangle}{(E_3 - E_i - 3\hbar\omega)(E_2 - E_i - 2\hbar\omega)(E_1 - E_i - \hbar\omega)} \right|^2. \quad (7.5)$$

In this expression i and f correspond to the initial bound state and the final state in the ionisation continuum, the indices 1, 2 and 3 indicate the intermediate first, second and third quantum states respectively, ω is the laser radiation frequency and V is the Hamiltonian of molecule–field interaction. The figures in the matrix elements denote the virtual states as one, two or three photons are absorbed. With varying laser frequency, resonances can arise at the fundamental, double and triple frequencies according to the position of molecular levels. The ionisation yield in this case increases greatly provided that the selection rules for multiphoton transitions are satisfied. As the electron absorption bands of most molecules begin in the UV spectral region, the recording of two- and three-photon resonance spectra is the most typical case when tunable dye laser radiation is in use. The method of multiphoton ionisation is widely used at present for excited state spectroscopy of polyatomic molecules. It is essential that the multiphoton selection rules differ from those of single-photon transitions. Thus, the spectroscopic data obtained in multiphoton ionisation complement those from ordinary single-photon spectroscopy. In some cases it is possible to study electronic molecular states not so far detected (see [7.3, 7.27]). An essential advantage of the method is that it allows investigation of the electronic states laying in the VUV spectral region as well as ionising almost any molecules using visible-range tunable lasers.

Figure 7.8 shows a typical spectrum of multiphoton molecular ionisation using benzene as an example [7.28]. The three-photon ionisation potential

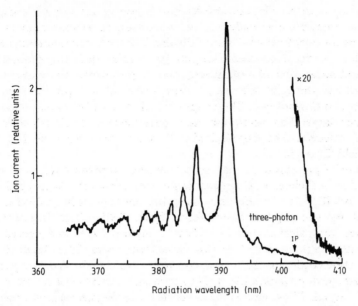

Figure 7.8 Photoionisation spectrum of multiphoton resonances of benzene [7.28].

corresponds to the 402 nm wavelength, and so a signal at shorter wavelengths corresponds to three-photon ionisation and, in the region of 400–500 nm, to four-photon ionisation. Studies of the three-photon ionisation spectrum show that the structure observed belongs to two-photon resonance with the $^1E_{1g}$ state prediced theoretically.

The three-photon resonance spectrum of benzene in the region of 400–500 nm corresponds almost completely to transitions to Rydberg states, the three-photon Rydberg structure being in fact identical to the single-photon absorption spectrum apart from the difference in intensities that points to a strong influence of a two-photon virtual state. Likewise the multiphoton ionisation spectra of a great number of polyatomic molecules have been studied [7.1–7.4, 7.27]. Despite the fact that the method of multiphoton molecular ionisation was developed only a short time ago, it is at present actually a standard technique and is widely used to investigate electronic molecular spectra.

7.2.3 Molecular fragmentation in strong laser fields

To achieve the maximum efficiency in two-step photoionisation it is necessary that the radiation absorption at each step should be close to the saturation defined by conditions (7.1) and (7.2). When $\phi \simeq \phi_{sat}$, the interaction of the molecules with the laser radiation becomes more complex in character: the molecular ions of two-step photoionisation as well as the neutral fragments formed by competitive photodissociation processes can absorb additional photons which in the general case gives rise to a large number of fragment ions.

Let us consider briefly the results obtained for the benzene molecule which provides a very convenient subject for studies into the processes of photoionisation and photofragmentation of polyatomic molecules [7.29–7.33] due to the great amount of thermochemical and spectroscopic data available for this molecule. The energy of the first excited electronic state S_1 of benzene is larger than half the ionisation potential ($I = 9.25$ eV) which enables excited molecules to be photoionised by the second photon in a one-frequency laser field ($2\hbar\omega > I$).

Figure 7.9 presents some mass spectra of benzene photoions formed under the action of excimer KrF laser radiation with $\hbar\omega = 5$ eV. At low radiation fluxes $\phi \lesssim 10^{-3}$ J cm^{-2} the mass spectrum contains only molecular ions formed by two-step photoionisation (figure 7.9(b)). The fragmentation becomes very strong with $\phi \gtrsim 5$ J cm^{-2} (figure 7.9(c)) and exceeds the fragmentation in the case of electron impact with energy 70 eV (figure 7.9(a)). Under these conditions the ion component C^+ is one of the strongest peaks in the mass spectrum (figure 7.9(c)). The appearance potential of atomic carbon ions C^+ equals 27 eV, and so the molecule (or its fragments) must absorb at

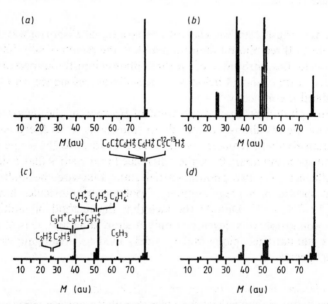

Figure 7.9 Mass spectra of the benzene molecule: (*a*) photoionisation mass spectrum with the radiation intensity $I = 10^5$ W cm^2 ($\lambda = 249$ nm, $\tau_{pulse} = 20$ ns); (*b*) with $I = 2 \times 10^6$ W cm^{-2} [7.32]; (*c*) with $I = 3 \times 10^8$ W cm^{-2} [7.30]; (*d*) electron impact mass spectrum, $E_{el} = 70$ eV, given for comparison.

least six photons to form such a fragment. The total ion yield is very high and approximates to 10% with a laser flux of about 10 J cm^{-2} [7.31]. Besides its high efficiency, the photoionisation of benzene by excimer laser radiation is selective: even at maximum radiation intensities the mass spectra of photoions are almost free of such atmospheric components as H_2, H_2O and others.

The lifetime of the benzene molecule in the intermediate excited state is longer than the laser pulse duration [7.30]. This brings about a high efficiency of formation of molecular ions when two UV photons are absorbed. Further formation of fragment ions will occur during a laser pulse in successive photodissociation of ions starting from molecular ones [7.31, 7.33]. The formation of fragment ions with laser intensity 10^7 to 10^9 W cm^{-2} (figure 7.9) cannot be caused by the absorption of photons by a neutral molecule above the ionisation limit with a transition to highly excited autoionisation states [7.34, 7.35] since the decay rate of autoionisation states, as a rule, is higher than 10^{12}–10^{13} s^{-1}, i.e. greatly exceeds the rate of laser-induced transitions. This is consistent with the absence of two-charge molecular ions in the photoionisation mass spectrum of benzene even though the threshold of their formation is lower than the appearance potential of C^+ ions [7.30].

Comparison of the fragmentation pattern in the cases of laser photoionisation and electron-impact ionisation (figure 7.9) shows in both cases that the ion fragments are almost the same but the degree of fragmentation in the first case may be much greater. In the second case the molecule acquires an energy of about 20 eV exceeding the ionisation potential and then its ionisation occurs with a subsequent consecutive decay of the over-excited ion into fragments. Despite the fact that the energy of ionising electrons is rather high (usually 70 to 100 eV), by virtue of the Franck–Condon principle just a small fraction of this energy is transferred to the ion at a vertical transition in the form of vibrational and electronic excitation. On multiple absorption of photons under the action of sufficiently powerful UV radiation this restriction is absent and the energy absorbed per molecule may greatly exceed 30 eV.

Experiments with many different molecules [7.36–7.43] show that in many cases the interaction of molecules with intense radiation may materially differ from that considered above. The degree of ion fragmentation can be much greater and the relative yield of molecular ions very small even at low laser radiation intensities. This is explained by the dissociation of molecules during their excitation to intermediate electronic states. In this case the main channel of ion formation can be photoionisation of neutral molecular fragments. In one of the first experiments [7.37] on acetaldehyde photoionisation the energy of the UV photons was lower than the molecular ionisation potential that led to the complete disappearance of molecular ions because of a high rate of the competitive process of dissociation from a highly excited state. The role of dissociation is also great when the benzene molecule is photoionised by ArF laser radiation ($\lambda = 193$ nm) through the intermediate electronic-excited state

$^1B_1$ [7.30]. The analysis of data on photoion yield in this case made it possible to estimate the lifetime of this state as equal to 10^{-11} s. The formation of the benzene ion $C_6H_6^+$ during benzaldehyde photoionisation via high-lying singlet states [7.31] was explained by an effective rearrangement of benzaldehyde into the benzene molecule in an intermediate state.

In these experiments the dissociation time of the molecules in intermediate excited states was much shorter than the laser pulse duration which conditioned a high quantum yield of dissociation and, as a result, effective suppression of the channel of direct stepwise molecular photoionisation. The resulting neutral molecular fragments which absorb at the laser radiation frequency can then be ionised during a laser pulse. The role of the mechanism of ionisation of neutral molecular fragments in the formation of a photoionisation mass spectrum of polyatomic molecules, however, may be great even in the presence of stable intermediate states. The transition of a molecule to a state over the ionisation limit from an electronic-excited state with a large store of vibrational energy, or with the nuclear configuration of this state differing materially from that of the ion, may result in a drastic increase of molecular fragmentation in competition with its ionisation [7.31]. Therefore, in order to reach a maximum efficiency of stepwise molecular photoionisation at minimum fragmentation it is necessary that the ionising photon energy should exceed the vertical potential of molecular ionisation from the excited state [7.31].

It is a very difficult task to carry out a detailed theoretical calculation of the fragmentation of polyatomic molecules in the field of powerful laser radiation, because it requires a knowledge of a great amount of spectroscopic and photochemical data for all intermediate photoproducts. So in [7.33, 7.44–7.46] much consideration was given to the development of approximate calculation methods.

In the model of absorption-multiple-fragmentation [7.44, 7.45] it is assumed that photoionisation followed by multiple photon absorption leads to energy-rich molecular ions which fragment in a series of consecutive and parallel unimolecular reactions. All absorbed energy is supposed to be uniformly distributed over vibrational degrees of molecular ions during the laser pulse. Then it is possible to calculate the complete fragmentation pattern by considering statistically the decay probability and the content of energy in all the fragments (by the RRKM theory [7.47]).

It is known for the case of benzene that the main fragmentation channels of initial $C_6H_6^+$ ions with excitation energies of 5–10 eV are the reactions [7.44]

$$C_6H_6^+ \rightarrow C_6H_5^+ \rightarrow C_6H_4^+$$

$$C_4H_4^+ \rightarrow C_4H_3^+ \rightarrow C_4H_2^+ \tag{7.6}$$

$$C_3H_3^+ \rightarrow C_3H_2^+.$$

The threshold of formation of the primary fragments $C_6H_5^+$, $C_4H_4^+$ and $C_3H_3^+$

lies around 4 eV. Further fragmentation is described by decays of the type

$$C_m H_n^+ \rightarrow C_k H_l^+ + C_{m-k} H_{n-l} \tag{7.7}$$

Figure 7.10(a) presents the results of such a calculation for the fragmentation of benzene by KrF laser radiation. The cross section of radiation absorption by $C_6 H_6^+$ ions was taken to be 2×10^{-18} cm². Their comparison with the results of measurements (figure 7.10(b)) shows a satisfactory agreement of the model calculation with experiment.

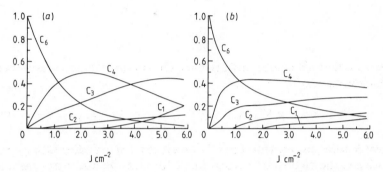

Figure 7.10 Dependence on laser flux of the relative abundance of different ion groups of C_n^+ in the case of benzene: (a) theoretical calculation [7.44]; (b) experiment [7.30].

Further refinement of the theoretical model for molecular multiple-photon fragmentation must include a more realistic consideration of the process of radiation energy absorption in an ionic system. In fact, the excitation energy of benzene molecular ions being just 7–8 eV, their dissociation rate calculated by RRKM is equal to about 10^{10} s^{-1} [7.33]. So during a nanosecond pulse different fragments are formed which, in their turn, can also absorb radiation, thus forming parallel fragmentation channels. The inclusion of parallel absorption–fragmentation channels ('ladder switching') in calculations [7.33] gives better agreement with experiment and allows the explanation of some characteristic features in laser mass spectra.

As noted above, the fragmentation of polyatomic molecules in the process of multiphoton ionisation depends materially on both laser radiation wavelength and intensity. Under similar conditions the photoionisation mass spectra of isomers also differ significantly [7.35]. While such a dependence of fragmentation on radiation parameters and molecular structure is of great interest for different applications, it is still important to make clear to what extent this dependence indicates the specific (dynamic) properties of the processes of photoexcitation and photofragmentation of individual molecules. To investigate this question in [7.46] a method was developed for calculating the most 'statistical' pattern of molecular fragmentation during multiphoton ionisation. According to [7.46], the fragmentation pattern is most statistical when maximum entropy is attained for a medium consisting of noninteracting particles: initial molecules and their neutral and ionised fragments. In this case the only parameter defining fragmentation is the average energy absorbed by one initial molecule.

Let χ_j be the number of molecules or radicals of the j type and χ_{ij} be the fraction of these particles in the quantum state i. Then

$$\sum_i \chi_{ij} = 1 \qquad (7.8)$$

$$\sum_j \chi_j = \chi \qquad (7.9)$$

where χ is the total number of fragments. The system entropy S is expressed in the following way

$$S = -R \sum_j \chi_j \ln \chi_j - R \sum_j \chi_j \sum_i \chi_{ij} \ln \chi_{ij} + R\chi \ln \chi \qquad (7.10)$$

where R is the gas constant. Thus the task is reduced to finding such χ_j and χ_{ij} which give the maximum of expression (7.10) under the following constraints: (i) the conservation of the average absorbed energy $\langle E \rangle$

$$\langle E \rangle = \sum_j \chi_j \sum_i \varepsilon_{ij} \chi_{ij} \qquad (7.11)$$

where ε_{ij} is the energy of the quantum state i for a particle of the j type; (ii) the conservation of matter and charge

$$\chi_k = \sum_j a_{kj} \chi_j \qquad (7.12)$$

where χ_k is the number of atoms of the kth element or the total charge and a_{kj} is the number of k atoms in a molecule of species j.

A satisfactory agreement between the results of calculation of fragmentation for some molecules, varying parameter $\langle E \rangle$, and the experimental data confirms a statistical character of fragmentation of polyatomic molecules [7.48, 7.49]. Of five molecules studied in [7.48] the largest disagreement with experiment is observed for the triethylenediamine molecule which seems to be related to a nonstatistical character of its fragmentation.

Despite the fact that the theories of molecular fragmentation under the action of powerful laser radiation satisfactorily describe the mass spectra of photoproducts, the theoretical estimations of kinetic energies of fragments are far from the experimental values. According to [7.50], in the case of benzene the measured value of kinetic energy of the C^+ ion, that equals 0.26 eV, agrees best with the value calculated by the theory of absorption-multiple-fragmentation (0.39 eV). The theory of parallel absorption-fragmentation ('ladder switching') and the entropy model give kinetic energies of C^+ as 0.66 and 0.76 eV [7.50].

7.3 Laser photoionisation mass spectrometry

The analysis of substances by the mass spectrometric method consists of transforming the molecules of a sample into ions with their subsequent separation and analysis by mass-to-charge ratio. The different ion abundances form a mass spectrum that characterises the sample. There are many methods for the formation of free ions of sample molecules and their mass analysis (see [7.51, 7.52]). The progress made in this direction makes the mass spectrometric method of analysis one of the most sensitive ones. Most commercial mass spectrometers have sensitivities at the level of pico- and femtogram of substance [7.53].

The efforts of scientists concerned with mass spectrometry have always been bent to designing ion sources with a high ionisation efficiency at moderate molecular fragmentation. One of the most widely used and effective methods of molecular ionisation in the mass spectrometer is ionisation by electrons with energies of about 100 eV. The characteristic efficiency of molecular ionisation in a standard ion source with electron impact ionisation (i.e. the ionisation probability of a molecule during its flight through the ionisation region) is far from ultimate and comes to a value of about 10^{-4} [7.51]. Molecular mass spectra, as a rule, consist of many fragment ions, and molecular ions are often absent. Despite the fact that the analysis of fragment pattern gives useful information on molecular structure, it is rather difficult to analyse a multicomponent mixture. Because of superposition of mass spectra it is particularly difficult to analyse molecular components with a low abundance. The difficulties connected with rather a low selectivity of the method are usually overcome by preliminary extraction of the desired components from a mixture in a chromatograph [7.52].

As compared with all known methods, resonant laser photoionisation has some unique characteristic features which make this method very attractive and effective in many applications even though it is not so universal with respect to molecules as ionisation by electrons.

Firstly, resonant laser photoionisation of molecules is selective since it includes excitation of discrete quantum states of molecules with the selectivity inherent in optical spectra. In some cases this allows detection of molecules exclusively on the basis of photoionisation spectra without using mass analysis [7.54–7.56]. In the general case with laser radiation frequency scanning, the mass spectrum of photoions is modulated with a corresponding optical spectrum of the molecule and the information obtained has the form of highly selective two-dimensional mass-optical spectra [7.57].

Secondly, the efficiency of laser photoionisation of molecules may be very high, up to 100%. This is due to the fact that the laser photon density can be increased by more orders than that of electrons without disturbing the ionic-optical properties of the ion source in the mass spectrometer.

Thirdly, the rate of laser photoionisation with powerful pico- and subpicosecond laser pulses may exceed the rate of electron impact ionisation by more than 10^{10} times. This opens up new possibilities for mass spectrometry in detecting unstable short-lived products and in studying high-rate nonstationary processes.

7.3.1 Experimental technique

Laser photoionisation can be used in any mass spectrometer but most often mass separation of ions is performed with a time-of-flight mass spectrometer [7.30, 7.32, 7.58]. This is mainly connected with the fact that laser photoionisation of molecules usually comes about during a short laser pulse, $\tau_{pul} = 10^{-8}$ s, and the time-of-flight mass spectrometer makes it possible to record the whole mass spectrum of photoions during a pulse. This essentially decreases the measuring time, especially if account is taken of the comparatively small repetition rate of laser pulses $v \lesssim 100$ Hz. Figure 7.11 shows a typical scheme of time-of-flight mass spectrometer with a laser photoionisation ion source. The photoions formed under laser radiation are pushed out by the electric pulse into the region of the accelerating field and then after mass separation in the field-free region of drift they fall onto the electron multiplier cathode (figure 7.11). The use of two accelerating electric fields and one field-free region is an optimal configuration of time-of-flight mass spectrometer without changes in ion trajectory [7.59]. In such a mass spectrometer unit mass resolution can be attained by using velocity focusing usually for masses of no more than 500 au. The time-of-flight mass spectrometer of the reflection type [7.60] is also used in experiments on laser photoionisation and gives a mass resolution $m/\Delta m = 3900$ at a mass of 78 au [7.61]. The signal from the electron multiplier in the form of a sequence of pulses corresponding to different mass peaks can be detected in digital form by a fast analogue-to-digital converter (figure 7.11).

When selective laser photoionisation is used in a mass spectrometer of any type, two essential conditions should be met. First, special measures must be taken to avoid nonselective molecular ionisation by the photoelectrons formed, as scattered light falls onto the surface, as well as emitted from the windows. In this connection it is desirable that the electrodes of the ion source should be kept at a small potential (see figure 7.11 and [7.32]). The delay of the ejecting electric pulse about the instant of photoion formation makes it possible to remove low-energy electrons from the space without their ionisation of the molecules.

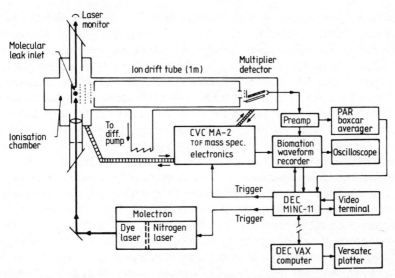

Figure 7.11 Scheme of time-of-flight mass spectrometer with laser photoionisation [7.58].

Another necessary condition is the use of oil-free evacuation of the mass spectrometer since the presence of oil vapour leads to an impermissible increase of the background ionisation level due to the high efficiency of ionisation of organic molecules by laser radiation [7.62].

The introduction of a sample is an important part of any mass spectrometric technique. It is promising to use molecular beams in the method of laser photoionisation. On the one hand, due to its high sensitivity, laser photoionisation is of interest for the diagnostics of molecular beams. On the other hand, the use of cooled molecular beams produced by free expansion of jets enables the photoionisation selectivity and efficiency to be increased considerably due to a considerable line narrowing in optical transition spectra.

7.3.2 Two-dimensional mass–optical spectrometry

One of the most interesting applications of laser photoionisation mass spectrometry combining optical and mass selectivities is detection and study of atomic and molecular clusters in supersonic free jets [7.63]. This approach permits optical absorption spectroscopy in molecular beams with the identification absorbing particles [7.64–7.69]. The problem of mass identification of clusters, however, becomes more complicated because the molecular ions formed by selective stepwise photoionisation can then undergo spontaneous fragmentation in the presence of certain vibrational excitations in the ion. This is favoured by the fact that, as in the case of benzene [7.64], a neutral cluster is bound by weak dipole-induced dipole dispersion forces while the parent ion of this cluster is bound much more strongly by monopole-induced dipole forces. In this case the Franck–Condon factors for photoionisation favour large change in vibrational quanta of the inter-molecular modes and a large store of vibrational energy in the complex ion.

Figure 7.12(a) illustrates the influence of fragmentation on the identification of benzene clusters [7.64]. Since the S_1 state of benzene and its clusters has an energy higher than one half of the ionisation potential, the laser radiation exciting the S_1 state also brings about subsequent photoionisation with the excess of photon energy over the ionisation limit being 2300 cm^{-1} for a monomer and more than 4000 cm^{-1} for different clusters. Despite the fact that the photoionisation spectrum is recorded monitoring the dimer mass (figure 7.12(a)), the basic structure in the spectrum belongs in fact to the trimer, which at the second ionisation step dissociates to the dimer. To eliminate the fragmentation of two-step ions it is necessary to apply a two-frequency scheme with the total photon energy being near the ionisation limit. In this case (figure 7.12(b)) the fragmentation of even higher clusters is almost absent and this allows unambiguous identification of optical spectra for individual clusters. Thus, the optical and mass selectivities complement each other: an optimal choice of an optical scheme for stepwise photoionisation makes it possible to attain a high mass selectivity which allows the recording of optical spectra of individual clusters in mixtures. A knowledge of absorption spectra of different clusters, in its turn, allows their detection in a beam even at a low mass

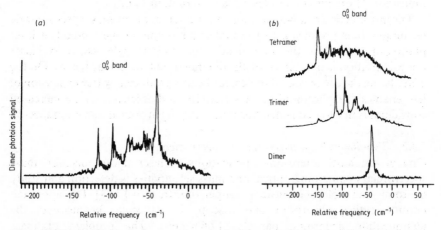

Figure 7.12 (a) Single-frequency photoionisation spectrum of benzene clusters in a cooled molecular beam with the ions recorded by dimer mass [7.64]. The zero of the frequency scale corresponds to the O–O electron transition of the monomer. The structure of the spectrum belongs mainly to the benzene trimer (see figure 7.12(b)). (b) Two-frequency photoionisation spectra of benzene clusters at a fixed wavelength of ionising radiation, $\lambda_2 = 2789 \text{ Å}$ [7.64].

selectivity since the ions of heavy clusters dissociate into lighter ones (figure 7.12(*a,b*)).

From the very beginning of the development of laser molecular photoionisation many experiments were concerned with the study of the mass spectra of the resulting photoions (see §7.2.3). As discussed above, the formation of fragment ions in the photoionisation mass spectrum is caused by the dissociation of molecular ions and/or the photoionisation of neutral photoproducts resulting from molecular excitation. In the first case the spectral dependence of fragment yield and composition is rather weak and is mainly connected with the variation of absorbed energy [7.14, 7.70]. The degree of molecular ion fragmentation in this case is mainly controlled by the variation of laser flux. It should be noted that, as the molecules are photoionised through intermediate multiphoton resonances, highly fragmented mass spectra are almost always observed since, when the radiation intensity is sufficient for a molecule to overcome the first step of multiphoton absorption, all the rest of the single-photon resonant transitions responsible for ion dissociation are close to saturation [7.34, 7.58].

In the second case the photoionisation mass spectrum greatly depends on the properties of the molecule in an intermediate excited state. The measurement of mass–optical spectra in this case gives much new information on excited molecules, their dissociation channels and products. In [7.31], for example, it has been found that at stepwise photoionisation of the benzaldehyde molecule (C_6H_5CHO) the peak of benzene ions $C_6H_6^+$ is very strong. It should be noted that benzene can be formed from benzaldehyde under collisionless conditions only through rearrangement dissociation of benzaldehyde with detachment of the aldehyde group CHO and transfer of the aldehyde hydrogen atom to the benzene ring [7.31]. Study of the spectral dependence of the yield of benzene ions [7.71] has shown that their appearance correlates with the absorption spectrum of benzene, and hence rearrangement dissociation occurs not in the ion system but in the electronic-excited state of the benzaldehyde molecule. Similar processes of rearrangement dissociation have been observed during multiphoton ionisation of the *p*-xylene molecule [7.42].

In many other cases [7.72–7.78] mass–optical molecular spectroscopy has proved to be an effective method for studying excited molecular states.

7.3.3 Detection of molecules and radicals

To get a maximum efficiency of multiphoton ionisation the laser flux should be great enough to saturate molecular optical transitions. Under this condition the molecule (or its fragments) can absorb many laser photons. As a result, a high-fragmentation mass spectrum is formed, and the information on original molecules may be lost. Study of the mechanism of formation of a photoionisation molecular mass spectrum under intense UV laser radiation has shown that ultimate ionisation efficiencies with moderate molecular fragmentation can be obtained under conditions when the intermediate electronic-excited state is stable during a laser pulse and the laser photon energy at the second step exceeds the vertical potential of molecular ionisation from an excited state. The last condition corresponds to a small probability of dissociation of the molecule, after it absorbs a second photon, as compared with ionisation. The maximum ionisation efficiency of a polyatomic molecule at a fixed degree of fragmentation with stable intermediate excited states depends on the ratio between the cross sections of molecular photoionisation and molecular ion photofragmentation. Molecular ions usually resonantly absorb laser radiation, too, since the electron terms of an ion-radical are placed, as a rule, lower than those of the neutral molecule. For estimation, let us assume that all the cross sections of resonant optical transitions in a polyatomic molecule and in an ion are equal and determine the efficiency of two-step ionisation in a single-frequency laser field as a function of degree of fragmentation. We shall assume, and this is valid for many molecules, that the absorption of one UV quantum by a molecular ion leads to its dissociation [7.33, 7.44]. The kinetic equations for the populations of ground state N_g, intermediate excited state N_{exc}, the numbers of molecular N_m and fragment ions N_f have the form

$$d/dt N_g = -N_g \sigma I \tag{7.13}$$

$$d/dt N_{exc} = N_g \sigma I - N_{exc} \sigma I \tag{7.14}$$

$$d/dt N_m = N_{exc} \sigma I - N_m \sigma I \tag{7.15}$$

$$d/dt N_f = N_m \sigma I \tag{7.16}$$

where σ denotes the cross sections of optical transitions, I is the laser radiation intensity (photons/cm^2 s). In the first equation we neglected the backward transition of the molecule from the excited to the ground state assuming the statistical weight of the excited state to be much larger than that of the ground state. Figure 7.13 presents some calculated dependences of the yields of molecular ions, fragment ions and the total number of photoions on laser flux. It can be seen that with photoionisation efficiencies $\lesssim 10\%$ the fraction of fragment ions is very small. But it increases quickly and, at an ionisation efficiency of 60%, the fractions of molecular and fragment ions become equal (figure 7.13). With further increase of laser flux the ionisation efficiency increases to 100% and the fraction of molecular ions in the mass spectrum decreases to zero (figure 7.13). Thus, the simplest calculation shows that it is possible to attain exclusively high efficiencies of laser resonant photoionisation for polyatomic molecules ($> 10\%$) at their moderate fragmentation ($< 50\%$). The experimental results are considered below.

In [7.79] it has been shown using the example of the naphthalene molecule that in an optimal case the use of selective laser photoionisation allows a 100% ionisation efficiency to be obtained and makes it possible to detect single

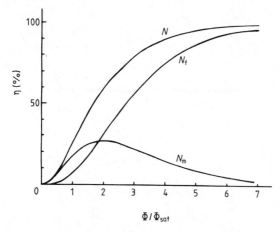

Figure 7.13 Dependences of the yields of molecular ion N_m, fragment ions N_f and the total number of photoions $N = N_m + N_f$ normalised to the initial number of molecules in the photoionisation region on the laser flux normalised to the saturation flux $\phi_{sat} = \hbar\omega/\sigma$.

molecules in the ionisation volume during a laser pulse. The molecules in the experiment were ionised by excimer KrF laser radiation ($\lambda = 249$ nm). The lifetime of the intermediate electronic excited state was similar to the laser pulse duration, 20 ns, and the energy of two laser photons exceeded by 2 eV the ionisation potential of the molecule (8.12 eV).

Figure 7.14 shows some dependences of an ionic signal summed over all masses (curve 1) and separately for a molecular ion (curve 2) on laser flux and the mass spectra of the resultant photoions. In the region of low laser fluxes the dependence of the ionisation yield of naphthalene is quadratic in character and the mass spectrum consists only of molecular ions (oscillogram (a) in figure 7.14). With laser fluxes higher than 0.05 J cm^{-2} the yields of both molecular and integral ion signals of naphthalene become saturated. In the region of 0.15 to 0.02 J cm^{-2} the dependence of the total signal tends to a plateau that is followed by an increasing fraction of fragments in the mass spectrum. A small drop in molecular ion yield occurs because of their subsequent photodissociation by laser radiation.

The quadratic character of the photoion yield in the region of low laser fluxes is explained by a two-step process of molecular photoionisation since the naphthalene potential ($I = 8.12$ eV) is lower than the energy of two laser photons ($2\hbar\omega = 10$ eV). The saturation of naphthalene ionisation in the region of 0.05 to 0.10 J cm^{-2} qualitatively agrees with the saturation of photon absorption at the first step. When the total ion signal comes to a plateau with pulse energies of 0.15 to 0.20 J cm^{-2}, this indicates that the photon absorption at the second step becomes saturated, too. Since the energy of two laser

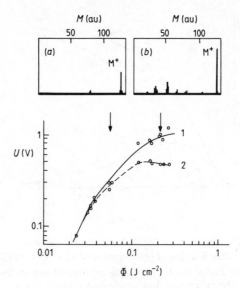

Figure 7.14 Dependences of the total photoion signal (curve 1) and the molecular ion signal of naphthalene (curve 2) on the laser flux ($\lambda = 249$ nm) with the mass spectra of resulting photoions at (a) $\phi = 0.06$ J cm^{-2} and (b) $\phi = 0.2$ J cm^{-2} [7.79].

photons significantly exceeds the molecular ionisation potential and the nonlocalised π electron detaches in the process of photoionisation, the competition of molecular dissociation relative to its ionisation, as it absorbs a second photon, can be neglected. Hence the photoionisation efficiency at such laser fluxes approximates to 100%. In this case 50% of the resultant molecular ions of naphthalene dissociate (figure 7.14) due to further absorption of laser photons. Under these conditions, as recording is performed at one molecular mass of naphthalene, the detection sensitivity of molecules per laser pulse is equal to four molecules in the photoionisation volume which corresponds to a partial naphthalene pressure of 10^{-14} Torr with its relative concentration in the air being 10^{-9} [7.79].

The results obtained are almost ultimate with respect to molecular detection efficiency in the photoionisation volume and correspond to the detection of single molecules in a laser pulse. It should be noted that it was the selectivity of laser stepwise photoionisation that enabled a high detection sensitivity for naphthalene molecules to be obtained. The mass spectrum of photoions was totally free from the basic components of air as simple molecules cannot absorb in the 250 nm region. Even such rather poor selectivity of laser ionisation considerably increases the analytical possibilities of mass spectrometry. Standard mass spectrometers, for example, have a constant background at all masses caused by scattered ions which limits detecting relative concentrations to the level of 10^{-6} to 10^{-8} [7.80].

Experiments on isotope-selective photoionisation of such polyatomic molecules as benzene [7.81] and aniline [7.82] demonstrate a possibility of essential improvement of the analytical characteristics of resonant laser photoionisation when narrow-band tunable radiation is used. The application of cooled supersonic molecular jets in this case is particularly attractive. In [7.83] the authors demonstrated a possibility of using supersonic molecular jets for cooling such complex molecules as phthalocyanine ($C_{32}N_8H_{18}$). The absorption spectrum of this molecule under deep cooling of vibrational and rotational degrees of freedom consisted of lines with width from 0.5 to 1 cm^{-1}. Study of the resonant photoionisation of aniline molecules cooled to the rotational temperature $T_{rot} \lesssim 0.2$ K shows that with the radiation intensity $I \lesssim 1$ MW cm^{-2} no broadening of optical transition lines is observed [7.84]. With $I = 10$–13 MW cm^{-2} a nonresonant substratum is formed and the lines are broadened up to 5 cm^{-1}. But at such radiation intensities the ionisation efficiency is near 100% [7.84].

As to laser ionisation efficiency naphthalene molecules are not unique. Table 7.1 contains experimental data on the two-step photoionisation efficiency for some molecules at a laser ionising·intensity of 10^7 W cm^{-2} [7.85]. With this radiation intensity there was almost no molecular fragmentation observed. It can be seen that when the laser radiation intensity approximates to the saturation intensity the molecular photoionisation efficiency is rather high and exceeds 10%. In experiments with metal carbonyls [7.86] and aniline [7.84] the laser photoionisation efficiencies were also close to 100%.

A much higher efficiency of resonant laser molecular photoionisation, as compared with that of electron impact (by 10^3 to 10^4 times) is due to the fact

Table 7.1 Experimental values of radiation absorption cross section at the first and second steps and two-step photoionisation efficiencies of some polyatomic molecules in a single-frequency laser field [7.85]. The radiation frequency corresponds to the onset of molecular electron absorption.

Molecules	$\langle \sigma_1 \rangle$ (cm^2)	$\langle \sigma_2 \rangle$ (cm^2)	I_{sat} (W cm^{-2})	Ionisation efficiency, $I = 10^7$ (W cm^{-2})
Benzene	2.7×10^{-17}	3.4×10^{-18}	1.5×10^8	3.7%
Thiophene	1.8×10^{-17}			0.01%
Toluene	6.0×10^{-18}	2.8×10^{-17}	8.3×10^7	6.2%
Naphthalene	4.8×10^{-18}	1.6×10^{-19}	2.8×10^9	0.05%
Aniline	4.5×10^{-17}	2.0×10^{-17}	2.3×10^7	25%

that very high photon fluxes can easily be attained in a laser pulse. In order that a molecular optical transition of characteristic cross section $\sigma = 10^{-17}$ cm^2 will be saturated, the photon flux in a pulse should be $\phi \simeq 1/\sigma = 10^{17}$ photons/cm^2 which corresponds to a radiation pulse energy of 10^{-3} J focused on an area of 10^{-2} cm^2. Tunable dye lasers can easily provide this energy in the visible and ultraviolet spectral regions $\lambda > 220$ nm. For average ion currents the difference between laser and electron beam ionisation is not so large because laser photoionisation is pulsed in character (usually $\tau_{\text{pulse}} = 10^{-8}$ s with a pulse repetition rate of 10 to 100 Hz) and ionisation by electron beam is continuous. If we assume that the efficiency of laser photoionisation is 10%, the efficiency of the electron beam ionisation is 10^{-2}%, the laser pulse repetition rate 100 Hz and the characteristic cross dimensions of the ionisation region 0.1 cm, it is easy to determine that the ratio of average ion currents $i_{\text{las}}/i_{\text{el}} \simeq 10$. It should be noted that, to calculate the effective duty cycle of pulsed photoionisation, it is necessary to use not the laser pulse duration as in [7.87] but the time in which the molecule flies through the ionisation region (in our case $\tau_{\text{fl}} \simeq 10^{-5}$ s).

Another field of application of laser photoionisation is related to the possibility of detecting molecules and radicals in single quantum states. In combination with a high detection sensitivity, time and spatial selectivity it gives researchers an effective means for studying IR and UV dissociation [7.88–7.94], reactions in molecular beams [7.95–7.96], etc. With the use of up-to-date tunable lasers the characteristic sensitivity of detection of molecules and radicals in one quantum state comes to 10^4 cm^{-3} [7.89–7.90]. The sensitivity of state selective two-step photoionisation of the hydrogen molecule with the transition $B^2\Sigma_u^+ (v' \leqslant 0, J') \leftarrow X^1\Sigma_u^+ (v'' = 0, J'')$ as an intermediate step equals 10^6 cm^{-3} when tunable VUV laser radiation is used [7.98]. Spatial selectivity made it possible in some experiments to detect photofragments with simultaneous recording of their angular and time-of-flight distribution [7.99].

It is often necessary to use independent tunable radiation at each step in order to reach maximum selectivity of multistep photoionisation. In the general case this leads to a significant decrease of selective photoionisation efficiency as compared with the one-frequency scheme. The point is that a multifrequency scheme may have other nonselective photoionisation channels which should be suppressed by limiting the radiation intensity. Indeed, in a typical case of two-step photoionisation $2\hbar\omega_1 < I < 2\hbar\omega_2$ where ω_1 and ω_2 are the laser radiation frequencies respectively at the first and second steps, I is the ionisation potential, two-step molecular photoionisation through high-lying electronic states by second-step radiation is possible (figure 7.15(c)). Besides, three-step photoionisation is possible when, in the presence of a resonance, the molecule absorbs in succession three photons of the laser with the ω_1 frequency (figure 7.15(b)). Let us compare the relative contributions of each of three channels of molecular photoionisation (figure 7.15). For estimation, all the cross sections of the optical transitions shown in figure 7.15 are assumed to

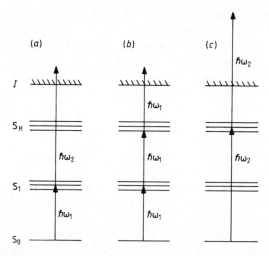

Figure 7.15 Schemes of molecular ionisation in a two-frequency laser field: (a) selective two-step photoionisation; (b) three-step photoionisation under the action of first-step laser radiation; (c) two-step photoionisation by second-step laser radiation.

be equal to each other and the lifetimes of all intermediate levels are assumed to exceed the laser pulse durations. The statistical weights of upper levels for all transitions will be considered to be much larger than those for lower ones. This enables the backward transitions under radiation to be neglected. It is clear that, when the laser flux approximates the saturation energy of transitions $\phi \simeq \phi_{sat} = \hbar\omega/\sigma$, all three channels make approximately the same contribution to molecular ionisation. Far from saturation the number of ions N_i formed by these channels will be

$$N_i^{(a)} = N_0 \sigma^2 \phi_1 \phi_2 / \hbar\omega_1 \hbar\omega_2 \qquad (7.17)$$

$$N_i^{(b)} = 1/6 N_0 \sigma^3 (\phi_1/\hbar\omega_1)^3 \qquad (7.18)$$

$$N_i^{(c)} = 1/2 N_0 \sigma^2 (\phi_2/\hbar\omega_2)^2 \qquad (7.19)$$

where ϕ_1, ϕ_2 are the first- and second-step laser fluxes and N_0 is the number of molecules in the photoionisation volume. It may be seen that the numbers of ions formed by channels (7.18) and (7.19) have higher power dependences on the laser fluxes ϕ_1 and ϕ_2. Therefore, as the laser pulse energies decrease, the ratio of the useful signal of two-step ionisation by channel (7.17) to the background (7.18) and (7.19) will grow. Taking, for example, signal-to-noise ratio $N_i^{(a)}/(N_i^{(b)} + N_i^{(c)}) = 10$ we have from (7.17)–(7.19) that the two-step photoionisation efficiency equals 0.01%. Nevertheless, the situation for real molecules may be more favourable. The main difference from the model consideration lines in the fact that in the highly excited electronic states which participate in multiphoton ionisation through channels (7.18) and (7.19) (figure 7.15(*b,c*)) the rates of relaxation processes may be so high that the photoionisation from these states will be suppressed. In the experiments on selective two-step photoionisation [7.13, 7.22], for example, there was no signal of two-step photoionisation observed as the molecules (formaldehyde, benzaldehyde, benzophenone, etc) were excited by vuv radiation with a photon energy of 7.7 eV.

7.3.4 *Analysis of molecules on a surface*

In all the above cases consideration is given to photoprocesses yielding ions as the molecules are irradiated in a low-pressure gas and when during their interaction with the laser radiation pulse they can be considered to be isolated from the collisions between themselves or with the walls. However, there are a great number of nonvolatile and thermally labile molecules which are hard to convert to the gas phase. So the production of free ions by direct irradiation of solid samples is of great interest. Generally speaking, mass spectrometry with laser desorption of ions (LDMS) has been developing for many years and covers a great number of experimental approaches and physical processes of free ion formation [7.52, 7.87, 7.100]. But it has lately been shown that in many cases the mechanism of of formation from solids is also related to the resonant laser photoionisation of molecules desorbed under the action of the same laser pulse [7.101]. Therefore this case is also discussed in this chapter.

In surface experiments the sample is usually placed directly into the ion source of the mass spectrometer [7.102, 7.103]. This allows mass analysis of the resulting photoions. In the simplest case the sample, which may be a trace amount of solid organic substance (molecular crystals) or adsorbed molecules, is irradiated by one uv radiation pulse. Since the formation of ions is preceded by thermal evaporation of the molecules from the surface, the ion signal N_i is proportional to the product of the desorption probability P_{des} and the multiphoton ionisation yield η

$$N_i = N_0 P_{des} \eta \tag{7.20}$$

where N_0 is the number of exposed molecules. The radiation absorption in the surface layer leads to its heating to the temperature ΔT which is easy to evaluate assuming that all the absorbed energy converts to heat and taking into account the fact that the heat has no time to penetrate into the sample during a laser pulse $\tau_{pul} \lesssim 10^{-8}$ s

$$\Delta T = \hbar \omega K \phi / c\rho \tag{7.21}$$

where ϕ is the laser flux, K is the absorption coefficient of the medium, c and ρ are respectively the weight and heat capacity of medium. The desorption rate W_{des} of molecules depends greatly on the surface temperature and is described by the Arrhenius-type formula

$$W_{des} = \Omega \exp[-E_{bind}/k(T_0 + \Delta T)] \tag{7.22}$$

where $\Omega = 10^{12}-10^{14}\,\text{s}^{-1}$ is the characteristic vibration frequency of molecules on the surface, E_{bind} is the energy of binding with the surface and T_0 is the initial surface temperature. The probability of molecular desorption is apparently equal to

$$P_{des} = 1 - \exp\left(-\int_0^t W_{des}(\tau)\,d\tau\right). \tag{7.23}$$

The value of heating ΔT at a fixed laser flux is proportional to the absorption coefficient $K(\lambda_{las})$ (see (7.21)). Thus the desorption probability P_{des}, according to (7.22) and (7.23), depends significantly on laser radiation wavelength and, together with the spectral dependence of multiphoton ionisation efficiency $\eta(\lambda_{las})$, governs the resonant nature of the ion yield.

Figure 7.16 presents the mass spectrum of photoions formed as molecular adenine crystals are irradiated by an excimer KrF laser [7.102]. The absence of adenine fragmentation is a characteristic feature of the mass spectrum. As well as molecular ions of adenine, the mass spectrum has peaks of alkali metals Na^+ and K^+ which present in the sample as impurities; it also has a weak peak of adenine–Na^+ clusters (figure 7.16). A small value, or the absence, of fragmentation in operation with nonvolatile organic and bio-organic molecules in this method is caused by the fact that both stages of laser desorption and photoionisation come about under 'soft conditions'. As for multiphoton ionisation of free molecules, the fragmentation, as discussed above, is minimum when the sum of laser photons slightly exceeds the molecular ionisation potential and intermediate excited molecular states are stable. The absence of molecular decomposition at the first step of evaporation is caused by the fact that, due to the very high efficiency of laser molecular photoionisation in the gas phase (up to 100%), the number of desorbed

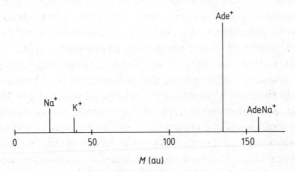

Figure 7.16 Mass spectrum of photoions formed by irradiating adenine crystals [7.102].

molecules may be very small. This means that the surface needs little heating. In the case of adenine, for example (figure 7.16), the signal of molecular ions can be reliably recorded with the laser flux on the surface of $\phi = 10^{-3}$ J cm^{-2}. This value results in heating the surface to ΔT no higher than 100 °C [7.104]. Such a low value of heating, together with its pulsed character ($\tau_{las} \lesssim 10^{-8}$ s) causes a polyatomic molecule to detach without decomposition. It should be stressed that it is very easy to record mass spectra of nonvolatile and thermally labile organic compounds by the method concerned. According to relations (7.20)–(7.23) and experimental studies [7.104–7.106], this takes place only in the case of sufficiently shortwave laser radiation that resonantly excites molecules both on the surface and in the gas phase.

To maximise the sensitivity of the analysis, it was proposed in [7.107] to apply a two-frequency IR–UV laser scheme where the function of radiation is separated relative to molecular evaporation and photoionisation. In this scheme under the action of the first IR laser pulse irradiating the surface the molecules desorb and are then resonantly ionised by the second laser pulse at a distance from the surface. Such separation allows, firstly, optimisation of the energy of the first desorbing laser pulse to reach a maximum yield of molecular desorption (overheating of the surface may bring about molecular decomposition) and the second ionising laser pulse to obtain a maximum photoionisation efficiency. Secondly, the use of IR radiation in molecular desorption makes it possible to get rid of unwanted photochemical reactions on the surface included by powerful UV radiation. This approach made it possible in [7.107] to obtain a detection sensitivity of naphthalene molecules adsorbed on the surface of graphite equal to 10^{-8}–10^{-10} monolayers.

7.4   Photoionisation detection of molecules in buffer gases

Mass spectrometric ion detection increases the selectivity and hence the analytical possibilities of resonant laser molecular photoionisation. In all types of mass spectrometers, however, mass separation should be provided with collisionless motion of ions in electric and magnetic fields. As a rule, gas pressure must not exceed 10^{-5} Torr. Thus, in analysing samples at the atmospheric pressure, for example, the concentration of molecules, when introduced into the mass spectrometer, has to be decreased by 10^8 or 10^5–10^6 times in the case of effective differential evacuation in the ion source. At low relative concentrations of the impurities to be investigated in the sample (less than 10^{-9}) this may cause no detectable molecule to be found in the photoionisation volume. So it is of great interest to develop selective laser photoionisation methods which would allow detection of molecules at atmospheric pressure directly in the air or in buffer gases.

7.4.1 Experiments in ionisation chambers

The analytical applications of ionisation chambers recording the total photocurrent are based on the selectivity of resonant laser molecular ionisation. Figure 7.17 schemetically shows a typical experimental set-up [7.55] which consists of an ionisation chamber comprising a cylindrical (sometimes planar [7.54, 7.108]) capacitor, a recording system and a tunable laser radiation source. The ionisation chamber can work under conditions of full photocharge collection and gas amplification. In the first case the completeness of photoion collection is controlled by the presence of a plateau in the dependence of the ion signal on the voltage across the electrodes of the chamber. The ion detection sensitivity in this case is governed by the noise of the amplifier and has a typical value of about 10^3 ions/pulse [7.56].

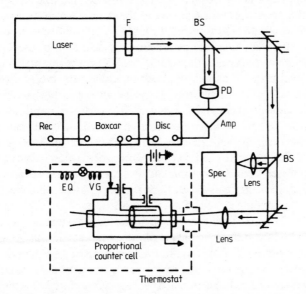

Figure 7.17 Scheme of the set-up for detection of molecules with an ionisation chamber [7.55]. BS, beam splitters; F, filter; PD, photodiode; EQ, equilibrium coil; VG, vapour generator coil.

In the second case the electric field strength must be chosen high enough so that an electron has time between the two collisions to acquire an energy sufficient to ionise the buffer gas molecules. Under these conditions one electron can initiate an avalanche with a duration of about 10^{-6} s and amplitude 10^{-14} C [7.55]. At low laser flux the recording is performed in a pulse (avalanche) count mode. At high laser fluxes, and hence high ionisation rates, the avalanches are overlapped in time and the average current is measured. Such a method of ion detection allows the formation of every electron–ion pair to be recorded [7.109].

The analytical possibilities of detection of low concentrations of polyatomic molecules in the air using a simple ionisation chamber were studied in [7.54] using the example of aniline. The energy of the first electronic-excited singlet state S_1 of aniline is more than half the ionisation potential, and so effective two-step molecular photoionisation is possible in a one-frequency laser field. The two-step photoionisation spectrum was recorded at a relative concentration of aniline in the air of 10^{-5} with the signal-to-noise ratio higher than 10 despite a comparatively low radiation peak intensity (2 W) and long laser pulse duration (10^{-7} s). As the calculations show [7.54], the use of a more powerful laser source with a pulse duration $\tau = 10^{-8}$ s would allow aniline to be detected in the air at the level of 10^{-11}. Under real conditions, however, the sensitivity might be limited by the background signal of ionisation of other organic impurities and by the multiphoton ionisation of basic air components.

Being a characteristic representative of the class of aromatic molecules the naphthalene molecule is well suited for laser photoionisation experiments. It was used to study the ultimate possibilities of detection of small molecular concentrations in a buffer gas (nitrogen) [7.56]. The measurements were taken under the conditions of complete photoion collection, the photoionisation volume being 0.1 cm^3. The photosignal value was linearly dependent on the naphthalene concentration in the region of eight orders of naphthalene partial pressure variation and the ultimate sensitivity was 5×10^4 mol cm^{-3} with the signal-to-noise ratio being equal to 2. The photoionisation of naphthalene molecules in this case was carried out using a two-step scheme, the pulse energy of dye laser second-harmonic radiation was 1 mJ, its duration 5 ns and the laser beam diameter was 1 mm [7.56]. The naphthalene molecule has an absorption maximum at the $S_1 \leftarrow S_0$ transition at $\lambda = 278.5$ nm and a minimum at 277.5 nm. In the experiment the naphthalene ionisation signal was expressed as the difference of two photosignals at the wavelengths of the absorption maximum and minimum. Further increase of sensitivity by 10 to 100 times can be obtained not only by increasing the laser pulse energy and the photoionisation volume but also when the ionisation chamber operates under avalanche conditions.

During photoionisation detection of molecules in an ionisation chamber laser ionisation selectivity is a very important factor and is actually the basic factor limiting the sensitivity of the technique. Therefore in some cases three-step photoionisation may turn out to be useful. On the one hand this enables us to make the most use of longwave radiation and thereby to reduce the background ionisation. On the other hand the presence of a second resonant transition between the bound states increases photoionisation selectivity. Such an approach was realised using the iodine molecule [7.110] and made it possible to reach a detection sensitivity of molecules in the air at the level of 10^{-9}.

7.4.2 Ion mobility spectrometer

The method of plasma chromatography is widely used for analytical purposes [7.111, 7.112]. In the standard plasma chromatograph buffer gas ions are formed first under the action of a radioactive source of ^{63}Ni with β decay. Then the chain of ion–molecular reactions develops ions of impurity molecules which enter the drift region (figure 7.18). In this region the ions move in the buffer gas with a superposed electric field and are separated in space and time according to their mobility.

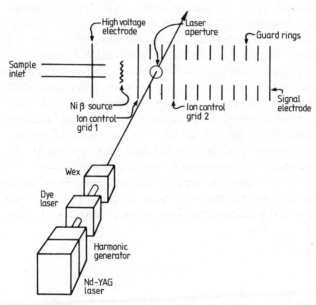

Figure 7.18 The ion mobility spectrometer with resonant laser molecular photoionisation on the basis of a plasma chromatograph [7.113].

The use of resonant laser photoionisation instead of the technique of ion production described allows us to overcome, in principle, many problems relating to the fact that ion–molecular reactions often result in many ion–molecular combinations thus complicating the analysis [7.113]. On the other hand, the mobility separation of the resultant photoions makes it possible to significantly increase the selectivity and hence the sensitivity of detection of impurity molecules in the air or in a buffer gas and thereby to overcome the basic problems of nonselective photoion detection in the ionisation chamber.

Figure 7.18 schematically shows an ion mobility spectrometer with laser photoionisation [7.113] developed on the basis of the commercial plasma chromatograph. The ions are formed under the action of resonant laser radiation in the gap between grids 1 and 2. When operating with laser ionisation, grid 1 is maintained at a constant potential and blocks the ions in the region of the radioactive source. Grid 2 is governed by a pulsed potential and ejects the photoions into the region of drift 8 cm long (figure 7.18). The ejecting pulse may vary from 0.05 to 0.5 ms. The shorter the ion injection, the higher the resolution. The sensitivity in this case is lower. The spectrometer works at a temperature of about 200 °C so that it is protected against the pollution caused by the substances under study. The sample was admitted through a quartz leak at 275 °C. Pure nitrogen was used as a buffer gas. Photoionisation was realised as usual by the second harmonic of tunable dye laser radiation or by the fourth harmonic of Nd–YAG laser radiation.

Figure 7.19 shows two-step photoionisation ion mobility spectrograms for naphthalene, benzene and toluene at two wavelengths (280 and 266 nm). The naphthalene concentration in nitrogen was equal to 2×10^{-8} (20 PPB). The molecular peaks are well resolved except for a residual background in the region of high drift times. In the case of longer-wave radiation, $\lambda = 280$ nm (figure 7.19(a)), there is mainly a naphthalene peak observed in the spectrum since the molecules of benzene and toluene cannot absorb at this wavelength and so cannot be resonantly ionised by laser radiation. This demonstrates the essence and the potentialities of the method consisting of a combination of optical and ion-chromatographic selectivities of detection of impurities. The

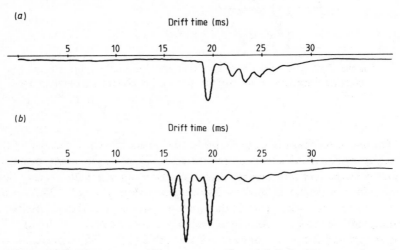

Figure 7.19 Photoion mobility spectra of naphthalene (concentration of 20 PPB) and benzene and toluene traces: (a) laser radiation wavelength $\lambda = 280$ nm, pulse energy $E = 1.5$ mJ; (b) $\lambda = 226$ nm, $E = 2$ mJ. The electric field strength is 171 V cm^{-1} [7.113].

maximum detection sensitivity for benzene in nitrogen was 10^{-9} and was limited by the presence of residual benzene in the spectrometer rather than by the sensitivity of the recording system.

7.5 Conclusions

Progress in the development of laser radiation sources in the UV spectrum has allowed the effect of resonant multiphoton molecular ionisation to be accessible for experimental studies and applications. Apart from the fact that these studies have improved our understanding of molecular and ion physics, the researches carried out in the past five to seven years have revealed a number of unique properties of resonant laser molecular photoionisation important for analytical applications. Among them are:

(i) selectivity—up to photoionisation and detection of molecules in separate quantum states;

(ii) efficiency—the probability of molecular ionisation in a laser field reaches 100%;

(iii) high time and spatial resolution.

It is sufficient to say that already, by early 1986, the methods of laser photoionisation analytics make it possible to detect single molecules in a time interval of 10^{-12} s.

Further investigation into the resonant interaction of laser radiation with molecules will no doubt induce the development of new possibilities of highly selective photoionisation. Of great interest, for example, is the resonant excitation of molecular vibrations with subsequent photoionisation of vibrationally excited molecules via intermediate electronic states [7.1, 7.114–7.116].

It seems quite possible that the development of laser photoionisation methods in combination with such present-day means of analysis as multiple mass spectrometry [7.52], gas and liquid chromatography and others will in the next few years give birth to a highly selective detector for trace quantities of molecules, the sensitivity of which will be several orders higher than that of all analytical methods available now. This will enable us to solve the old problem of experimental physics of detection by physical methods of 'odours' with a sensitivity attainable today only by the olfactory organs of animals.

References

7.1   Antonov V S and Letokhov V S 1981 *Appl. Phys.* **24** 89
7.2   Antonov V S, Letokhov V S and Shibanov A N 1984 *Usp. Fiz. Nauk* **142** 177 (in Russian)

7.3   Johnson P M 1980 *Acc. Chem. Res.* **13** 20

7.4   Boesl U, Neusser H J and Schlag E W 1983 *Acc. Chem. Res.* **16** 355

7.5   Letokhov V S and Chebotayev V P 1977 *Nonlinear Laser Spectroscopy* (Springer series in Optical Sciences, vol 4) (Berlin: Springer)

7.6   Rothberg L J, Gerrity D P and Vaida V 1981 *J. Chem. Phys.* **75** 4403

7.7   Rhodes C K (ed) 1984 *Excimer Lasers* (Springer Series 'Topics in Applied Physics', vol 30) (Berlin: Springer)

7.8   Woodworth J R and Rice J K 1978 *J. Chem. Phys.* **69** 2500

7.9   Berkowitz J 1979 *Photoabsorption, Photoionization, and Photoelectron Spectroscopy* (New York: Academic)

7.10   Long S R, Meek J T and Reilly J P 1982 *J. Chem. Phys.* **79** 3206; *J. Phys. Chem.* **86** 2809

7.11   Duncan M A, Dietz T G and Smalley R E 1981 *J. Chem. Phys.* **75** 2118

7.12   Cooper D E, Frueholz R P, Klimcak C M and Wessel J E 1982 *J. Phys. Chem.* **86** 4892

7.13   Antonov V S, Knyazev I N, Letokhov V S, Matiuk V M and Potapov V K 1978 *Opt. Lett.* **3** 37

7.14   Antonov V S and Shibanov A N 1982 *Opt. Spectrosc. (USSR)* **52** 234

7.15   Brophy J H and Rettner C T 1979 *Chem. Phys. Lett.* **67** 351

7.16   Colson S D, Cheung M Y and Glownia J H 1980 *Chem. Phys. Lett.* **76** 515

7.17   Fujii M, Ebata T, Mikami N and Ito M 1983 *Chem. Phys. Lett.* **101** 578

7.18   Antonov V S, Knyazev I N, Letokhov V S, Matiuk V M, Movshev V G and Potapov V K 1977 *Pis'ma Zh. Tekhn. Fiz.* **3** 1278 (in Russian)

7.19   Antonov V S 1982 *Opt. Spectrosc. (USSR)* **52** 5

7.20   Ackerhalt J R and Shore B W 1977 *Phys. Rev.* A **16** 277

7.21   Ackerhalt J R and Eberly J H 1976 *Phys. Rev.* A **14** 1705

7.22   Andreyev S V, Antonov V S, Knyazev I N and Letokhov V S 1977 **45** 166

7.23   Green B I and Farrow R C 1983 *J. Chem. Phys.* **78** 3336

7.24   Perry J W, Scherer N F and Zewail A H 1983 *Chem. Phys. Lett.* **103** 1

7.25   Knee J L and Johnson P M 1984 *J. Chem. Phys.* **80** 13

7.26   Dietz T G, Duncan M A and Smalley R E 1982 *J. Chem. Phys.* **76** 1227

7.27   Parker D H, Berg J O and El-Sayed M A 1978 in *Advances in Laser Chemistry* ed A H Zewail (Berlin: Springer) p. 320

7.28   Johnson P M 1976 *J. Chem. Phys.* **64** 4143

7.29   Zandee L and Bernstein R B 1979 *J. Chem. Phys.* **70** 2574

7.30   Reilly J P and Kompa K L 1980 *J. Chem. Phys.* **73** 5468

7.31   Antonov V S, Letokhov V S and Shibanov A N 1980 *Sov. Phys.–JETP* **51** 1113

7.32   Antonov V S, Letokhov V S and Shibanov A N 1980 *Appl. Phys.* **22** 293

7.33   Dietz W, Neusser H J, Boesl U, Schlag E W and Lin S H 1982 *Chem. Phys.* **66** 105

7.34   Zandee L and Bernstein R B 1979 *J. Chem. Phys.* **71** 1359

7.35   Lubman D M, Naaman R and Zare R N 1980 *J. Chem. Phys.* **72** 3034

7.36   Seaver M, Hudgens J W and De Corpo J J 1980 *Int. J. Mass Spectrosc. Ion Phys.* **34** 159

7.37   Reilly J P and Kompa K L 1980 *Adv. Mass Spectrosc.* **8B** 1800

7.38   Dietz T G, Duncan M A, Liverman M G and Smalley R E 1980 *J. Chem. Phys.* **73** 4816

7.39   Leutwyler S and Even U 1981 *Chem. Phys. Lett.* **84** 188; *Chem. Phys.* **58** 409

7.40   Fisanick G J, Gedanken A, Eichelberger T S IV, Kuebler N A and Robin M B 1981 *J. Chem. Phys.* **75** 5215

7.41   Martin T P 1982 *J. Chem. Phys.* **77** 3815

7.42   Takenoshita Y, Shinohara H, Umemoto M and Nishi N 1982 *Chem. Phys. Lett.* **87** 566

7.43   Baba M, Shinohara H, Nishi N and Hirota N 1984 *Chem. Phys.* **83** 221

7.44   Rebentrost F, Kompa K L and Ben-Shaul A 1981 *Chem. Phys. Lett.* **77** 394

7.45   Rebentrost F and Ben-Shaul A 1981 *J. Chem. Phys.* **74** 3255

7.46   Silberstein J and Levine R D 1980 *Chem. Phys. Lett.* **74** 6

7.47   Andlaner B and Ottinger C 1971 *J. Chem. Phys.* **55** 293

7.48   Silberstein J and Levine R D 1981 *J. Chem. Phys.* **75** 5735

7.49   Lichtin D A, Bernstein R B and Newton K R 1981 *J. Chem. Phys.* **75** 5728

7.50   Carney T E and Baer T 1982 *J. Chem. Phys.* **76** 5968

7.51   Ligon W V 1979 *Science* **205** 151

7.52   Burlingame A L, Dell A and Russel D H 1982 *Anal. Chem.* **54** 363R

7.53   Millard B J 1979 *Quantitative Mass Spectrometry* (London: Heyden) p. 120

7.54   Brophy J H and Rettner C T 1979 *Opt. Lett.* **4** 337

7.55   Frueholz R, Wessel J and Whealtley E 1980 *Anal. Chem.* **52** 281

7.56   Klimcak C and Wessel J 1980 *Appl. Phys. Lett.* **37** 138

7.57   Letokhov V S 1976 *Usp. Fiz. Nauk* **118** 197 (in Russian)

7.58   Bernstein R B 1982 *J. Phys. Chem.* **86** 1178

7.59   Wiley W C and McLaren I H 1955 *Rev. Sci. Instrum.* **26** 1150

7.60   Mamyrin B A, Karataev V I, Shmikk P V and Zagulin V A 1973 *Zh. Exp. Teor. Fiz.* **64** 82 (in Russian)

7.61   Boesl U, Neusser H J, Weinkauf R and Schlag E W 1982 *J. Phys. Chem.* **86** 4857

7.62   Rockwood S, Reilly J P, Hohla K and Kompa K L 1979 *Opt. Commun.* **28** 175

7.63   Anderson J B 1974 in *Molecular Beams and Low Density Gas-Dynamics* ed P P Wegener (New York: Dekker) p. 1

7.64   Hopkins J B, Powers D E and Smalley R E 1981 *J. Phys. Chem.* **85** 3739

7.65   Herrmann A, Leutwyler S, Schumacher E and Wöste L 1977 *Chem. Phys. Lett.* **52** 418

7.66   Shinohara H and Nishi N 1982 *Chem. Phys. Lett.* **87** 561

7.67   Leutwyler S, Even U and Jortner J 1983 *J. Chem. Phys.* **79** 5769

7.68   Choo K Y, Shinohara H and Nishi N 1983 *Chem. Phys. Lett.* **95** 102

7.69   Rademann K, Brutschy B and Baumgartel H 1983 *Chem. Phys.* **80** 129

7.70   Lichtin D A, Datta-Ghosh S and Newton K R 1980 *Chem. Phys. Lett.* **75** 214

7.71   Long S R, Meek J T, Harrington P J and Reilly J P 1983 *J. Chem. Phys.* **78** 3341

7.72   Fisanick G J, Eichelberger T S IV, Heath B A and Robin M B 1980 *J. Chem. Phys.* **72** 5571

7.73   Cooper C D, Williamson A D, Miller J C and Compton R N 1980 *J. Chem. Phys.* **73** 1527

7.74   Proch D, Rider D M and Zare R N 1981 *J. Photochem.* **17** 249

7.75   Fisanick G J and Eichelberger T S IV 1981 *J. Chem. Phys.* **74** 5962

7.76   Rettner C T and Brophy J H 1981 *Chem. Phys.* **56** 53

7.77   Seaver M, Hudgens J W and De Corpo J J 1982 *Chem. Phys.* **70** 63

7.78   Lichtin D A, Squire D W, Winnik M A and Bernstein R B 1983 *J. Am. Chem. Soc.* **105** 2109

7.79   Antonov V S, Letokhov V S and Shibanov A N 1981 *Opt. Comm.* **38** 182

7.80   Middleditch B S (ed) 1979 *Practical Mass Spectroscopy* (New York: Plenum)
7.81   Boesl U, Neusser H J and Schlag E W 1979 in *Laser Spectroscopy IV* ed H Walther and K W Rothe (Springer Series in Optical Science, vol 21) (Berlin: Springer) p. 164
7.82   Leutwyler S and Even U 1981 *Chem. Phys. Lett.* **81** 578
7.83   Fitch P S H, Haynam C A and Levy D H 1980 *J. Chem. Phys.* **73** 1064
7.84   Dietz T G, Duncan M A, Liverman M G and Smalley R E 1980 *Chem. Phys. Lett.* **70** 246
7.85   Boesl U, Neusser H J and Schlag E W 1981 *Chem. Phys.* **55** 193
7.86   Duncan M A, Dietz T G and Smalley R E 1979 *Chem. Phys.* **44** 415
7.87   Hunt D F 1982 *Int. J. Mass Spectrosc. Ion Phys.* **45** 111
7.88   Zacharias H, Schmiedl R and Welge K H 1980 *Appl. Phys.* **21** 127
7.89   Zacharias H, Rottke H and Welge K H 1981 *Appl. Phys.* **24** 23
7.90   Danon J, Zacharias H, Rottke H and Welge K H 1982 *J. Chem. Phys.* **76** 2399
7.91   Rockney B H and Grant E R 1981 *Chem. Phys. Lett.* **79** 15
7.92   Morrison R J S and Grant E R 1982 *J. Chem. Phys.* **77** 5994
7.93   Dulgnan M T, Hudgens J W and W,yatt J R 1982 *J. Phys. Chem.* **86** 4156
7.94   Stuke M, Reisler H and Wittig C 1981 *Appl. Phys. Lett.* **39** 201
7.95   Deldman D L, Lengel R K and Zare R N 1977 *Chem. Phys. Lett.* **52** 413
7.96   Goldstein N, Greenblatt G D and Wiesenfeld J R 1983 **96** 410
7.97   Hayden J S and Diebold G J 1982 *J. Chem. Phys.* **77** 4767
7.98   Rottke H and Welge K H 1983 *Chem. Phys. Lett.* **99** 456
7.99   Rockney B H and Grant E R 1982 *J. Chem. Phys.* **77** 4257
7.100   Conzemius R J and Capellen J M 1980 *Int. J. Mass Spectrosc. Ion Phys.* **34** 197
7.101   Egorov S E, Letokhov V S and Shibanov A N 1983 in *Surface Studies with Lasers* ed F R Aussenegg *et al* (Springer Series in Chemical Physics, vol 33)
7.102   Antonov V S, Letokhov V S and Shibanov A N 1980 *Pis'ma Zh. Exp. Teor. Fiz.* **31** 471 (in Russian)
7.103   Chai J-W and Reilly J P 1984 *Opt. Commun.* **49** 51
7.104   Antonov V S, Letokhov V S and Shibanov A N 1981 *Appl. Phys.* **25** 71
7.105   Lubman D M and Naaman R 1983 *Chem. Phys. Lett.* **95** 325
7.106   Opsal R B and Reilly J P 1983 *Chem. Phys. Lett.* **99** 461
7.107   Antonov V S, Egorov S E, Letokhov V S and Shibanov A N 1983 *Pis'ma Zh. Exp. Teor. Fiz.* **38** 185 (in Russian)
7.108   Andreyev S V, Antonov V S, Knyazev I N and Letokhov V S 1975 *Phys. Lett.* **54A** 91
7.109   Williams A and Sara R I 1962 *Int. J. Appl. Radiat. Isot.* **13** 229
7.110   Darznek S A, Zverev M M and Kopit C P 1983 *Kvant. Electron.* **10** 1270 (in Russian)
7.111   Karasek F W, Kim S H and Hill H H Jnr 1976 *Anal. Chem.* **48** 1133
7.112   Hagen D F 1979 *Anal. Chem.* **51** 870
7.113   Lubman D M and Kronick M N 1982 *Anal. Chem.* **54** 1546
7.114   Esherick P and Anderson R J M 1980 *Chem. Phys. Lett.* **70** 621
7.115   Glownie J H, Romero R J and Sander R K 1982 *Chem. Phys. Lett.* **88** 292
7.116   Haas Y, Reisler H and Wittig C 1982 *Chem. Phys. Lett.* **92** 109

8 Laser Desorption Mass Spectrometry of Nonvolatile Organic Molecules

A N Shibanov

8.1 Introduction

Mass spectral analysis is one of the most powerful analytical methods characterised by both high sensitivity and high information capability [8.1, 8.2]. In most types of spectrometer electron impact ionisation and chemical ionisation methods are used to obtain ions from the sample molecules. These methods have been comprehensively studied and many experimental results have been accumulated. For a large number of molecules mass spectra obtained under standard conditions are summarised in special atlases.

To apply the methods of electron impact and chemical ionisation, it is necessary that the molecules of the sample under study should be in the ion source of the mass spectrometer in the gas phase. In the case of gases and volatile compounds it is not difficult to introduce the substance into the ion source and this is usually done through a standard capillary inlet system. For substances having a low vapour pressure under conventional conditions ($P < 10^{-6}$ Torr), the direct inlet system is used [8.1]. In this case the sample is placed into a tube, one end of which is sealed off and the open end enters the ion source of the mass spectrometer. As the tube is heated to several hundreds of degrees centigrade, a relatively high vapour pressure is set up in it, and the molecular density at the end of the tube is quite sufficient for analysis. But the direct inlet technique is inapplicable for a number of organic and bio-organic compounds due to their low vapour pressure and thermal instability.

It is evident that researchers could not accept the situation when they were unable to use mass spectral analysis for quite a number of extremely important compounds. The first step taken to solve this problem was to develop methods for chemical modification of sample molecules in order to increase their volatility [8.3, 8.4]. But these methods can only be applied to a limited class of molecules and make the procedure of analysis more complicated.

At the end of the 1960s different methods were developed to produce ions for nonvolatile and thermally unstable molecules without their preliminary chemical modification. At present there are quite a number of methods which allow the production of ions of almost any organic and bio-organic molecules with masses of up to several thousands of atomic units (see [8.5]). Each of these methods, however, has particular shortcomings. Today many laboratories are engaged in both improving the suggested methods and in seeking new ones.

The common feature of all methods known now is that the solid sample is placed directly in the mass spectrometer source. The characteristic features of a particular method are defined by the physical action on the sample that leads to ion formation. Therefore, the methods of mass spectrometric analysis of nonvolatile compounds are distinguished by the methods of ion production.

The following nonlaser methods are the most intensively investigated and used in mass spectrometry.

(1) Field desorption mass spectrometry, FDMS [8.6].

(2) Mass spectrometry with the formation of ions due to bombardment of the sample by the decay products of Cf-252 nuclei (there are two names used in literature: plasma desorption mass spectrometry, PDMS, and fission-fragment induced desorption mass spectroscopy, FFIDMS) [8.7, 8.8].

(3) Secondary ion mass spectrometry, SIMS [8.9].

(4) Fast atom bombardment mass spectrometry, FABMS [8.94].

(5) Desorption chemical ionisation mass spectrometry, DCIMS [8.10, 8.11].

Along with these methods, laser methods were developed to produce ions of nonvolatile organic and bio-organic compounds (see [8.12, 8.13]). Now it is customary to unite the laser ionisation mass spectrometric methods by the common name laser desorption mass spectrometry, LDMS.

8.1.1 Pioneer work on laser desorption mass spectrometry

The first papers concerned with LDMS were published in 1966–8 (F J Vastola, A J Pirone and R O Mumma [8.14–8.18]). In [8.14, 8.15] it was found that pulsed laser pyrolysis of polymers could produce not only neutral species but also ions. Later, in [8.16–8.18], Vastola and co-workers were able to obtain mass spectra of the ions formed by the action of pulsed laser radiation on some polyaromatic compounds and organic salts.

Figure 8.1 schematically shows the set-up used in [8.16–8.18]. The investigated sample was fixed onto a holder and placed into the source of a time-of-flight mass spectrometer. The pulsed radiation of a ruby laser

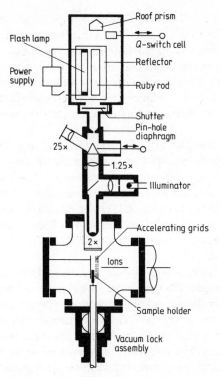

Figure 8.1 Experimental arrangement for mass spectrometry of nonvolatile compounds with a laser ion source [8.18].

($\lambda = 694.3$ nm) with a pulse duration of 800 μs and energy of 3.8 J was focused onto the sample with a spot of 0.8 mm. The action of laser radiation on the sample gave rise to both ions and neutral molecules. The neutral molecules were detected through their ionisation by an electron beam.

In [8.16] three classes of molecules were studied: conjugated aromatic compounds, alkyl-substituted aromatic compounds and the amino acid leucine. The common feature of all the mass spectra was the presence of a rather intense signal of the molecular ions M^+. The mass spectrum of conjugated aromatic compounds consisted almost entirely of one molecular ion. The mass spectrum of the second group, alkyl-substituted aromatic compounds exhibit not only molecular ions but also heavier ions and light fragment ions. Alkyl group ions were not observed, however. Finally, the mass spectrum of leucine consisted of, besides a molecular ion, a large number of fragment ions among which light alkyl ions could be observed. It is worthy of note that the electron-impact ionisation mass spectrum of leucine (the energy of electrons was 25 eV) did not exhibit molecular ions.

Quite a different mass spectrum was observed in [8.17, 8.18] where the formation of ions due to laser irradiation of organic salts was investigated. Mass spectra of sodium and potassium salts of sulphanates, sulphates and thiosulphates were obtained. In most of the mass spectra the only organic ion was that formed due to an alkali cation attachment to the parent molecule $(M + alkali)^+$. Besides, the mass spectrum contained a number of inorganic ions. In some cases molecular dimer ions of the sample $(2M + alkali)^+$ were observed. Neither molecular nor fragment ions were observed in the mass spectrum.

The paper by Kistemaker *et al* [8.19] published in 1978 was of special significance in the development of LDMS as applied to analysis of nonvolatile and thermally unstable bio-organic molecules. The mass spectra of oligosaccharides, glycosides, peptides and some other biological molecules were obtained with the use of pulsed CO_2 laser radiation ($\lambda = 10.6$ μm). All the mass spectra exhibited peaks of the ions $(M + Na)^+$ and $(M + K)^+$. For example, quasi-molecular ions of digitonin $(C_{56}H_{92}O_{29} + Na)^+$ with mass of 1251 au were observed.

Later on greater interest was shown in LDMS. Many experiments were carried out to investigate the mass spectra of some compounds of specific interest for biology, medicine and pharmacology. Special investigations have been performed to ascertain the mechanisms of ion formation. In the technical aspect of development the method of LDMS has graduated from unique laboratory installations to a commercially available instrument [8.20].

In recent years many experiments have been performed with the use of different laser and mass spectrometer equipment. Pulsed sources of radiation were a CO_2 laser, a ruby laser at its fundamental frequency and its second harmonic, an Nd laser at the first, second, third and fourth harmonics, an N_2 laser and an excimer laser, and a dye laser. The ions of nonvolatile organic compounds have been obtained using CW CO_2 laser radiation [8.21].

Thus, the wavelength scale of radiation used in the experiments on LDMS ranges from the middle IR spectrum to the near UV one. The laser radiation intensity at which ions can be formed lies in the range 1 to 10^{11} W cm^{-2}.

8.1.2 Mechanism of ion formation

It is unlikely that the formation of ions under the action of laser radiation with its parameters lying over a wide range can be explained by one mechanism. One should most likely assume that there are several mechanisms of ion formation. The role of each of them depends on both the laser radiation parameters and the physical properties of the sample. But until recently our ideas of the specific mechanisms of ion formation have been rather vauge. And only in investigations carried out recently [8.22–8.28] has it been made possible to elucidate the role of thermal molecular desorption, gas-phase ion–molecular reactions and photoionisation processes in the formation of ions.

Due to these investigations it became possible to analyse and make some systematisations of the available experimental results.

As far as the experimental conditions and dominating physical processes of ion formation are concerned, all the experiments can be divided into three groups.

The first group comprises the experiments where IR laser radiation was used. The mass spectra in this case reveal a high abundance of quasi-molecular ions $(M + C)^+$, where M is the intact molecule and C is the metal cation, and a low degree of fragmentation of intact molecules. The main mechanism of ion formation in this case is as follows [8.22–8.26]. When the sample is heated by laser radiation, the molecules and cations of metals (in most cases alkali ones) are subjected to desorption. The subsequent gas-phase ion–molecular reactions define the character of the mass spectrum. The reaction of attachment of the alkali cation to the intact molecule gives a quasi-molecular ion $(M + \text{alkali})^+$.

As well as this mechanism, the formation of ions in some classes of molecules occurs due to thermodesorption of molecular cations from the sample surface [8.29–8.32] or dissociative ionisation of clusters on a hot surface [8.25–8.31].

The second group comprises the experiments where UV laser radiation or a combination of UV and IR lasers were used. The incident radiation intensity does not exceed 10^6 W cm^{-2} with a pulse duration of 10^{-8} s. The characteristic features of the mass spectra in this case is that the molecular ions dominate in them and fragmentation is either small or absent.

The mechanism of ion formation is as follows [8.27, 8.28]. The heating of the sample surface by laser radiation leads to molecular desorption. The desorbed molecules are ionised by UV laser radiation due to resonant absorption of two or three quanta by a molecule.

Laser microanalytical mass spectrometry forms the third group of experiments. This technique is used for high-spatial resolution mass spectrometry. In this case the laser radiation (usually in the UV range) is focused onto the sample in a very small spot ($d \simeq 1\ \mu$m). The radiation intensity ranges from 10^8 to 10^{11} W cm^{-2} and the physical conditions differ from those used in the first two groups: the small volume of the sample under laser radiation and the high UV radiation intensities, at which nonresonant multiphoton absorption is possible. The characteristic features of the mass spectra [8.13] are the presence of both positive and negative ions, a high abundance of protonated and deprotonated ions, $(M + H)^+$ and $(M + H)^-$, as well as cationised ions $(M + C)^+$.

Unfortunately, the ion formation mechanisms under these conditions are still not clearly understood. Therefore, the experiments were placed into this group on a methodical basis which is tentative in character. As will be shown below, despite the use of the laser microanalytical technique, it was possible in some experiments to realise the conditions of ion formation characteristic of the first or the second groups. But there are conditions which can, in our opinion, be realised only by using strongly focused laser radiation. The mechanism of mass spectra formation in experiments of the third type must differ significantly from the others.

Thus, it can be said rather conventionally that three physically different methods of laser mass spectrometry of nonvolatile and thermally unstable organic molecules were studied in the three groups of experiments. It should be noted at once that the degree of understanding of the mechanisms of ion formation and the level of technical development of these methods differ greatly. For example, the elementary processes in the method of laser desorption and uv laser photoionisation are clearly understood at present (see §8.3). At the same time, the mass spectrometric data obtained by this method are rather scarce. The situation is quite different for the microanalytical method. Now the West German firm of Leybold Heraeus is producing a commercial instrument LAMMA-500 (laser microprobe mass analyser). A rich volume of mass spectroscopic information has been obtained for organic and inorganic molecules. However, there are still many unsolved questions connected with the mechanism of ion formation under strongly focused intense uv laser radiation.

Below we will consider in detail some experiments on ion formation as laser radiation acts on organic compounds as well as the mass spectra of different classes of organic and bio-organic molecules.

8.2   Laser desorption mass spectrometry with ir radiation

All the experiments discussed below were performed with ir lasers ($\lambda \geq 1\,\mu$m), and this section considers mechanisms of ion formation in which the heating of a sample is the primary effect of absorption of laser radiation. Later sections concern cases when laser radiation absorption results in electronic transitions in the molecules.

8.2.1   Experimental technique
Mass spectrometers which permit recording of the whole mass spectrum or a large part of it after a single laser pulse are preferred to scanning mass spectrometers. In the second case the sample must be subjected to multiple irradiation in order to record the mass spectrum. This may cause a change in sample composition and, accordingly, the mass spectrum can be distorted.

The time-of-flight mass spectrometer is one of the most convenient instruments for detecting the ions formed under pulsed laser radiation. This is the type of the mass spectrometer that was used in the first experiments on LDMS [8.16, 8.17] (figure 8.1).

Modern time-of-flight mass spectrometers designed as a 'mass reflection' [8.33, 8.34] give mass resolution $m/\Delta m$ of some several thousands (see [8.36]).

Until now the signal from the secondary electron multiplier of the mass spectrometer has been recorded with an oscilloscope. This made it impossible to realise fully the possibilities of the instrument, i.e. to carry out simultaneous high-resolution detection of a wide range of the mass spectrum. When transient recorders appeared this problem was solved. The time-of-flight mass spectrometer's advantage of highly efficient detection of the ions formed in the ion source is of prime importance.

In [8.19, 8.36] Kistemaker *et al* used another method to solve the problem of detecting the wide range of the mass spectrum after a single laser pulse. For this purpose a home-made magnetic sector mass spectrometer was used (figure 8.2) in which a part of the mass spectrum is projected on a chevron CEMA (channeltron electron multiplier array) detector. The secondary electron output of the CEMA was proximity focused on a phosphorescent screen, coated on a fibre-optic window. Through the window the mass spectrum could be observed as a series of luminescent lines. This image was detected and digitised with a Vidicon camera coupled to a 500-channel optical analyser.

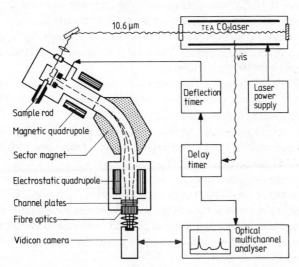

Figure 8.2 Experimental arrangement of an IR laser desorption mass spectrometer with a simultaneous ion detection capability [8.19].

A magnetic quadrupole lens was placed between the magnet and the ion source and provided a sharp image of mass spectral lines on the ion detector. The electrostatic quadrupole lens located between the magnet and the ion detector made it possible to change the mass spectrometer dispersion.

This design allowed the detection of a section of the mass spectrum in the mass range m_1–m_2 with the ratio $m_1/m_2 = 1.2$. Due to some deterioration of resolution the mass ratio m_1/m_2 could be extended up to 1.6. Most of the mass spectra given below were obtained using the set-up described.

A quadrupole mass spectrometer was used in [8.21] to investigate ion formation under CW CO_2 laser radiation.

The samples in the experiments discussed below were prepared by drying the solution on the surface of the sample holder. In this case the substance to be investigated was dissolved in methanol or water. Several microlitres of the solution with a concentration of 0.1 to 1.0 mg ml^{-1} were put on the surface of the sample holder and dried in air or in a vacuum. This method allows the production of a relatively uniform film of the substance, with microcrystalline structure and several thousands of angstroms thick. The technique of methanol solution electrospraying was used to produce more uniform films. The sample holder was usually made of stainless steel or quartz.

CW and pulsed lasers were used as radiation sources. Most experiments were performed with pulsed CO_2 lasers ($\lambda = 10.6$ μm). The duration of the main part of the laser pulse was 100 ns [8.19] or 40 ns [8.26]. The main part of the laser pulse was followed by a 'tail' with a length of several hundreds of nanoseconds and an intensity 10–100 times less than the main pulse. In some experiments Nd–YAG laser ($\lambda = 1.06$ μm) pulses of about 10^{-4} s were used.

The laser radiation was usually focused onto the sample in a spot several tenths of a millimetre in diameter. The radiation intensity at which ions were formed ranged from 10^6 to 10^8 W cm^{-2}.

In [8.21, 8.37] ion formation under CW CO_2 laser radiation was investigated. In the absence of focusing the incident radiation intensity was 10–30 W cm^{-2} and in the presence of focusing it was 10^2–10^4 W cm^{-2}.

Interesting studies were carried out by Heresch *et al* in [8.38], where they investigated the possibility of combining a scanning mass spectrometer with a pulsed Nd–YAG laser in LDMS. The mass spectra were detected with a commercial magnetic sector instrument (MS-902, Kratos-AEI) the ion source of which was equipped with optical windows for laser radiation. The Nd–YAG laser was used in the open-Q mode (pulsewidth 80 ms) with a pulse repetition rate of 50 Hz. Laser pulses were synchronised with the mass spectrometer scanning. In this way the mass spectrum could be recorded under pulse repetition mode.

To record the mass spectrum in this case the sample must undergo multiple irradiation, it must have at least one laser pulse per point of the mass spectrum. Therefore stringent requirements are placed upon the ion mass spectrum being constant from pulse to pulse. It is necessary that, first of all, the absolute value of the ion signal should remain constant, and, secondly, that the qualitative composition of the sample should not change under laser irradiation.

According to [8.38], when the irradiated sample surface was smooth (a film of sucrose produced by drying the solution), the signal of the ions $(M + Na)^+$ was reduced by a thousand times after several laser pulses (figure 8.3). The best results were obtained for the samples with mechanically produced notches

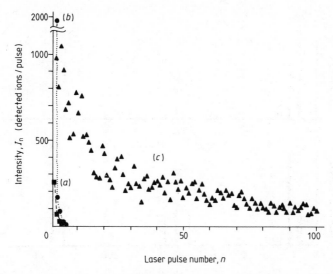

Figure 8.3 Plot of laser pulse number, n, versus intensity, I_n, of the corresponding ion pulse of sucrose, $(M + Na)^+$ for typical ion pulse series at (a) smooth sample surface, (b) isolated particle of sample, and (c) pretreated sample spot (notch) [8.38].

about 0.1 mm wide. The half-amplitude decrease of the signal in this case took place after 10 to 15 laser pulses.

8.2.2 Mass spectra of different classes of organic compounds

In this section we shall consider mass spectra of different classes of organic nonvolatile compounds obtained with the use of the IR–LDMS technique. Much attention was given in the development of LDMS to the mass spectra of bio-organic molecules for two reasons. Firstly rapid progress in biological sciences requires more and more reliable analytical methods—such methods must have high absolute and relative sensitivity. Secondly, the low volatility and thermal instability of many important biological compounds make it impossible to apply conventional ionisation methods.

Here we are going to consider the most important characteristics of the mass spectra of different organic and bio-organic compounds. More detailed information can be found in the papers cited below. Most of the mass spectra presented below were obtained by Kistemaker *et al* [8.19, 8.37].

(a) Oligosaccharides

This class is an example of nonvolatile compounds. In the mass spectra of sucrose produced by thermal evaporation of samples and electron-impact ionisation light fragmentary ions prevail. The signal at the molecular ion mass is very weak or absent [8.39].

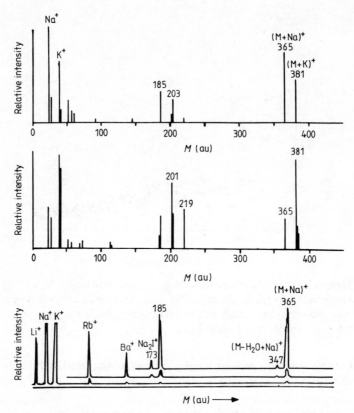

Figure 8.4   IR-LDMS of sucrose ($C_{12}H_{22}O_{11}$, $M = 342$ au) obtained with (*a*) pulsed CO_2 laser, (*b*) pulsed Nd laser [8.19], and (*c*) cw CO_2 laser [8.21].

Figure 8.4 presents sucrose mass spectra ($M = 342$ au) obtained with pulses of CO_2 ($\lambda = 10.6$ μm) and Nd ($\lambda = 1.06$ μm) laser radiation (8.4(*b*)) [8.19] as well as with cw CO_2 laser radiation (8.4(*c*)) [8.21]. Attention must be drawn to the intense peaks of mass quasi-molecular ions, $M/Z = 365$ and 381, formed due to attachment of alkali metal cations to the sucrose molecule $(M + Na)^+$ and $(M + K)^+$. It should be noted that the mass spectra are rather simple. In the region of 200 au there is a group of peaks, each corresponding to the cationised fragments of a sucrose molecule. These fragments must be due to cleavage at the glycosidic oxygen accompanied by hydrogen transfer. The peaks with $M/Z = 185$ and 203 correspond to the ions $(C_6H_{10}O_5 + Na)^+$ and $(C_6H_{12}O_6 + Na)^+$, and those with $M/Z = 201$ and 219 are their potassium analogues. In the low-mass range the following ions are observed: Na^+, K^+, H_2ONa^+ and H_2OK^+.

As seen from figure 8.4(*a,b*) the mass spectra produced when the sample is irradiated by pulses with $\lambda = 10.6$ μm and 1.06 μm, are very similar whereas the absorption of sucrose at $\lambda = 10.6$ μm is much higher than at $\lambda = 1.06$ μm. This fact points to the importance of heating just the substrate rather than the direct heating of the organic molecules [8.19]. The role of the substrate is more obvious in experiments with cw CO_2 lasers [8.37]. There were no ions formed when a thin layer (0.1 μm) of sucrose applied on a stainless steel substrate was exposed to cw radiation ($I \simeq 10^3$ W cm^{-3}). When the substrate was made of quartz, which absorbs IR radiation well, both alkali metal ions and quasi-molecular ions of sucrose $(M + \text{alkali})^+$ could be observed.

Stoll and Röllgen [8.21] could observe the formation of quasi-molecular ions of sucrose with 20 W cm^{-2} of cw CO_2 laser radiation. In this case they irradiated a thick layer (0.5 mm) of sucrose mixed with NaI.

The detachment of H_2O from the sucrose molecule characteristic of oligosaccharide pyrolysis was observed only in [8.21] (figure 8.4(c)) and [8.38], where ions $(M–H_2O + \text{alkali})^+$ could be seen in mass spectra. In [8.38] the mass spectrum was recorded under pulse repetition mode. Thus long heating of the sample by cw radiation or pulsed radiation leads to a stronger thermal fragmentation of the sucrose molecules compared with the case of monopulse irradiation.

In recording the mass spectrum under the pulse repetition mode Heresch [8.40] observed artifacts, i.e. ions with masses $(342 + n162 + \text{Na})^+$ and $(162 + n164 + \text{Na})^+$, where $n = 1$–7. In [8.41] it was shown that the formation of these ions in the case of sucrose was caused by the polymerisation of the prolysis products of intact molecules in the sample.

Zakett *et al* reported [8.49] that, as the sucrose sample on a silver substratum was irradiated by a pulsed Nd–YAG laser, the ion $(M + Ag)^+$ was dominant in the mass spectrum. Besides sucrose some other unsubstituted oligosaccharides have been studied in [8.19, 8.42]. It may be seen from table 8.1 that the quasi-molecular ions $(M + \text{alkali})^+$ are almost always dominant. Besides, the mass spectra always have peaks which correspond to the cationised molecular fragments formed by the cleavage of glycoside bonds.

Table 8.1 Peak intensities of cationised molecules and their fragments in laser desorption mass spectra of unsubstituted oligosaccharides [8.19].

	M (au)	M	M-18	M-162	M-180	M-324	M-342	M-486	M-504	M-522
Glucose	180	100	0	—	—	—	—	—	—	—
Sucrose	342	100	0	32	24	—	—	—	—	—
Gentiobiose	342	100	6	15	6	—	—	—	—	—
Raffinose	504	98	15	100	25	28	26	—	—	—
Stachyose	666	81	4	100	30	64	34	33	92	68

Note: if the generalised formula for oligosaccharides is written as HO(SO)$_n$H, the fragment (M-18) will have the structure of (SO)$_n$, (M-162) – HO(SO)$_{n-1}$H, etc.

Thus, the mass spectrum makes it possible to determine the molecular weight of a specific oligosaccharide as well as the number of glycoside links.

(b) Glycosides

This class of biologically active compounds, which is of great interest for pharmacology, was investigated in [8.19, 8.37, 8.42]. For example, figure 8.5 shows the digoxin mass spectrum obtained by irradiating the sample by CO_2 laser pulses [8.19]. Similar oligosaccharides, the mass spectra of all the investigated glycosides have rather intense signals from the quasimolecular ions $(M + alkali)^+$, up to the pentaglycoside digitonin $(C_{56}H_{92}O_{29}$, $M = 1228$ au). Besides the ions $(M + alkali)^+$, one can observe many fragment ions in the form of cationised fragments of the intact molecules. In the high-mass range there are fragments due to the expulsion of one or more glycosidic units. In the lower range one can observe glycosidic units, the possible fragments of these units and of aglycone.

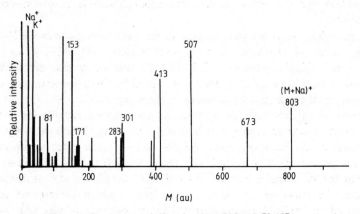

Figure 8.5   IR-LDMS of digoxin $(M = 780$ au) [8.19].

The mass spectrum of glycosides provides enough information to determine the structure of the molecules with some uncertainty due to isobaric glycosidic units.

(c) Nucleotides

This class is one of the most interesting of nonvolatile bio-organic compounds for mass spectrometry. Because of their extremely low volatility electron ionisation mass spectra [8.4] have been obtained for derivative compounds only.

Figure 8.6 shows the mass spectra of adenosine, adenosine-5-monophosphate acid (AMP) and its sodium salt (AMP–Na$_2$) [8.19]. Unlike oligosaccharides and glycosides, one can observe in the mass spectra of

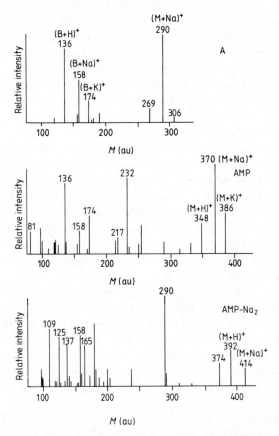

Figure 8.6 IR-LDMS of adenosine (A), adenosine-5′-monophosphoric acid (AMP), and adenosine-5′-monophosphate disodium salt (AMP–Na$_2$) [8.19].

nucleosides and nucleotides ions formed by proton transfer as well as ions containing alkali metal cations. For example, the quasi-molecular ions are represented by $(M+Na)^+$, $(M+K)^+$ and $(M+H)^+$. Among the cationised and protonated base $(B+Na)^+ - M/Z = 158$, $(B+K)^+ - M/Z = 174$ and $(B+H)^+ - M/Z = 136$ which are present in all three mass spectra (figure 8.6). The intense peaks among the fragment ions in the AMP and AMP–Na$_2$ mass spectra correspond to (adenosine + Na) and phosphoric acid residues: HPO_3H^+ ($M/Z = 81$) for AMP and $NaPO_2Na^+$ ($M/Z = 109$), $NaPO_3Na^+$ ($M/Z = 125$) and $Na_2PO_4Na^+$ ($M/Z = 165$) for AMP–Na$_2$.

When nucleosides are irradiated by CW CO_2 or pulsed Nd lasers it is possible to obtain mass spectra with intense peaks of the quasi-molecular ions [8.37]. For nucleotides the situation is somewhat more complicated. According to [8.37] quasi-molecular ions can be produced only for sodium salts of nucleic acids. The mass spectrum of AMP–Na$_2$, for example, has a peak for the ion (AMP–Na$_2$ + H)$^+$ – M/Z = 392, but the signal is very low and the reproducibility is poor. At the same time, in [8.38] the spectrum of AMP–Na$_2$ was produced by using a pulsed Nd–YAG laser. Along with the peak M/Z = 392, the quasi-molecular ions ((AMP–Na$_2$) + Na)$^+$ and ((AMP–Na$_2$) + K)$^+$ were observed in the mass spectrum.

(d)   *Amino acids and oligopeptides*

The mass spectra for unsubstituted amino acids and oligopeptides have been studied for a long time. For most amino acids it is possible to obtain mass spectra containing molecular ions by evaporating the sample directly in the ionisation chamber with electron or chemical ionisation. An exception to this is arginine for which it is impossible to produce molecular ions under these conditions [8.43, 8.44]. Good results can be obtained by pulsed evaporation of the sample from a Teflon surface. Under these conditions chemical ionisation can be applied to obtain mass spectra containing molecular ions for quite a number of oligopeptides [8.45].

Mass spectra of arginine, cystine and methionine have been obtained by Kistemaker and co-workers [8.19] using a pulsed CO$_2$ laser. The mass spectra of all the amino acids investigated are characterised by an intense signal of quasi-molecular ions as well as by a small degree of fragmentation. The mass spectrum of arginine, for example, (figure 8.7), besides ions with M/Z = 175, 197 and 213 corresponding to (M + H)$^+$, (M + Na)$^+$ and (M + K)$^+$, contains weak peaks of ions with M/Z = 158, 180 and 196 formed due to lack of ammonia.

The mass spectra of oligopeptides obtained with a pulsed CO$_2$ laser [8.19] contain only the quasi-molecular ions (M + Na)$^+$ and (M + K)$^+$. A small degree of fragmentation of the intact molecules does not allow structural

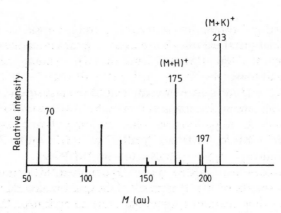

Figure 8.7   IR-LDMS of arginine (M = 174 au) [8.19].

analysis of oligopeptides. In this case additional measures should be taken to ensure the required level of fragmentation.

It is impossible to produce quasi-molecular ions by irradiating oligopeptides with a pulsed Nd laser or a cw CO_2 laser [8.37]. In this case the mass spectra contain only weak peaks of fragment ions. In the case of arginine, on the contrary, the use of any laser makes it possible to produce the quasi-molecular ions $(M + Na)^+$ and $(M + K)^+$. It is worthy of note that, when the sample was irradiated by a cw CO_2 laser, no protonated quasi-molecular ions were observed in the amino acids' mass spectra [8.21, 8.37].

(e) *Quaternary ammonium salts*

This class of compounds presents serious problems for electron ionisation mass spectrometry. As is shown in [8.45], the main contribution to the decomposition of the cation R_4N^+ is made by the process of electron-impact ionisation rather than by sample heating and evaporation. The large store of vibrational energy in the cations R_4N^+ leads to their fast dissociation yielding fragment ions $(R_3N)^+$, $(R_3NH)^+$ and others. Nevertheless, in [8.47, 8.48] cations R_4N^+ were observed in electron ionisation mass spectra.

The formation of ions by the action of IR laser radiation on quaternary ammonium salts was investigated in [8.21, 8.32, 8.42]. When a sample of tetrabutylammonium iodide $(C_4H_9)_4NI$ is irradiated by a cw CO_2 laser with $I \simeq 30$ W cm^{-2}, its mass spectrum contains only the cation $(C_4H_9)_4N^+$. As the IR radiation intensity is increased to 400 W cm^{-2}, the fragment ions $M/Z = 184, 142, 100$ au appear in mass spectra. It should be noted that with $I \simeq 100$ W cm^{-2} the time of observation of $(C_4H_9)_4N^+$ cations was equal to several minutes while for other molecules (sucrose and organic acids) it did not exceed several seconds.

The mass spectrum of $(C_4H_9)_4NI$ was also obtained with a pulsed Nd–YAG laser [8.42]. In this case one could observe an intense signal of the cation $(C_4H_9)_4N^+$ and a fragment ion, $M/Z = 142$.

The formation of ions in the case of tetramethyl ammonium chloride irradiated by CO_2 laser pulses was investigated in [8.32]. The main peak in the mass spectrum corresponded to the cation $(CH_3)_4N^+$; there were no fragment ions. In the higher mass range there was a peak corresponding to the cluster $([(CH_3)_4N]_2Cl)^+$. In this experiment it was found that the efficiency of cation formation was independent of the halogen of the salt.

(f) *Negative ions*

The formation of negative ions in IR laser desorption has not been studied. There are only two works [8.21, 8.37] where negative ions were observed. In [8.21] when citric acid was irradiated with cw CO_2 laser a strong negative ion signal was observed and identified as a deprotonated intact molecular ion (M–H)$^-$. The mass spectrum of negative ions of citric acid had a number of fragmentary ions and an ion with $M/Z = 85$ identified by the authors [8.21] as a double-charged ion $(M-2H)^{2-}$. The mass spectrum of negative ions produced by irradiating a mixture of sucrose and LiI contained only I^- and LiI_2^-. No organic negative ions were observed.

In reference [8.37], where samples of adenosinemonophosphate (AMP) and peptide were irradiated by Nd laser pulses, quasi-molecular ions $(M-H)^-$ were observed.

(g)   Conclusions

The foregoing makes it possible to draw the following conclusions. When nonvolatile organic molecules are exposed to IR laser radiation it is possible to produce quasi-molecular ions giving information on the mass of the intact molecules. In most cases the mass spectrum contains a number of specific fragment ions which give information on the molecular structure of the substance.

The mass spectra produced by the IR–LDMS method have the following characteristic features.

(1) The overwhelming majority of ions are formed as a result of attachment of the alkali cations to the molecules or its fragments.

(2) For some classes of compounds ions of protonated molecules are observed along with ions of cationised molecules.

(3) The degree of intact molecule fragmentation for different classes of compounds is from moderate to low. Quasi-molecular ions are very abundant in the mass spectra.

(4) The best results are obtained with a pulsed CO_2 laser.

8.2.3   Mechanisms of ion formation in the IR–LDMS method

The absorption of IR laser radiation by a solid leads directly to its heating. So it is obvious that the formation of organic molecular ions under IR laser radiation must be caused by the heating of the sample. Nevertheless, in early works it was suggested that there should be nonthermal effects of ion formation [8.21]. The assumption resulted from observing abnormally long times of emission of quaternary ammonium salt ions. The prevalence of quasi-molecular ions of thermally unstable molecules in the mass spectra cannot be easily explained within ordinary thermal ionisation.

Nevertheless, some facts already mentioned above indicate conclusively that the primary cause of ion formation is the heating of the sample. Ions can be observed under both pulsed radiation with $I \simeq 10^6$ W cm^{-2} and cw radiation with $I \simeq 10$ W cm^{-2}. An important condition of ion formation is the absorption of radiation either by a thick layer of organic sample [8.21] or by the substrate [8.37]. If an optically thin layer of substance is applied on a slightly absorbing substrate ions are not formed [8.37]. Finally, in an optically thin layer of substance the efficiency of ion formation is almost independent of radiation wavelength. This fact confirms again the significant part played by radiation absorption by the substrate [8.19].

Thus, some questions connected with the mechanism of ion formation arise. What processes caused by the laser heating of the sample give rise to molecular ions? Are the ions formed on the sample surface and then thermally desorbed, or formed in the gas phase from the thermally desorbed molecules? Why are quasi-molecular ions formed from molecules which can be easily dissociated by heating, and for which it is almost impossible to produce their molecular ions with electron ionisation.

Some works carried out over the past three years [8.21–8.26] have provided the answers to all these questions. It has been found that there are several processes giving rise to ions, each of them being characteristic of a certain class of molecules.

(a) Alkali metals

Alkali ions are always found in LDMS mass spectra as the very intense peaks. Their formation is related to ordinary thermal ionisation resulting from sufficiently strong heating of the surface [8.50]. In [8.22, 8.23] it was shown that the ions Na^+ and K^+ resulted from the heating of the sample surface by laser radiation to a temperature above 700–800 degrees centigrade. The measurements of the kinetic energy of sodium ions [8.51] show that the velocity of laser desorbed ions is near the thermal value.

(b) Non-ionic polar molecules

Since the formation of quasi-molecular ions is always followed by intense emission of alkali ions, it was proposed [8.19, 8.37] that $(M + alkali)^+$ ions were formed as a result of attachment of an alkali cation to the intact molecule in the gas phase. This hypothesis was supported by Kistemaker and co-workers [8.22, 8.23, 8.25], Stoll and Röllgen [8.25] and Cotter *et al* [8.26].

First of all, it has been found that the formation of alkali ions always precedes the formation of cationised molecules $(M + alkali)^+$. Figure 8.8 shows the time dependence of the ion signals K^+ and $(M + K)^+$ as a sucrose sample on a NiCr–Ni thermocouple is irradiated by a CW CO_2 laser. The desorption of neutral molecules is controlled by electron beam ionisation giving fragment ions with $M/Z = 126$ [8.22]. At low temperatures of the substrate ($T < 700$ K) the sucrose molecules only are evaporated intensively. As the temperature rises, first the K^+ ion and then $(M + K)^+$ and fragment ions appear. A similar correlation for the appearance of $alkali^+$ and $(M + alkali)^+$ was observed for the case of a pulsed CO_2 laser [8.25].

Assume that $(M + K)^+$ ions are emitted from a hot surface. Then we must accept that the quasi-molecular ions survive through heating over the temperature range from 800 to 1000 K which is unlikely. On the other hand, at temperatures insufficient for the formation of $alkali^+$ ions quasi-molecular ions are not observed. From this it follows that high surface temperatures are needed to form $alkali^+$ ions rather than $(N + alkali)^+$. $(M + alkali)^+$ ions can be formed either in the gas phase or on colder sections of the surface when

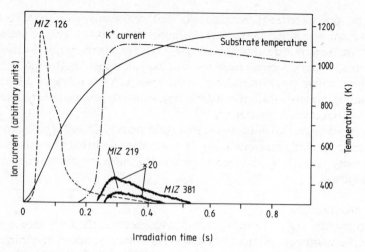

Figure 8.8 Substrate temperature and ion intensities as a function of irradiation time of a sucrose sample by continuous wave CO_2 laser $(10^4 \, W \, cm^{-2})$ [8.22].

alkali$^+$ ions fall onto them. Both mechanisms require that there should be two different temperature areas.

In [8.22, 8.24, 8.25] a situation was simulated in which two adjacent surface areas with different temperatures were created. A layer of sucrose was applied on the colder surface, and K$^+$ ions were thermally emitted from the hot surface. In [8.27] two wires resistively heated to different temperatures were used (figure 8.9(a)). In [8.24] a resistively heated wire was used as an emitter of K$^+$, and a thermocouple with sucrose applied and heated by cw CO_2 laser radiation served as an emitter of sucrose molecules (figure 8.9(b)). In [8.25] a wire with sucrose placed above the silica gel substrate with potassium iodide was heated by one cw CO_2 laser beam and simulated the situation in which there are two different temperature areas (figure 8.9(c)). In all three experiments effective formation of $(M+K)^+$ ions was observed under the following conditions: K$^+$ ions were emitted from the hot surface ($T \simeq 1000 \, K$) and sucrose molecules were emitted from the colder surface ($T = 400–500 \, K$) [8.24]. It is shown in [8.25] that a sucrose sample of 5 ng was sufficient for measurement of the mass spectrum. It is of interest that the addition of NaI to sucrose [8.25] did not give rise to $(M+Na)^+$ ions and the mass spectrum contained only $(M+K)^+$ ions and their fragments. From this it can be concluded that there is no desorption of cationised molecules from the surface of the sample containing sucrose.

However, the other possibility of surface effects remains. The attachment of a cation, in principle, can occur as a sucrose molecule from the gas phase strikes the hot surface without absorption, and hence without strong heating

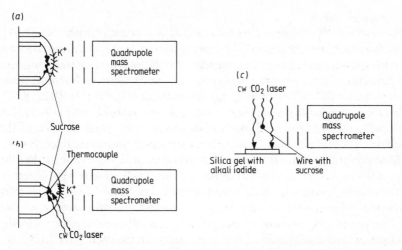

Figure 8.9 Schematic experimental set-up in three experiments with spatially separated sources of sucrose molecules and alkali ions: (*a*) molecules and ions emitted due to resistive heating; (*b*) sucrose sample heated by cw CO_2 laser irradiation, the alkali ion source is heated resistively; (*c*) sucrose molecules and alkali ions emitted due to heating by cw CO_2 laser irradiation.

of the molecule itself. This possibility was studied in [8.25] by applying a potential of 60 V to the potassium ion source and 30 V to the substrate with sucrose. Then the integral energy distribution of the resulting ions $(M+K)^+$ and K^+ was measured. The measurement showed that the formation of $(M+K)^+$ ions took place near the wire with sucrose with 30 V applied to it.

Thus, the conclusion is that the only possibility of $(M+K)^+$ ion formation is the attachment of the cation to the molecule in the gas phase. The possibility of such ionisation was proved by the following experiment [8.24]. A tungsten ribbon with sucrose on its surface was resistively heated to $T=460$ or 457 K, the control was performed with a thermocouple. The beam of K^+ ions produced with a commercial source travelled near the ribbon so that the K^+ ions did not impinge on the sucrose substrate. When the ribbon had attained a temperature of 460 K, only a signal of $(M+K)^+$ ions, $M/Z=381$, could be observed in the mass spectrum. The heating of the ribbon by a further 15 K resulted in the appearance of fragment ions $(M-H_2O+K)^+$. This fact shows once again that the sucrose molecule is thermally unstable.

The above results unambiguously prove that in the case of polar molecules like sucrose the attachment of metal cations in the gas phase is the main mechanism of ion formation. A similar conclusion can also be drawn for organic acids [8.25].

(c) *Organic salts*

When benzo (1,5)-crown-5 ether was heated in a mixture with NaI [8.31], $(M + Na)^+$ ions were observed. The mass spectrum, however, did not contain Na^+ ions which seemed to be inconsistent with the above stated mechanism of ion formation. The different behaviour of the crown ether results from strong complexation of Na^+ ion forming the salt molecule $(M + Na)^+ I^-$ [8.25, 8.52].

Ion formation in the case of organic salts has been investigated in [8.21, 8.25, 8.30, 8.32, 8.42] and several mechanisms have been revealed. One of the mechanisms is thermal desorption of preformed ions from the surface. For the salt the preformed ions are the anion and cation of it. It is obvious that the ratio of ion to neutral molecule desorption efficiency depends on the difference of adsorption energy of the ion and the neutral molecule. For a molecular cation with a large radius this difference may be small. In this case it is possible to observe not only thermal evaporation of the salt molecules but also the desorption of the salt cation. The quaternary ammonium salts $R_4N^+X^-$, where R is the alkyl group and X is the halogen, are a characteristic example of salts with a large-radius cation. The desorption of R_4N^- ions was observed in conventional heating of the sample in [8.25, 8.30, 8.53, 8.54] and under IR laser radiation in [8.21, 8.32, 8.42].

Table 8.2 presents temperature values for a number of quaternary ammonium salts at which neutral molecules (T_1) and their cations (T_2) are evaporated [8.53]. The fact that the temperatures T_1 and T_2 differ slightly shows that the desorption energies of the molecules and the cation are similar. The dependence of the desorption temperature of $(R_4N)^+$ cations on their radius is clearly defined here. In [8.32, 8.30, 8.31] the essential role of the energy of the bond of the cations with the crystal lattice in the formation of salt ions was pointed out.

If the salt cation radius is small and the energy of its bond with the crystal lattice is accordingly high, very high temperatures are needed for the thermodesorption of the cation from the surface. In this case very strong evaporation of neutral molecules and clusters occurs first. Such a situation occurs for organic salts of alkali metals. In such a molecule the alkali metal ion

Table 8.2 Lowest temperatures at which the evaporation of neutral molecules (T_1) and cations $(R_4N)^+$ (T_2) of quaternary ammonium salts [8.54] was detectable.

Salt	T_1 (°C)	T_2 (°C)	Salt	T_1 (°C)	T_2 (°C)
Me_4NI	264	†	n-Bu_4NI	83	138
Et_4NI	260	350	n-Bu_4NOAc	70	250
n-Pr_4NI	175	278	n-$Bu_4N(Ph_4B)$	178	240

† Not determined.

is a cation, the desorption of which calls for very high temperatures. All the potential mechanisms of ion formation in this case were investigated in [8.25, 8.54].

In [8.25] it was shown for the case of CH_3CO_2Na that cationised molecules could be formed in three processes. A schematic diagram of the experiments is presented in figure 8.10(c). Sodium acetate was applied on a wire near the surface of silica gel with potassium iodide. When the wire and silica gel were exposed to cw CO_2 laser radiation, there were three types of ions observed in the mass spectrum: $((CH_3CO_2Na)_nK)^+$, where $n = 1, 2, 3, 4$; $(CH_3CO_2Na_2)^+$ and $(CH_3CO_2K_2)^+$. The main peak in the mass spectrum corresponded to the ion $(CH_3CO_2NaK)^+$.

The formation of $((CH_3CO_2Na)_n + K)^+$ ions results from the attachment of the K^+ cation desorbed from silica gel to the molecule or the cluster of sodium acetate evaporated from the wire.

$(CH_3CO_2Na_2)^+$ ions are formed as a result of dissociative ionisation when dimers or clusters $(CH_3CO_2Na)_n$ collide with the hot surface of the silica gel. The possibility of such ionisation was demonstrated in [8.54].

Finally, the formation of $(CH_3CO_2K_2)^+$ ions probably occurs during a cation-exchange reaction on the surface yielding CH_3CO_2K molecules with subsequent attachment of the K^+ cation.

The primary cause of ion formation in all the mechanisms under discussion is the heating of the sample. Of course, the question may arise as to why the signal of quasi-molecular ions of thermally unstable molecules is rather intense.

The process of evaporation of nonvolatile organic compounds under laser heating has so far not been investigated in detail. In some experiments [8.5, 8.45], however, it has been found that the heating rate is an important factor that defines the efficiency of evaporation of intact molecules from the surface without their dissociation. The faster the heating of the sample, the higher the probability of evaporation of a molecule as a whole. When the heating rate is very high, it is possible that a shock wave arising in the sample may cause the desorption of molecules from the surface.

Another essential factor defining the stability of the resulting complex $(M + \text{alkali})^+$ is low exothermicity of the reaction of attachment of the alkali metal cation to the molecule. This case differs greatly from electron ionisation when the resulting ions have a large store of vibrational energy (up to 10 eV) that leads to their strong fragmentation.

8.3 Laser desorption–photoionisation mass spectrometry

When the surface of organic crystals is exposed to uv laser radiation, the spectrum of physical processes is wider than in the case of ir radiation.

The primary result of absorption of uv radiation quanta is the excitation of molecular electrons. In this case quite a number of processes may occur. These include the formation of excitons, electrons and holes in the conduction band of crystals, and emission of electrons from the surface. As a result of some relaxation processes, the electron excitation energy is transformed to thermal motion. In the analysis of mechanisms of ion formation one should take into account that the neutral molecules which leave the surface can also interact with UV radiation which brings about their excitation, ionisation or fragmentation. At a sufficiently high intensity of laser radiation the physical picture becomes more complicated due to nonlinear photoprocesses which may take place both in a crystal and a gas.

Here we are going to consider the formation of ions in the case when the sample surface is exposed to pulsed UV radiation of moderate intensity, $I < 10^6$ W cm^{-2}. The absorption of radiation by an organic crystal in this case is connected with the resonant single-photon transitions in the molecules, and the resulting heating of the surface is not large enough to bring about phase transitions ($T_{surf} < T_{melt}$).

The formation of ions, when crystals of nucleic acid bases and anthracene are exposed to KrF laser radiation ($\lambda = 249$ nm, $\tau = 20$ ns) with $I = 10^4$–10^6 W cm^{-2} ($\phi = 10^{-4}$–10^{-2} J cm^{-2}), was observed in [8.54]. As seen from figure 8.10, in all the mass spectra the molecular ions M^+ are the most abundant. Fragment ions $M/Z = 84$ and 69 are present only in the mass spectra of thymine and uracil respectively. In addition, the mass spectra contain Na$^+$ and K$^+$ ions as well as cationised molecules $(M + Na)^+$.

A more complex ion mass spectrum can be observed when crystals of protacted tripeptide AcOHTrpAla–GlyOB^{t}u are irradiated [8.56, 8.57] (figure 8.11). As well as the quasi-molecular ion $(TrpAlaGlyOB^tu)^+$, the mass spectrum contains a number of fragment ions formed as a result of bond cleavage along the peptide chain with the charge being localised within the indole group of tryptophan. The lightest fragmentary ion is 3-methylene indole.

In [8.58, 8.59] it was assumed that the role of sample surface heating by laser radiation in the mechanism of ion formation cannot be essential. The estimations show that in the case of adenine at the least detectable ion signal the surface heating ΔT did not exceed 100 °C (the melting point of adenine $T_{melt} = 360$ °C). The high kinetic energies of ions, $E_{kin} \simeq 1$ eV, seemed to show that the role of thermal processes was not significant.

In subsequent experiments [8.27], however, it was found that the signal decreased as the adenine crystals were cooled, i.e. the efficiency of ion formation depended on the initial temperature of the crystals. In [8.60, 8.61] desorption of neutral molecules was observed, along with the formation of molecular ions. The number of desorbed neutral molecules in this case was larger than that of ions [8.27].

These facts allow us to assume that the formation of ions comes about as a result of laser photoionisation of the molecules thermally desorbed from the

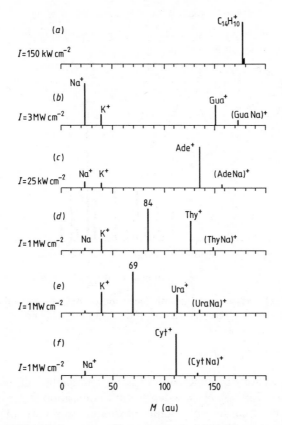

Figure 8.10 Mass spectra of ions obtained with KrF laser irradiation of molecular crystals of anthracene (*a*), guanine (*b*), adenine (*c*), thymine (*d*), uracil (*e*) and cytosine (*f*) [8.55].

surface. The surface was heated by the same uv pulse. The results of subsequent experiments on the formation of ions of the host molecules of organic crystals and the organic molecules adsorbed on the surface were consistent with theoretical estimations within the suggested model [8.62, 8.27]. The only fact that seems to be inconsistent with the suggested mechanism is the high kinetic energies of the ions, but this can be explained too. In [8.63] it was shown that the ions formed near the crystal surface could attain kinetic energies of several electron volts as a result of acceleration in the electric field of the surface charge. The latter arose as the surface was irradiated by uv laser pulses due to the photoeffect.

Since in the case being discussed the mechanism of ion formation consists of two elementary physical processes—molecular desorption under pulsed laser heating of the surface and molecular photoionisation above the surface—it is

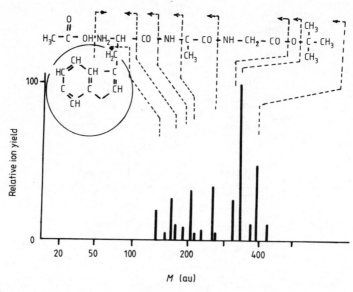

Figure 8.11 Mass spectrum of the peptide AcOHTrpAlaGlyOB'u obtained with KrF laser irradiation of the sample ($\lambda = 249$ nm, $\tau = 20$ ns, $\phi = 10$ mJ cm^{-2}) [8.56].

reasonable to separate the functions of heating and photoionisation between two laser pulses. This will allow independent optimisation of desorption and photoionisation in order to obtain maximum sensitivity of molecular detection. The sensitivity of detection of naphthalene molecules adsorbed on graphite with the use of such a two-laser technique was as high as 10^5 mol cm^{-2} or 10^{-9} of a molecular layer [8.64].

Below we shall consider in detail a phenomenological model of formation of molecular ions as the crystal surface is exposed to UV laser radiation of moderate intensity as well as present the results of experiments in which the basic physical characteristics of this process were investigated.

8.3.1 Phenomenological model of ion formation

In this section we shall consider in more detail a model of ion formation as the surface of organic crystals is exposed to UV laser radiation. The sequence of processes is as follows: absorption of UV radiation by the crystal; energy thermolisation; thermal desorption of molecules from the surface; and laser photoionisation of molecules in the gas phase.

We shall consider the case where the laser radiation frequency falls within the electron absorption band of the crystal. This condition is almost always fulfilled in the experiments discussed below. The characteristic values of absorption of crystals lie, in this case, in the range 10^3–10^5 cm^{-1}. The UV radiation falling on the crystal surface is absorbed in a layer with a width of $l \approx 1/\kappa = 0.1$–10 μm, and results in electron excitations, excitons, in this layer. If the quantum energy is larger than the width of forbidden zone of the crystal, $\hbar\omega > E_g$, electrons and holes may be found in the corresponding conduction bands with a probability $\eta_\pm$. The value $\eta_\pm$ depends on the absorbed photon energy. For adenine, for example, this value ranges from 3×10^{-3} to 3×10^{-2} [8.66]. Some of the electrons can escape the crystal surface (photoeffect).

Further energy transformation in the crystal occurs due to relaxation processes. Due to nonradiative transitions the exciton energy is transformed into vibrational molecular energy. The characteristic time of this process ranges from 10^{-9} to 10^{-12} s. If the exciton density is sufficiently high, the role of exciton–exciton annihilation, which also results in transformation of electronic to vibrational energy, becomes essential [8.65]. For example, if adenine crystals are irradiated by KrF laser pulses ($\lambda = 249$ nm, $\tau = 20$ ns, $\phi = 1$ mJ cm^{-2}) the density of the excitons excited during a laser pulse will be $n_0 = \kappa\phi/\hbar\omega \simeq 10^{20}$ cm^{-3}. The maximum instantaneous exciton density n_{max} is defined by the ratio of the excitation rate n_0/τ to the rate of relaxation caused by biexciton annihilation γn^2: $n_{max} = (n_0/\tau\gamma)^{1/2}$. If γ has its characteristic value 10^{-8} cm^{-3} s^{-1} [8.67], n_{max} equals 10^{18} cm^{-3}.

Nearly maximum exciton density can be obtained at the very beginning of the laser pulse when about $n_{max}/n_0 = 10^{-3}$ of the pulse energy is absorbed. The effective exciton lifetime in this case will be $\tau_{eff} = 1/n\gamma \simeq 10^{-10}$ s.

Thus, the absorbed UV radiation energy of nanosecond laser pulses is believed to be instantaneously transformed to thermal molecular motion. So the standard heat equation can be applied to describe the crystal surface heating dynamics

$$\frac{\partial I}{\partial t} = \chi \frac{\partial^2 I}{\partial X^2} + \frac{\kappa \, e^{-\kappa x}}{c\rho} I(t) \tag{8.1}$$

where $I(t)$ is the laser pulse intensity, $\int_0^\infty I(t)\, dt = \phi$, χ is the thermal diffusivity, c is the specific heat, ρ is the density and κ is the absorption factor of the sample.

Figure 8.12 (curve A) shows the calculated temperature of an adenine surface subjected to pulsed KrF laser irradiation with $\phi = 4$ mJ cm^{-2}. It can be seen that the cooling of the surface due to heat conduction during a laser pulse is essential even for pulses with their duration of 2×10^{-8} s. The temperature reaches its maximum at the end of the pulse.

As the sample temperature increases, the rate of evaporation of the molecules from the surface increases drastically:

$$dN_g/dt = N_s\Omega \exp(-E/kT) \tag{8.2}$$

where $\Omega = 10^{12}$–10^{14} s^{-1} is the characteristic frequency of molecular vibrations on the surface, E is the binding energy of molecule on the surface, N_g and N_s denote respectively the numbers of molecules over the surface and on the surface. The value $W_{des} = \Omega \exp(-E/kT)$ is the molecular desorption rate.

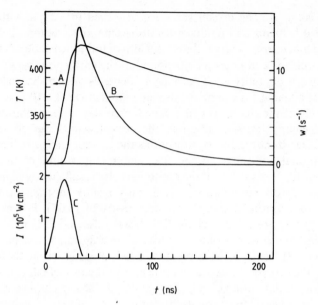

Figure 8.12   Numerical solution of equations (8.1) and (8.2) for adenine evaporation with KrF laser irradiation of the sample. A, surface temperature, T; B, molecule evaporation rate, W; C, laser radiation intensity, I [8.28].

Figure 8.12 presents the results of numerical integration of equations (8.1), (8.2) for the case of adenine crystals. The following parameters were used in the calculations: $\kappa = 3.6 \times 10^3\ \mathrm{cm}^{-1}$ [8.66], $c = 1.05\ \mathrm{J\,g}^{-1}\,°\mathrm{C}^{-1}$, $\rho = 1.2\ \mathrm{g\,cm}^{-3}$, $\phi = 4\ \mathrm{mJ\,cm}^{-2}$, $\tau = 20$ ns and $\Omega = 10^{13}\ \mathrm{s}^{-1}$. The binding energy E of molecules on the surface of adenine crystals is unknown. Moreover, this energy differs for molecules in different positions. Therefore, in estimating the evaporation kinetics the value of E was taken to be 1 eV, which was close to the crystal lattice energy per molecule. The calculations show that the evaporation rate increases by 10^5 times by the end of the laser pulse and decreases by an order of magnitude about 80 ns after the pulse is over.

Similar calculations of the desorption kinetics can be carried out for the case when molecules of different sorts are adsorbed on the crystal surface. Since the number of molecules on the surface is finite in this case, one can observe full desorption of molecules from the surface at a sufficiently high energy of laser pulse. Such calculations were performed in [8.27] for the case of adsorption of naphthalene molecules on a rhodamine 6 G film (figure 8.13).

The following parameters were used in the calculations: $\hbar\omega = 5$ eV, $\kappa = 1.3 \times 10^5\ \mathrm{cm}^{-1}$, $\chi = 0.01\ \mathrm{cm}^2\,\mathrm{s}^{-1}$, $T_0 = 200$ K, $\rho = 1.2\ \mathrm{g\,cm}^{-3}$, $E = 0.5$ eV [8.62], $c = 1.26\ \mathrm{J\,g}^{-1}\,°\mathrm{C}^{-1}$, $\tau = 20$ ns, $\Omega = 10^{13}\ \mathrm{s}^{-1}$, $\phi = 15.7\ \mathrm{J\,cm}^{-2}$. With these parameters the desorption of all naphthalene molecules occurs during a

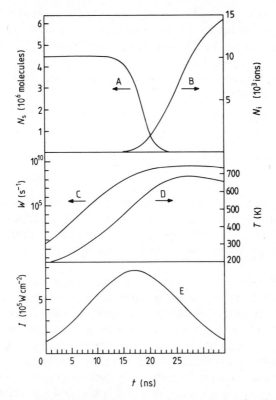

Figure 8.13 Numerical solution of equations (8.1)–(8.4) for the case of naphthalene molecules adsorbed on a rhodamine 6 G film. A, quantity of naphthalene molecules on a surface, N_s; B, quantity of naphthalene photoions N_i; C, desorption rate, W; D, surface temperature, T; E, laser radiation intensity, I [8.28].

laser pulse. Figure 8.13(c) shows that a noticeable desorption rate is attained at a substrate temperature of about 500 K, when the lifespan of naphthalene molecules on the surface $\tau_{des} = 1/W_{des}$ is of the order of a laser pulse duration. Increase of the surface temperature leads to a sharp decrease in τ_{des}.

The molecules which leave the crystal surface during a laser pulse can be ionised by the same radiation due to absorption of two or more quanta. Since the process of molecular photoionisation in the gas phase was discussed in detail in the previous chapter, we shall restrict ourselves here to general remarks only.

The ionisation potential of most organic molecules is higher than 7 eV, and in the case of single-photon ionisation one should use vacuum ultraviolet radiation. The radiation used in the experiments carried out so far had a quantum energy less than or equal to 5 eV. In this case the process of photoionisation is two-step in character. After absorbing the first quantum the molecule goes from the ground electronic state to one of the excited electronic states. The absorption of the second quantum results in ionisation of the electronic-excited molecules.

The process of two-step molecular photoionisation is described by the following system of equations

$$\frac{dN_{exc}}{dt} = \frac{I(t)}{\hbar\omega}\,\sigma_1 N_g - \frac{I(t)}{\hbar\omega}\,\sigma_2 N_{exc} - \frac{I(t)}{\hbar\omega}\,\sigma_i N_{exc} - N_{exc}/\tau \qquad (8.3)$$

$$\frac{dN_i}{dt} = \frac{I(t)}{\hbar\omega}\,\sigma_i N_{exc} \qquad (8.4)$$

where N_{exc} is the number of electronic-excited molecules, N_i is the number of ions, σ_1, σ_2 are the cross sections of photon stimulated transitions from the ground state into the excited one (1) and from excited into the ground state (2), σ_i is the photoionisation cross section from the excited state and τ is the excited state lifetime.

Figure 8.13 presents the results of numerical integration of the system of equations (8.1)–(8.4) describing ion formation as the surface of a rhodamine 6 G film with adsorbed naphthalene molecules is subjected to pulsed KrF laser irradiation. The laser pulse energy was chosen so that the molecules could desorb completely from the surface during a pulse. But the calculations show that this energy is not sufficient for effective molecular ionisation. The low ion yield can be explained by the fact that only the tail of the laser pulse takes part in photoionisation. So the laser pulse energy should be increased in order to increase the photoionisation yield. The temperature of effective molecular desorption in this case, 500 K, is attained earlier. But the surface temperature by the end of laser pulse may turn out to be too high, and this will lead to unwanted effects, for example intense evaporation of the substrate.

Thus, the analysis carried out shows that the process of ion formation can be optimised by separating the functions of desorption and photoionisation with the use of two lasers. Since the ionising pulse may be time-delayed with respect to the desorbing one, the requirements on desorption rate and hence surface heating are considerably reduced. It is advisable to use radiation with a low quantum energy, for example IR radiation, for molecular desorption, which will reduce the probability of unwanted photochemical processes on the surface.

As was shown in Chapter 7, with optimal radiation parameters it is possible to attain near 100% probability of molecular ionisation in the gas phase. Because of this the use of the two-pulse method of ion production in mass spectrometry can ensure a very high sensitivity of molecular detection.

8.3.2   Experimental technique
Before discussing the major physical characteristics of the method we must consider the experimental technique used in our studies (figure 8.14).

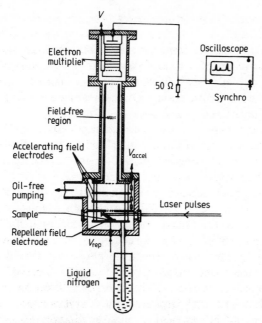

Figure 8.14 Experimental set-up for UV-LDMS.

The main element of the set-up is a time-of-flight mass spectrometer. The sample under study is placed on a metal substrate arranged on the ejecting electrode in the ionisation chamber of the time-of-flight mass spectrometer. The laser radiation is directed through the side quartz window into the ion source of the mass spectrometer and after reflection from the mirror it falls on the sample at 45°. The ions formed by the irradiation of the sample with an electric pulse ($E = 125$ V cm^{-1}) are ejected into the acceleration region and then separated with respect to their masses in the field-free region 40 cm long. The ions are detected with a secondary electron multiplier (SEM) and with a memory oscilloscope.

The samples used were in the form of fine-dispersed crystalline powders or in the form of polycrystalline film produced by drying the solution of the substance in a vacuum. Anthracene was used as a monocrystalline film. The electrode on which the sample was placed could be cooled to $T = 200$ K.

The sample was irradiated by a KrF excimer laser ($\tau = 20$ ns, $\lambda = 249$ nm), a second-harmonic Nd–YAG laser ($\tau = 10$ ns, $\lambda = 531$ nm), a XeCl excimer laser ($\tau = 15$ ns, $\lambda = 308$ nm) and a CO$_2$ laser ($\tau = 100$ ns, $\lambda = 9.6$ μm).

The time-of-flight mass spectrometer resolution was 200 at half-amplitude level. The transit time of ions to the SEM cathode was measured from the leading edge of the ejecting pulse with a precision-delay pulse generator. The ejecting pulse could be applied to the electrode with a different delay τ_d with respect to the laser pulse which made it possible to focus ions of different masses relative to their initial velocities [8.68].

8.3.3   *Ion formation of host molecules of organic crystals*

As has been already said, the mass spectrum of the ions formed as the molecular crystals of nucleic acid bases and anthracene are irradiated by UV pulses consists mainly of one molecular ion. This is consistent with the desorption model at weak heating of the surface and molecular photoionisation at moderate intensity of laser radiation when the probability of ion photofragmentation is not large (see Chapter 7).

The heating of the surface by laser radiation estimated for the case of adenine shows that under the conditions of the experiments described below the value of heating ΔT ranges from 100 to 300 °C. The estimations were performed on the assumption that the absorption κ of the crystal was independent of laser pulse energy. To check the validity of this assumption, the transmission coefficient of a thin film of adenine deposited on a quartz substrate was measured at different radiation intensities with $\lambda = 249$ nm. The transmission coefficient of this film turned out to be constant over a range of fluxes from low values where the absorption was obviously linear to $\phi = 15$ mJ cm^{-2} when destruction of the film was observed. From this it follows that the role of nonlinear absorption of UV radiation with frequency falling within the electron absorption band of crystals is not essential.

The dependence of the ion yield N_i on laser radiation flux ϕ is very sharp [8.58]. For adenine this dependence is approximated as $N_i \propto \phi^{6.9}$ for the initial crystal temperature $T_0 = 300$ K and as $N_i \propto \phi^{7.4}$ for $T_0 = 200$ K [8.27]. As the initial temperature is reduced, the ion yield decreases drastically. These facts can be explained well within the model discussed in §8.3.1 and is related to the exponential dependence of the molecular desorption rate on crystal surface temperature.

The dependence of the ion yield on wavelength $N_i(\lambda)$ is defined by two factors: the dependence of the crystal absorption factor on wavelength, $\kappa(\lambda)$, since κ is responsible for the heating of the surface and the probability of molecular desorption P_{des}; and the dependence of the efficiency of molecular ionisation in the gas on wavelength, $\eta_i(\lambda)$. Thus, $N_i \propto N_S P_{des}(\phi, \kappa(\lambda))\eta_i(\lambda)$.

In reference [8.58] consideration is given to ion formation of adenine irradiated by three pulsed lasers: a KrF laser ($\lambda = 249$ nm), a XeCl laser ($\lambda = 308$ nm) and a N$_2$ laser ($\lambda = 337$ nm) (figure 8.15). The radiation wavelength of a KrF laser falls within the electron absorption band of both the crystal and the molecule of adenine. Therefore, under the action of these laser pulses one can observe effective formation of molecular ions of adenine even with $\phi = 2$ mJ cm^{-2}. In the case of the XeCl laser the energy flux was increased to 12 mJ cm^{-2} so that adenine ions could be observed. The signal of adenine molecular ions in this case was 40 times smaller than in the case of the KrF laser. This is due to the fact that the wavelength of the XeCl laser falls within the absorption band edge of the crystal and the molecule of adenine. This in its turn causes the efficiency of surface heating and the photoionisation efficiency of desorbed molecules to be reduced. Finally, the N$_2$ laser radiation does not

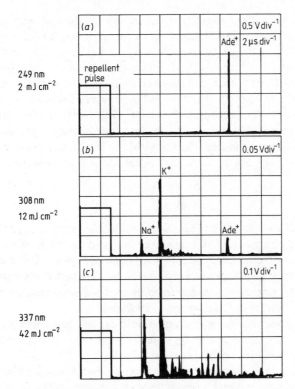

Figure 8.15 UV-LDMS of adenine [8.58].

coincide with the exciton absorption band of the crystal and is not absorbed by the adenine molecule. Figure 8.15 shows that the mass spectrum in this case can be observed if radiation with a much higher energy flux, $\phi = 42$ mJ cm^{-2}, is used when the crystal heating and the molecular ionisation occur due to multiphoton absorption. In this case strong fragmentation of the molecules takes place and the molecular ion M$^+$ is not observed.

Desorption of neutral molecules can take place when adenine crystals are irradiated with picosecond pulses of the fourth-harmonic Nd–YAG laser [8.60, 8.61] and with nanosecond pulses of KrF laser [8.27, 8.28]. The neutral molecules desorbed from the surface were detected using pulses of the KrF laser which was used to accomplish molecular ionisation in the gas phase.

In [8.60, 8.61] the ionising pulse radiation covered a rather long region of the sample surface and so just a small number of desorbed molecules came into the ionisation area. It was therefore difficult to determine the relation between the number of the desorbed molecules and the ions formed under the action of the first pulse. This problem was solved in an experiment [8.27] with two independent KrF lasers. The pulse of the second, probing, laser with a time delay $\tau_d = 2$ μs was directed along the sample surface so that all the molecules which escaped the surface after the first pulse entered the ionisation area of the probing pulse. The measurements were taken with the first pulse flux $\phi_1 = 4$ mJ cm^{-2} and the second pulse flux $\phi_2 = 2.3$ mJ cm^{-2}. When the sample was only exposed to the second pulse there was no ion signal. The measurements showed that the number of ions N_2 formed from neutral molecules under the action of the second pulse was 11 times higher than the number of ions N_1 formed under the action of the first pulse on the adenine surface. As adenine molecules can be photoionised in the gas phase as a result of absorption of two photons, the ion yield for the second pulse depends as $N_2 \sim \phi_2^2$. Thus if the energy fluxes are equal, the ion ratio will be

$$\frac{N_2}{N_1} \left(\frac{\phi_1}{\phi_2}\right)^2 = 36.$$

This result shows that if all the desorbed molecules could escape the surface at the very beginning of the laser pulse, the ion yield would be 36 times higher than the one observed in the experiment with one laser pulse. This is consistent with the numerical calculations carried out in §8.3.1, where it is shown that the desorption rate can reach its maximum at the end of the laser pulse and most of the molecules desorb in the absence of radiation.

The results presented show that one should use two laser pulses, desorbing and ionising, to increase the ion yield. In this case it is possible to independently optimise both processes.

The initial velocity distribution of ions can be studied using a time-of-flight mass spectrometer operating without velocity focusing of ions. In this case the time of arrival of ions at the electron multiplier cathode is defined by the relation $t = t_0 - \alpha V$, where t_0 means the time of arrival of ions with a zero initial velocity. Thus, the signal from the electron multiplier for ions of the same mass is a function of the initial velocity distribution of ions.

The shape of a signal from the electron multiplier for adenine ions is given in figure 8.16 [8.58]. Below is the scale of initial ion velocities. The pulse shape in figure 8.16(a) was obtained by averaging the signal over a large number of oscillograms since the signal value in this case corresponded to several ions per pulse. The pulse shapes in figure 8.16(b,c) were obtained by averaging over four oscillograms. The function of the initial velocity distribution of ions has a maximum with $V = 10^5$ cm s^{-1} that corresponds to an ion kinetic energy of 0.7 eV.

These results seem to be inconsistent with the model of ion formation proposed in §8.3.1. However, in reference [8.63], it was shown that such high velocities could be attributed to ion acceleration in the field of surface charges arising in organic crystals under pulsed UV radiation. It should be noted that, as a finely dispersed powder with a crystal size of about a micron is irradiated, the characteristic dimensions of local nonuniform electric fields are of the same order. The ion acceleration time in these fields comes to several nanoseconds. That is why it is almost impossible to distinguish between the real initial

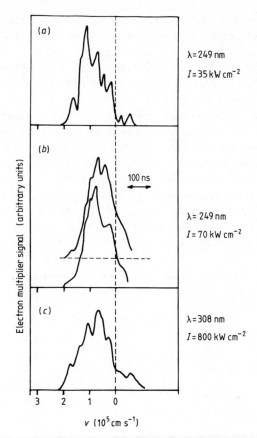

Figure 8.16 The adenine ion signal time dependence (the scale of ion velocities is at the bottom) [8.58].

velocity of ions and the velocity acquired by acceleration in the field of charged crystals. The presence of surface charges was proved in an experiment with a monocrystalline film of anthracene under homogeneous irradiation of a surface with area $S = 0.3 \times 3$ mm^2 [8.63].

A wide velocity distribution of ions deteriorate the mass spectrometer resolution. This effect can be overcome by applying the two-pulse method. In this case it is necessary to use longwave radiation, which does not lead to the photoeffect, for molecular desorption.

8.3.4 Formation of ions of the molecules adsorbed on the surface

Under thermal surface desorption of molecules with subsequent photoionisation in the gas phase it is quite possible, in principle, to observe the ions of the molecules adsorbed on the surface if they are subjected to multiphoton ionisation. This effect was observed in experiments with adenine crystals if there was anthracene vapour in the mass spectrometer cavity [8.26, 8.62]. As the adenine crystals were exposed to UV pulses, the mass spectrum contained ions of both adenine and anthracene. When the space above the surface was irradiated only anthracene ions were observed, but with $\phi = 6$ mJ cm^{-2} the signal was weaker by a factor of 2 or 3. Cooling of the ejecting electrode with the sample to $T = 200$ K led to a strong decrease of the ion signal connected with the anthracene in the gas phase and an increase of the signal from the surface. From this it follows that anthracene ions are formed by the photoionisation of the anthracene molecules adsorbed on the adenine surface. With $\phi = 10$ mJ cm^{-2} the signal from the surface exceeded the signal from the gas phase by more than 200 times.

Figure 8.17 shows the dependence of ion signals N_i on laser radiation flux ϕ for anthracene adsorbed on the surface of adenine and a film of rhodamine 6 G. For anthracene adsorbed on adenine (figure 8.17(a)) with $T = 200$ K the dependence is approximated as $N_i \sim \phi^8$, with $T = 300$ K it is similar to the

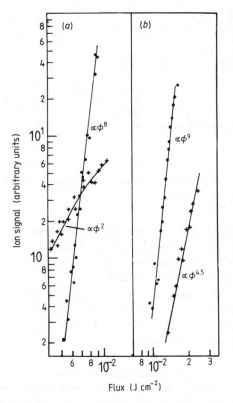

Figure 8.17 The dependence of the anthracene ion signal on the flux of KrF laser pulse irradiating a crystalline powder of adenine (a), rhodamine 6 G film (b). Dots, $T = 200$ K; crosses, $T = 300$ K.

dependence of anthracene vapour ionisation yield on energy flux $N_i(\phi)$. This fact can be explained thus: at a low temperature of the sample the number of resulting ions is defined by the number of molecules desorbed during a pulse and their ionisation efficiency. With $T = 300$ K all the molecules desorb at the beginning of the laser pulse and the ion yield is defined only by the ionisation efficiency of the molecules in the gas phase.

For the case of adsorption of anthracene on rhodamine 6 G (figure 8.17(b)) the corresponding curves are steeper and vary between $N_i \sim \phi^9$ with $T = 200$ K and $N_i \sim \phi^{4.5}$ with $T = 300$ K. In this case the signal from the surface is much stronger than the signal from the gas phase. For example, with $T = 200$ K and $\phi = 16$ mJ cm^{-2} the ratio of these signals was higher than 10^3.

Some experiments were performed with two laser pulses. A polycrystalline film of rhodamine 6 G ($T = 200$ K) with naphthalene molecules adsorbed on it was exposed to pulsed radiation with $\lambda = 531$ nm and $\phi = 4.3$ mJ cm^{-2}. There were no ions formed under the action of this pulse. Then after a time delay of $3\ \mu$s the sample was irradiated with UV pulses with $\lambda = 249$ nm and $\phi = 1$ mJ cm^{-2}. A signal from naphthalene ions was observed. This did not depend on the energy flux of the first pulse in the range from 4 to 6 mJ cm^{-2} which pointed to full desorption of the naphthalene molecules from the surface. The irradiation of the space above the surface of the sample by KrF laser pulses only gave rise to naphthalene ions but the signal was 2×10^3 times less than the signal from the surface irradiated by the two pulses. When the surface was irradiated by one KrF laser pulse, the dependence could be approximated as $N_i \sim \phi^{7.6}$. With $\phi = 16$ mJ cm^{-2} the N_s/N_g ratio was equal to 17. A further increase in energy flux resulted in intense evaporation of the rhodamine 6 G film which made it difficult to detect naphthalene ions.

Thus, the two-pulse technique used for detecting molecules adsorbed on a surface was more effective as it was also in the case of ions formed from the host molecules of organic crystals.

8.3.5 Detection of small numbers of molecules

The method of pulsed laser desorption, that allows the molecules of the sample to be desorbed without being fragmented, combined with effective laser photoionisation must, in principle, have a very high sensitivity. The possibilities of this method and the factors limiting its ultimate sensitivity were studied in reference [8.64] where anthracene and naphthalene molecules adsorbed on the surface of graphite were detected.

In this experiment a graphite substrate was placed on the ejecting electrode in the ion source of the time-of-flight mass spectrometer (figure 8.14) and was cooled to 200 K. The mass spectrometer contained naphthalene and anthracene vapour with partial pressures of 10^{-12} to 10^{-13} Torr estimated from photoionisation measurements. The molecules of naphthalene and anthracene were adsorbed on the cold surface of graphite and were in equilibrium with the gas phase.

The molecules were desorbed by heating the graphite surface with pulsed CO_2 laser radiation. To photoionise the desorbed molecules an excimer KrF laser beam was directed parallel to the substrate surface. The delay between the desorbing and ionising pulses was 0.7 to 1.0 μs.

According to [8.69], with a KrF laser energy flux of 0.2–0.3 J cm^{-2} the ionisation efficiency of naphthalene molecules in the gas phase reached 100%. But, since the photoion signal was too strong and beyond the range of the SEM linearity the molecules were ionised by pulses with $\phi = 0.015$ J cm^{-2} which provided an ionisation efficiency of about 2% [8.69].

Figure 8.18 shows the dependence of the photoionisation signals of naphthalene and anthracene ions on the energy flux of CO_2 laser pulses ϕ_{CO_2}.

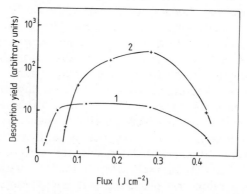

Figure 8.18 The dependence of desorption efficiency of naphthalene molecules (1) and anthracene molecules (2) on CO_2 laser pulse flux, ϕ_1. The flux of the KrF laser pulse, $\phi_2 = 1.5 \times 10^{-2}$ J cm^{-2} [8.64].

With $\phi_{CO_2} > 0.06$ J cm^{-2} the dependence $N_i(\phi_{CO_2})$ for naphthalene comes to saturation which corresponds to full desorption of naphthalene molecules from the surface before the arrival of an ionising pulse. With IR pulse energy fluxes $\phi_{CO_2} \geqslant 0.3$ J cm^{-2} the signal of naphthalene molecules drops which is apparently explained by the thermal dissociation of the molecules. In this case fragment ions and substrate ions C^+ and C_2^+ appeared in the mass spectrum. Also a glow in the substrate could be observed which indicated strong heating of its surface.

In the dependence $N_i(\phi_{CO_2})$ for anthracene there is not a well pronounced region of saturation. This is due to the fact that at temperatures at which the anthracene molecules are fully desorbed from the surface a larger fraction of molecules undergoes thermal dissociation.

It is easy to evaluate the detection sensitivity of naphthalene molecules on the surface. The observed ion signal, 10^3 ions per pulse at full desorption of molecules corresponds to a surface concentration of $10^6 \, mol \, cm^{-2}$ or about 10^{-8} of a molecular layer. In the calculation it was taken into account that the photoionisation efficiency was 2% and the ion detection efficiency was 50% [8.69]. If the photoionising pulse energy is optimal, i.e. $0.2-0.3 \, J \, cm^{-2}$, the detection threshold of adsorbed molecules for a signal of 1 ion/pulse will be 50 to 100 molecules/cm^2.

Thus, the results of [8.64] show that under optimised desorption and photoionisation conditions it is possible to detect molecules of $10^{-8}-10^{-10}$ of a monolayer. For molecules with a high adsorption energy the main factor limiting the sensitivity is thermal dissociation of the molecules on the surface.

8.4 Laser microanalytical mass spectrometry of organic and bio-organic molecules (LAMMA method)

In studying objects with a complex spatial structure it is necessary to carry out mass spectral analysis of their separate parts. Such a task has to be solved in studying tissue sections, separate cells or individual parts of a cell. Such objects have to be investigated with micron or submicron resolution. The sensitivity of the method must be high because the amount of substance to be studied is usually about $10^{-10} \, g$ or less. Some methods of microanalytical mass spectrometry were elaborated [8.71] to solve these tasks. The laser method is one of the most promising.

As mentioned in §8.1, the method of laser microanalytical mass spectrometry has found successful realisation in the LAMMA-500 instrument [8.72, 8.73] (see also [8.74]). New models have also been developed lately: LAMMA-1000 [8.75] and LIMA (Cambridge Consultants Ltd, Great Britain) [8.76]. However, all the experiments on laser microanalytical mass spectrometry of organic compounds have only been performed so far with the LAMMA-500 instrument, with the exception of [8.75]. That is why below we will use for simplicity the terms 'LAMMA' instrument and 'LAMMA method'.

Initially the LAMMA method was applied to elementary analysis of biological samples with a high spatial resolution [8.72, 8.77]. In [8.73, 8.78] it was shown that the LAMMA method permits the study of the content of many elements (Na, K, Cs, Rb, Mg, Ca, Cr, Pb, etc) in different biological objects. It was also shown that the method would make it possible to investigate different tissues (retina, muscles, etc) at cell and subcell resolution as well as single bacteria and dust particles. The spatial resolution achieved was equal to $0.5 \, \mu m$. The absolute sensitivity for different elements lay in the range from $10^{-18} \, g$ to $10^{-20} \, g$ with a relative sensitivity of 10^{-4} weight per cent.

The first experiments [8.78, 8.79] revealed that, along with ions of different elements, the mass spectra contained ions which were of organic origin. In the LAMMA spectra of crystals of sulphanilic acid and nicotinic acids the quasi-molecular ions $(M+H)^+$ and $(M-H)^-$ were the most abundant. In the subsequent experiments mass spectra of many classes of organic and biorganic compounds were obtained. Despite the fact that the LAMMA method is now being successfully applied to solve specific analytical tasks, the mechanisms of formation of ions of organic molecules are still not clearly understood [8.13]. This fact, of course, hampers the further development of the method and its wider application. Thus much consideration has recently been given to the investigation of the process of ion formation in the LAMMA method [8.83, 8.89, 8.91].

8.4.1   LAMMA-500, technical parameters

The LAMMA-500 instrument consists of two basic components: a mass spectrometer and a laser. The instrument underwent changes as the mass spectrometric and laser techniques developed. In the first experiments an ordinary time-of-flight mass spectrometer was used [8.72, 8.77], and then it was replaced by a mass reflectron [8.73]. In all these experiments different laser sources were used: the first and second harmonics of a Q-switched ruby laser, the first, second, third and fourth harmonics of a Q-switched Nd–YAG laser and the N_2 laser.

Here we shall consider in more detail the last version of LAMMA-500 (figure 8.19). The instrument design is described in detail in [8.73, 8.78] and in [8.74].

Figure 8.19(*a*) depicts the specimen chamber of the instrument, and figure 8.19(*b*) shows an enlarged diagram of the location of the sample. The sample, in the form of a thin semitransparent section 0.3–1.0 μm thick or as microcrystals, is mounted on a copper grid for electron microscopy. The grid is sometimes covered with an organic film of collodion or formvar. The grid with the sample is located in the vacuum chamber of the instrument in the immediate vicinity of the 0.15 mm thick quartz window. On the other side opposite the window there is a microscope objective which focuses the laser radiation on the sample.

The ions formed under the action of laser radiation are collected and accelerated by an electrostatic ion optical system and directed into the drift tube of the mass spectrometer. The mass spectrometer is designed according to the scheme of the mass reflectron [8.33, 8.34] (figure 8.19(*c*)). The accelerated ions drift in the first field-free region at the end of which is placed an electrostatic ion mirror. On reaching the mirror the ions change their motion by about 180° and enter the second field-free region. At the end of the second field-free region a secondary electron multiplier is located. If the potentials of the mass spectrometer electrodes are optimal, it provides a first-order compensation of the effect of the initial velocities of ions on the drift time and achieves a mass resolution $M/\Delta M = 800$–1000. Choosing the correct polarity of the potentials of the electrodes makes it possible to detect both positive and negative ions.

The signal from the multiplier is amplified by a preamplifier and is fed to the

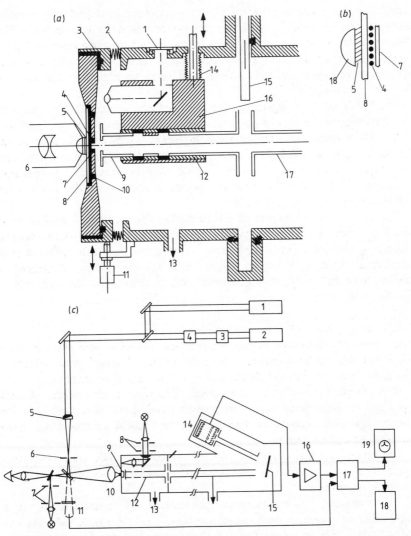

Figure 8.19 LAMMA-500. (*a*) sample chamber and (*b*) enlarged view of the specimen arrangement. Key for parts (*a*) and (*b*): 1, window for transmitted light; 2, 14, bellows; 3, *x*, *y* movable vacuum flange; 4, copper grid; 5, immersion liquid; 6, microscope objective; 7, sample; 8, cover slide; 9, ion lens; 10, ring; 11, micrometer, *x–y* manipulator; 12, plastic washer; 13, two vacuum pumps; 15, pendulum gate; 16, movable fixture for light transmission and ion optical system; 17, time-of-flight tube; 18, front lens of the microscope objective.

Key for part (*c*): general schematic diagram of LAMMA-500. 1, He–Ne-laser; 2, Nd–YAG laser; 3, frequency converter; 4, filter; 5, achromatic lens; 6, diaphragm; 7, illuminator; 8, condensor; 9, sample; 10, objective; 11, photodiode; 12, ion lens; 13, two vacuum pumps; 14, electron multiplier; 15, ion reflector; 16, pre-amplifier; 17, transient recorder; 18, strip-chart recorder; 19, CRT display.

input of a transient recorder which performs analogue-to-digital conversion of the signal and storage in a buffer memory with a capacity of 2048 bytes. The shortest sampling time of the signal is 10 ns. The information from the buffer memory can be converted to an analogue signal and fed to an oscilloscope or a strip-chart recorder and also transmitted into the computer memory for further processing.

The Nd–YAG laser with a Q switch is used as a source of radiation with $\tau = 15$ ns. Most often the samples are irradiated by the fourth harmonic of laser radiation obtained with a frequency multiplier at $\lambda = 264$ nm. The achromatic lens focuses the laser radiation onto a diaphragm 60–70 μm in diameter, and its image is transferred onto the sample with an immersed microscope objective. All this allows a radiation spot to be focused on the sample of a diffraction limited size of 0.5 μm.

The flange on which the sample is mounted, can be manually shifted in the X and Y directions via two micrometer screws and thereby the desired part of the sample is selected. The sample can be observed in both reflected and transmitted light by using an appropriate lighting system. Alignment is realised with an He–Ne laser the beam path of which coincides with that of the Nd–YAG laser.

8.4.2   Mass spectra of various compounds

The volume of mass spectrometric information which has been obtained with the LAMMA method exceeds the information obtained with all other methods. This is explained by the use of a commercial instrument.

Here we shall consider the most essential characteristic features of mass spectra of various organic compounds which enable us to reveal their common features and differences of ion formation in the LAMMA method and in the methods described above.

The most essential characteristics of molecules which affect the mass spectrum are the presence of polar groups or ion bonds and the presence of aromatic groups.

For non-polar molecules, such as unsaturated hydrocarbons, it is impossible to produce quasi-molecular ions like $(M+H)^+$, $(M-H)^-$ or $(M+\text{alkali})^+$. The mass spectra of both positive and negative ions contain many fragment ions [8.80].

(a)   Carbohydrates

The mass spectra of positive ions for oligosaccharides, glycosides and other carbohydrates have dominant $(M+\text{alkali})^+$ ions. Figure 8.20 shows the mass spectrum of stachyose [8.81]. Besides the quasi-molecular ion, it contains a number of peaks corresponding to the cationised fragments formed as a result of detachment of one or more glycosidic units from the intact molecule. The mass spectra of glycosides, along with the quasi-molecular ion $(M+\text{alkali})^+$, always have cationised fragments with a smaller number of glycosidic units

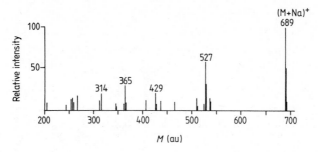

Figure 8.20 LAMMA spectrum of stachyose with NaCl (weight ratio 5:1) [8.81].

[8.81]. The mass spectra of carbohydrates obtained with the LAMMA instrument have many features in common with the spectra obtained with CO_2 laser pulses [8.19].

(b) *Organic acids*

The basic peak in the mass spectra of negative ions of organic acids always corresponds to the quasi-molecular ion $(M-H)^-$ [8.73, 8.80, 8.81], and at near-threshold intensities it is sometimes the only one. The mass spectra of positive ions always contain intense signals of fragment ions. Nevertheless, the quasi-molecular ion $(M+H)^+$, as a rule, is represented by a more or less intense peak. The same character of the mass spectrum is observed for molecules which do not contain the carboxyl COOH group, such as for example ascorbic acid [8.81]. The intensity of the signal of quasi-molecular ions $(M+alkali)^+$ is small or there is no signal at all. This is an essential difference between these spectra and the mass spectra of organic acids obtained with IR lasers.

(c) *Amino acids and micropeptides*

The mass spectra of quite a number of amino acids—glycine, alanine, valine, leucine, isoleucine, tyrosine and phenylalanine—were studied in [8.82, 8.84]. The mass spectra of positive and negative ions of valine are shown in figure 8.21. In the mass spectra of negative ions, as in the case of ordinary organic acids, the basic peak corresponds to deprotonated molecules $(M-H)^+$. The most intense signal for positive ions is observed for the mass $(M-45)$ which corresponds to the $(M-COOH)^+$ ion. The $(M+H)^+$ ions are also abundant in the mass spectrum. Cationised molecular ions are observed as weak signals. In [8.84] it is noted that for aromatic amino acids an intense signal can be observed for the ions of the amino acid residual R^+. As a whole, consideration must be given to an essential difference between the mass spectra of amino acids irradiated by IR laser pulses and measured by the LAMMA instrument.

The mass spectra of dipeptides given in figure 8.22 were investigated in

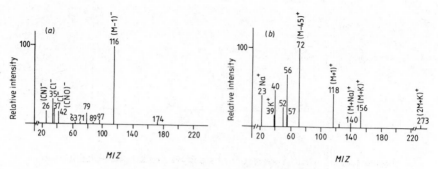

Figure 8.21 LAMMA spectra of valine [8.84].

[8.82, 8.84]. It is noted that the mass spectra of dipeptides and an equimolar mixture of corresponding amino acids differ greatly. In the latter case the mass spectrum is a sum of mass spectra of amino acids. The mass spectra of dipeptides contain rather intense signals for the $(M+H)^+$ and $(M-H)^-$ ions. Besides, they contain decarboxylated amino acids which points to a break in the peptide bond. At the same time, the decarboxylated molecular peptide ion $(M-45)^+$ is not observed. It is worthy of note that one can observe deprotonated molecular ions of the amino acids constituting peptides which cannot be formed by a simple cleavage of the peptide bond. In accordance with [8.84], this fact points to a significant role of the chemical reactions taking place during evaporation of the sample and ion formation.

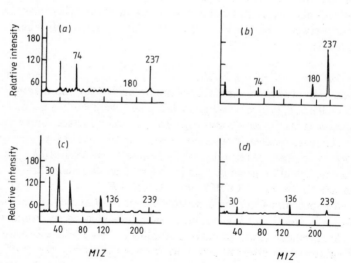

Figure 8.22 Negative (a, c) and positive (b, d) LAMMA spectra of glycine-tyrosine (a, b) and tyril-glycine (c, d) [8.84].

(d) Bases

As far as the mass spectra of positive and negative ions are concerned, bases are complementary to acids. In the case of bases the protonated molecular ions $(M+H)^+$ are dominant whereas the signal of $(M-H)^-$ ions is often rather weak. A great number of fragment ions, however, can be observed both in positive and negative mass spectra [8.80, 8.81, 8.90].

(e) Organic salts

This class of compounds differs greatly from those described above. The basic peaks in the mass spectra of positive and negative ions correspond to the cation and the anion of the salt. Quasi-molecular ions are not observed [8.81]. If the salt contains two cations $A^{2-}C_2^+$ or two anions $A_2^-C^{2+}$, the mass spectrum contains single-charged ions $(A^{2-}C^+)^-$ and (A^-C^{2+}), doubly charged anions and cations are not observed [8.80, 8.81]. When the intensity is far above the threshold of ion formation, the mass spectrum has a great number of fragment ions formed by the dissociation of the cation and the anion of the salt.

The mass spectra of quaternary ammonium salts [8.85] look like their corresponding mass spectra obtained with the IR laser. The salt cation $(R_4N)^+$ is the most abundant in the mass spectra. The detachment of R radicals gives rise to fragmentary ions.

(f) Aromatic compounds

This is the only class of molecules for which it is possible to produce the molecular radical ion M^+. Molecules such as anthracene, phenanthrene, dibenzothiophene and dibenzofuran yield abundant M^+ ions and few fragment ions [8.80]. The negative ion mass spectrum of these compounds is not specific and consists of a sequence of peaks of $(C_nM_m)^-$ with a well pronounced periodic structure of $C-12$.

In the presence of polar groups the mass spectrum of aromatic molecules contains not only the M^+ ions but also $(M+H)^+$, the ratio of their signals depending on laser radiation intensity. For example, in the mass spectrum of triphenylphosphine [8.86] at low laser radiation intensities M^+ ions prevail, whereas at high intensities the signal of the $(M+H)^+$ ion becomes predominant.

The presence of molecular M^+ ions and a low degree of fragmentation are general features of the mass spectra of aromatic compounds obtained by the LAMMA method and in the case of nonfocused moderate-intensity UV radiation (see §8.3).

(g) Metallo-organic complexes

The mass spectra of metallo-organic complexes were studied in [8.81, 8.87]. In [8.87] a thin polymer film of polyvinylbutynate (PVB) was used with molecules of metallo-organic complexes dissolved in it. Complexes of the metals As, Sb, Bi, Ge and Cd with aromatic and alicyclic ligands were investigated. The quasi-molecular ion $(M + H)^+$ was observed only in the case of $(C_6H_5)_3As$. The mass spectra of triphenyl arsine contain fragment ions formed in the process of successive detachment of the ligands $(M - 77)^+$, $(M - 2 \times 77)^+$, as well as the ions of $(C_6H_5)^+$ and the central atom. The peak intensity of ion of the central atom in the series As, Sb, Bi increases with increasing ionic radius of the atom. Along with the ions listed, the mass spectra contain rather a large number of fragment ions in the low-mass region.

(h) *High-molecular-weight compounds*

As a rule, with an increase in molecular weight the volatility and temperature stability of organic compounds decrease. Therefore, it has always been a problem for mass spectrometry to produce ions of heavy molecules. The potentialities of the LAMMA technique in this respect were demonstrated in [8.88, 8.89]. In [8.88] mass spectra of positive and negative ions for six cobalamines with masses ranging from 1000 to 1600 au were obtained. Quasi-molecular ions $(M + H)^+$ were obtained for all six compounds. In addition, the mass spectra have rather intense peaks of fragment ions which give information on the structure of the parent molecules.

In [8.89] the mass spectra of glycoside (digitonin) ($M = 1228$ au) and a synthetic lipid were studied, with sodium, potassium and caesium chlorides added to the sample. For both compounds there were intense signals of quasi-molecular ions. The quasi-molecular ions were obtained when the crystals of the sample were larger than the laser beam diameter. If the diameter of the crystals was equal to that of the laser spot, only light ions were observed in the mass spectrum.

As the results presented above show, the mass spectra of different compounds obtained by the LAMMA instrument vary greatly as compared with those obtained by the methods of IR–LDMS and LDMS with moderate-intensity UV radiation. Such differences are probably connected with the presence of several mechanisms of ion formation. The dominant role of a certain mechanism is defined by the physico-chemical properties of the sample. The mass spectra taken by the LAMMA and IR–LDMS methods are very alike for some compounds such as carbohydrates and quaternary ammonium salts. In the case of aromatic compounds this similarity is observed with moderate-intensity UV readiation LDMS.

8.4.3 *Mechanisms of ion formation in the LAMMA method*

When organic crystals are exposed to strongly focused intense UV laser radiation, rather a complicated physical phenomenon arises.

(1) At high intensities of UV laser pulses (10^8–10^{11} W cm^{-2}) multiphoton absorption in the crystal and in the molecules in the gas phase occur. For organic molecules containing heteroatoms O, N, S and others multiphoton

absorption is very effective because all these molecules have a resonant absorption (though weak) at $\lambda = 266$ nm.

(2) Strong heating of the surface can lead to phase transition and intense evaporation of the sample. Photoionisation processes in a vapour cloud give rise to plasma since the energy of two radiation quanta, $2\hbar\omega = 9.4$ eV, is sufficient to ionise most organic molecules.

Some authors state that with $I = 10^9$ W cm^{-2} neither intense evaporation nor plasma formation occur. The main argument in this case is the absence of destruction of the crystal or film surface detectable by an electron microscope. But it can easily be estimated that under the action of a laser pulse ($\tau = 10$ ns) focused into a spot of diameter 1 μm a vapour cloud of several cubic microns is formed with a density of 10^{18} molecules/cm^3 due to evaporation of a few molecular layers. If we take a small absorption cross section $\sigma = 10^{-20}$ cm^{-2}, the probability of two-photon ionisation with $I = 10^8$ W cm^{-2} ($\phi = 1$ J cm^{-2}) is about one per cent, i.e. the charged-particle density in the cloud will be 10^{16} ions/cm^3. The free path for gas-kinetic collisions in this case will be about 1 μm. It is obvious that under such conditions molecular and ion–molecular reactions must be rather effective.

(3) Strong radiation focusing results in the temperature over the crystal being very nonuniform. The region, where the conditions are considered to be uniform (~ 1 μm), is of the same order as or smaller than the region with high temperature gradients.

(4) The conditions under which processes of ion formation occur are essentially nonstationary. The time of plasma cloud formation (10 ns) and the time of its recombination and expansion can be shorter than the time of establishment of equilibrium parameters. This does not allow simple thermodynamic relations to be applied to analysis.

A shortage of experimental results does not make it possible at present to indicate conclusively which of the mechanisms of ion formation is basic in each specific case. Hercules *et al* [8.86] points to four spatial regions where different processes of ion formation may take place (figure 8.23).

The authors of [8.86] believe that in the region of direct action of radiation on the sample (1) it is possible to reach temperatures from 3000 to 6000 K. So emission of atomic ions and ions of light molecular fragments must take place in this region. Near region (1) there is a region of high thermal gradient (2). This region is subjected to the action of the shock wave from both the laser impact and from collapsing plasma. From this region heavier neutral fragments of molecules and their ions will be emitted. The formation of heavy ions, including quasi-molecular ones, is assumed to take place in region (3) as a result of surface ionisation and in (4) as a result of gas-phase ion–molecular reactions.

The main source of information on mechanisms of ion formation at present is mass spectra of different organic compounds obtained with the LAMMA

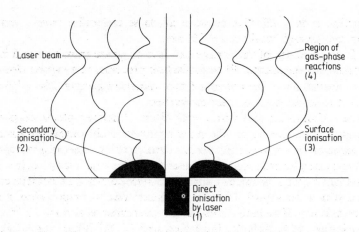

Figure 8.23 Diagram showing important phenomena in laser volatilisation and ionisation [8.86].

instrument. Their comparison with the mass spectra obtained with the IR–LDMS and UV–LDMS methods as well as the results of recent experiments [8.89, 8.81, 8.83] make it possible to define the mechanism of ion formation for some classes of molecules.

The mass spectra of carbohydrates (oligosaccharides, glycosides) obtained by the LAMMA and IR–LDMS methods are very alike. This suggests that the mechanism of ion formation in these two cases is the same. The results of [8.91, 8.89] confirm that the formation of cationised carbohydrates occurs by the attachment of a metal cation to the intact molecule in the gas phase [8.22–8.26]. The process of formation of quasi-molecular ions $(M + \text{alkali})^+$ of sucrose was studied as applied to some parameters of the sample [8.91].

It was found that the substitution of a silver film for the slightly absorbing polymer film on which the sample was placed resulted in a sharp increase of quasi-molecular ion yield. Thus, the heating of the substrate plays a significant role in the formation of ions (compare with [8.19, 8.23]).

The character of the mass spectra depends greatly on the dimension of the sucrose crystal under irradiation. The intensity of the $(M + \text{Na})^+$ ion is decreased by 10 times as the dimension of the crystal is reduced from 6.2 to 3.7 μm. No quasi-molecular ions are observed when the crystal is about 1 μm in size. Similar results were obtained in [8.89], where the mass spectra of such heavy molecules as digitonin and synthetic lipid were investigated. These results show that $(M + \text{alkali})^+$ ions are not formed directly in the region of UV radiation action but in peripheral regions. The important role of gas-phase ion–molecule reactions was confirmed in experiments [8.91] where $(\text{sucrose} + \text{Li})^+$ ions were observed in spite of spatially separated sources of sucrose molecules and Li^+ ions.

All the above listed facts show that cationised molecules are formed according to the mechanism suggested in [8.22–8.25], i.e. emission of alkali metal cations from the hot central region of the irradiation spot, desorption of molecules from the colder peripheral regions and gas-phase reaction of attachment of a metal cation to the molecule with the formation of $(M + alkali)^+$ ions.

In [8.92, 8.93] it was shown for a large number of organic molecules that, if a sample contained a compound of such metals as Ag, Cu, Pb or Cd, $(M + metal)^+$ ions could also be observed. It must be said that the process of cationisation of organic molecules is characteristic of the LAMMA and IR–LDMS methods.

The molecular ions M^+ of aromatic compounds are apparently formed by two-step photoionisation of the molecules evaporating from the surface under laser heating as in the case of UV–LDMS (see §8.3), since all aromatic molecules have a large absorption cross section at $\lambda = 266$ nm and the ionisation potential is lower than $2\hbar\omega = 9.4$ eV.

Interesting results are presented in [8.86] which point to peculiarities of strong focusing of UV radiation. At low radiation intensities the molecular ion M^+ is dominant in the mass spectrum of triphenylphosphine. At high intensities the $(M + H)^+$ ion becomes dominant. This points to the fact that at sufficiently high radiation intensities strong evaporation of the sample leads to a dense cloud formation in which ion–molecular reactions proceed effectively.

It is noted in [8.83] that the role of ion–molecular gas-phase reactions can be essential in the formation of negative quasi-molecular ions $(M + 15)^-$ of nitroaromatic compounds. In this case the following chain of reactions is assumed

$$AromH + NO_2^- \rightarrow \begin{bmatrix} & H \\ Arom & \diagup \\ & \diagdown \\ & NO_2 \end{bmatrix}^- \rightarrow AromO^- + HNO.$$

The mass spectra obtained by the IR–LDMS and LAMMA methods differ greatly for acids and bases. In the latter case the $(M + H)^+$ and $(M - H)^-$ ions are the most abundant.

In [8.83] it is shown for the case of amino acids that the formation of quasi-molecular ions occurs due to dissociation of the dimer with transfer of the proton. This assumption is based on the fact that in a crystal the molecules of amino acids are in the form of dimers. As they evaporate from the surface, UV dissociation is possible

$$
\begin{array}{ccc}
NH_3{}^+ \ldots {}^-OOC & NH_3{}^+ & NH_2 \\
\diagup & \diagdown \quad \diagup & \diagup \\
R-CH & CH-R \xrightarrow{h\nu} R-CH & + R-CH \\
\diagdown & \diagup \quad \diagdown & \diagdown \\
COO^- \ldots {}^+NH_3 & COOH & COO^-
\end{array}
$$

with the formation of a pair of ions $(M + H)^+$ and $(M - H)^-$.

The mass spectra obtained for deuterated valine show that the transfer of the proton occurs with the participation of hydrogen from the amino group and not from the aliphatic chain.

The role of attachment of a proton to an amino acid in the gas phase is not essential. As is shown in [8.83], as a valine sample deuterated on the amino group is irradiated by intense radiation, the mass spectrum contains the H^+ and D^+ ions, but the quasi-molecular ions are represented only by the peak $(M+D)^+$.

Since the mass spectra of quaternary ammonium salts obtained by the LAMMA method are almost similar to the mass spectra obtained by the IR–LDMS method and by resistive heating of the sample, it is obvious that in the LAMMA method the thermal desorption of salt cations $(R_4N)^+$ can be observed. It is not unlikely that in the case of other organic salts the main mechanism of ion formation is thermal desorption of salt anions and cations.

In conclusion it should be noted again that under the conditions of the LAMMA method quite a number of physico-chemical processes can take place and give rise to different ions. In some cases it is possible to determine more or less convincingly the dominant mechanism of ion formation. But there are still many unsolved questions, for example on the role of heterophase reactions in the formation of ions, which call for further studies.

8.5 Conclusions

This chapter has been concerned with different laser methods of mass spectrometry of nonvolatile thermally unstable organic and bio-organic molecules. Each of these methods has its characteristic features, its pros and cons.

The most sensitive is the two-pulse method of IR–UV–LDMS which allows detection of almost single molecules on a surface. But such high parameters can be so far attained only for aromatic compounds which can be effectively ionised in the gas phase. The development of lasers in the vacuum ultraviolet range will enable us to expand the class of molecules which can be detected with a high sensitivity by the IR–UV–LDMS method. The advantage of this method is that it provides selective ionisation of molecules. Although the electron absorption bands of most organic molecules are rather wide, in some cases it is possible to realise preferential ionisation of a definite type of molecule through proper choice of laser radiation frequency.

The IR–LDMS and LAMMA methods are more universal but their sensitivity is lower than in the case of IR–UV–LDMS. The main shortcoming of IR–LDMS and LAMMA is that the central part of the radiation spot is a source of metal cations or other ions which participate in secondary ion reactions giving rise to quasi-molecular ions. Inevitably, a considerable part of the sample in this case dissociates. This shortcoming can be eliminated by using the two-pulse method. If the regions where metal cations are formed and the molecules of the sample desorb are spaced apart and two laser pulses are used, it is possible to optimise independently the processes of molecular desorption and metal cation formation. This will enable the sensitivity of the cationisation technique to be increased significantly. It should be noted that the two-pulse method in this case is a logical development of the model experiments discussed in [8.22–8.25].

References

8.1 Beynon J H 1960 *Mass Spectrometry and Its Application to Organic Chemistry* (Amsterdam: Elsevier)

8.2 Middleditch B S (ed) 1979 *Practical Mass Spectrometry. A Contemporary Introduction* (New York: Plenum)

8.3 Johnstone R A W 1972 *Mass Spectrometry for Organic Chemists*

8.4 Lawson A M, Stillwell R N, Tacker M M, Tsybogama K and McCloskey J A 1971 *J. Am. Chem. Soc.* **93** 1014

8.5 Daves G S Jnr 1979 *Acc. Chem. Res.* **12** 359

8.6 Beckey H D 1969 *Int. J. Mass. Spectrom. Ion Phys.* **2** 500

8.7 Torgerson D F, Skowronski R P and Macfarlane R D 1974 *Biochem. Biophys. Res. Commun.* 616

8.8 Macfarlane R D and Torgerson D F 1976 *Science* **191** 920

8.9 Benninghoven A, Jaspers D and Sichtermann W 1976 *Appl. Phys.* **11** 35

8.10 Baldwin M A and McLafferty F W 1973 *Org. Mass Spectrom.* **7** 1353

8.11 Hansen C and Munson B 1980 *Anal. Chem.* **52** 245

8.12 Conzemius R J and Capellen J M 1980 *Int. J. Mass Spectrom. Ion Phys.* **34** 197

8.13 Hillenkamp F 1983 in *Ion Formation from Organic Solids* ed A Benninghoven (Springer Series in Chemical Physics, vol 25) (Berlin: Springer) p. 190

8.14 Vastola F J, Pirone A J and Knox B E 1966 in *Proc. 14th Ann. Conf. on Mass Spectrom. and Allied Topics, Dallas, Texas, May 22–7 1966* p. 78

8.15 Vastola F J and Pirone A J 1966 *Preprint* 10C-53 (American Chemical Society Division of Fuel Chemistry)

8.16 Vastola F J and Pirone A J 1968 *Adv. Mass Spectrom.* **4** 107

8.17 Vastola F J, Mumma R O and Pirone A J 1970 *Org. Mass Spectrom.* **3** 101

8.18 Mumma R O and Vastola F J 1972 *Org. Mass Spectrom.* **6** 1373

8.19 Posthumus M A, Kistemaker P G, Meuzelaar H L C and Ten Voever de Brauw M C 1978 *Anal. Chem.* **50** 985

8.20 Kaufmann R, Hillenkamp F, Wechsung R J, Heinen W J and Schurmann M 1979 *Scan. Electron. Microsc.* **2** 279

8.21 Stoll R and Röllgen F W 1979 *Org. Mass Spectrom.* **14** 642

8.22 Van der Peyl G J Q, Isa K, Haverkamp J and Kistemaker P G 1981 *Org. Mass Spectrom.* **16** 416

8.23 Van der Peyl G J Q, Haverkamp J and Kistemaker P G 1982 *Int. J. Mass Spectrom. Ion Phys.* **42** 125

8.24 Van der Peyl G J Q, Isa K, Haverkamp J and Kistemaker P G 1982 *Nucl. Instrum. Meth.* **198** 125

8.25 Stoll R and Röllgen F W 1982 *Z. Naturf.* **37A** 9

8.26 Cotter R J, Snow M and Colvin M 1983 in *Ion Formation from Organic Solids* ed A Benninghoven (Springer Series in Chemical Physics, vol 25) (Berlin: Springer) p. 206

8.27 Letokhov V S, Shibanov A N and Yegorov S E 1983 in *Surface Studies with Lasers* eds F R Aussenegg, A Leitner and M E Lippitsch (Springer Series in Chemical Physics, vol 33) (Berlin: Springer) p. 156

8.28 Egorov S E, Letokhov V S and Shibanov A N 1984 *Kvantov. Elektron.* **11** 1397

8.29 Stoll R and Röllgen F W 1980 *J. Chem. Soc. Chem. Commun.* 789

8.30 Cotter R J and Yergey A L 1981 *J. Am. Chem. Soc.* **103** 1596

8.31 Stoll R and Röllgen F W 1981 *Org. Mass Spectrom.* **16** 72

8.32 Van Breemen R B, Snow M and Cotter R J 1983 *Int. J. Mass Spectrom. Ion Phys.* **49** 35

8.33 Karataev V I, Mamyrin B A and Shmyk D V 1972 *Sov. Phys.–Tech. Phys.* **16** 1177

8.34 Mamyrin B A, Karataev V I, Shmyk D V and Zagulin V A 1973 *Zh. Eksp. Teor. Fiz.* **64** 82 (Engl. transl. *Sov. Phys.–JETP* **37** 45)

8.35 Baesl U, Neusser H J, Weinkauf R and Schlag E W 1982 *J. Phys. Chem.* **86** 4857

8.36 Tuithof H H, Boerboom A J H and Meuzelaar H L C 1975 *Int. J. Mass Spectrom. Ion Phys.* **17** 299

8.37 Kistemaker P G, Lens M M, Van der Peyl G J Q and Boerboom A J H 1980 *Adv. Mass Spectrom.* **8A** 928

8.38 Heresch F, Schmid E R and Huber J F K 1980 *Anal. Chem.* **52** 1803

8.39 Joy W K and Reuben B G 1970 *Dyn. Mass. Spectrom.* **1** 183

8.40 Heresch F 1983 *Int. J. Mass Spectrom. Ion Phys.* **47** 27

8.41 Heresch F 1983 in *Ion Formation from Organic Solids* ed A Benninghoven (Springer Series in Chemical Physics, vol 25) (Berlin: Springer) p. 217

8.42 Heresch F, Schmid E R and Huber J F K 1980 *Proc. 7th Int. Symp. on Mass Spectrom. in Biochem, Medicine and Environmental Research, Milan, 16–18 June 1980*

8.43 Junk G and Svec H 1963 *J. Am. Chem. Soc.* **85** 839

8.44 Leclercq P A and Desiderio D M 1973 *Org. Mass. Spectrom.* **7** 515

8.45 Beuhler R J, Flanigan E, Greene L J and Friedman L F 1974 *J. Am. Soc.* **96** 3990

8.46 Ohushi M, Barron R P and Benson W R 1981 *J. Am. Chem. Soc.* **103** 3943

8.47 Lee T D, Anderson W R Jnr and Daves G D Jnr 1981 *Anal. Chem.* **53** 304

8.48 Isa K and Yamada Y 1983 *Org. Mass Spectrom.* **18** 229

8.49 Zakett D, Schoen A E, Cooks R G and Hemberger P H 1981 *J. Am. Chem. Soc.* **103** 1295

8.50 Ready J F 1971 *Effects of High-power Laser Radiation* (New York: Academic)

8.51 Van der Peyl G J Q, Van der Zande W J, Bederski K, Boerboom A J H and Kistemaker P J 1983 *Int. J. Mass Spectrom. Ion Phys.* **47** 7

8.52 Röllgen F W, Giessmann U and Stoll F 1983 *Nucl. Instrum. Meth.* **201** 78

8.53 Cotter R J and Yergey A L 1981 *Anal. Chem.* **53** 1306

8.54 Schode U, Stoll R and Röllgen F W 1981 *Org. Mass Spectrom.* **16** 441

8.55 Antonov V S, Letokhov V S and Shibanov A N 1980 *Pis'ma Zh. Eksp. Teor. Fiz.* **31** 471 (in Russian)

8.56 Antonov V S, Egorov S E, Letokhov V S, Matveetz Yu A and Shibanov A N 1982 *Pis'ma Zh. Eksp. Teor. Fiz.* **36** 29 (in Russian)

8.57 Antonov V S, Khoroshilova E B, Kuzmina N P, Letokhov V S, Matveets Yu A, Shibanov A N and Yegorov S E 1982 in *Picosecond Phenomena III* (Springer Series in Chemical Physics, vol 23) (Berlin: Springer) p. 310

8.58 Antonov V S, Letokhov V S and Shibanov A N 1981 *Appl. Phys.* **25** 71

8.59 Antonov V S, Letokhov V S and Shibanov A N 1982 *Appl. Phys.* **28B** 245

8.60 Antonov V S, Letokhov V S, Matveyets Yu A and Shibanov A N 1982 *Laser Chem.* **1** 37

8.61 Antonov V S, Letokhov V S, Matveets Yu A and Shibanov A N 1982 *Poverkhnost* **1** 54 (in Russian)

8.62 Yegorov S E, Letokhov V S and Shibanov A N 1984 *Chem. Phys.* **85** 349

8.63 Egorov S E and Shibanov A N 1984 *Zh. Tekh. Fiz.* **54** 2270 (in Russian)

8.64 Antonov V S, Egorov S E, Letokhov V S and Shibanov A N 1983 *Pis'ma Zh. Eksp. Teor. Fiz.* **38** 185 (in Russian)

8.65 Agranovich V M and Galanin M D 1978 *Electron Excitation Energy Transfer in Condensed Media* (Moscow: Nauka)

8.66 Cherkasov Yu A, Kiseleva M N and Dodonova N Ya 1978 *Opt. Spektrosk.* **45** 1126

8.67 Bergman A, Levine M and Jortner J 1967 *Phys. Rev. Lett.* **18** 593

8.68 Wiley W C and McLaren I H 1955 *Rev. Sci. Instrum.* **26** 1150

8.69 Antonov V S, Letokhov V S and Shibanov A N 1981 *Opt. Commun.* **38** 182

8.70 Antonov V S, Letokhov V S and Shibanov A N 1984 *Usp. Fiz. Nauk.* **142** 177

8.71 Anderson C A 1973 *Microprobe Analysis* (New York: Wiley)

8.72 Hillenkamp F, Unsöld E, Kaufmann R and Nitsche R 1975 *Appl. Phys.* **8** 341

8.73 Kaufmann R, Hillenkamp F and Wechsung R 1979 *Med. Prog. Technol.* **6** 109

8.74 Denoyer E, Van Gieken R, Adams F and Nautsch D F S 1982 *Anal. Chem.* **54** 26a

8.75 Heinen H J, Meier S and Vogt H 1983 in *Ion Formation from Organic Solids* ed A Benninghoven (Springer Series in Chemical Physics, vol 25) (Berlin: Springer) p. 229

8.76 Dingle T, Griffiths B W and Ruckman J C 1981 *Vacuum* **31** 571

8.77 Unsöld E, Hillenkamp F and Nitsche R 1976 *Analysis* **4** 115

8.78 Kaufmann R, Hillenkamp F, Wechsung R, Heinen H J and Schurmann M 1979 *Proc. Scanning Electron Microscopy Symp., Washington, DC, April 16–22 1979* vol 29 p. 279

8.79 Nitsche R, Kaufmann R, Hillenkamp F, Unsöld E, Vogt H and Wechsung R 1978 *Israel J. Chem.* **17** 181

8.80 Heinen H J 1981 *Int. J. Mass Spectrom. Ion Phys.* **38** 309

8.81 Heinen H J, Meier S, Vogt H and Wechsung R 1980 *Adv. Mass Spectrom.* **8A** 942

8.82 Kupka K, Hillenkamp F and Schiller C 1980 *Adv. Mass Spectrom.* **8A** 935

8.83 Hercules D M, Parker C D, Balasanmugam K and Visvandaham S K 1983 in *Ion Formation from Organic Solids* ed A Benninghoven (Springer Series in Chemical Physics, vol 25) (Berlin: Springer) p. 222

8.84 Hillenkamp F 1981 *Kvantov. Elektron.* **8** 2655

8.85 Schuler B and Kruger F R 1979 *Org. Mass Spectrom.* **14** 439

8.86 Hercules D M, Day R J, Balasanmugam K, Dang T A and Li C P 1982 *Anal. Chem.* **54** 280A

8.87 Ollmann B, Kupka K-D and Hillenkamp F 1983 *Int. J. Mass Spectrom. Ion Phys.* **47** 31

8.88 Graham S W, Dowd P and Hercules D M 1982 *Anal. Chem.* **54** 649

8.89 Seydel U and Lindner B 1983 in *Ion Formation from Organic Solids* ed A Benninghoven (Springer Series in Chemical Physics, vol 25) (Berlin: Springer) p. 240

8.90 Schueler E and Krueger F R 1980 *Org. Mass Spectrom.* **15** 295

8.91 Wieser P and Wurster R 1983 in *Ion Formation from Organic Solids* ed A Benninghoven (Springer Series in Chemical Physics, vol 25) (Berlin: Springer) p. 235

8.92 Schueler B, Fiegl P and Hillenkamp F 1981 *Org. Mass Spectrom.* **16** 502

8.93 Balasanmugam K, Dang T A, Day R J and Hercules D M 1981 *Anal. Chem.* **53** 2296

8.94 Barber M, Bordoli R S, Sedgwick R D and Tyler A N 1981 *J. Chem. Soc. Chem. Commun.* 325

Index